The vibrational spectroscopy of polymers

Cambridge Solid State Science Series

EDITORS

Professor R. W. Cahn
*Department of Materials Science and Metallurgy,
University of Cambridge*

Professor E. A. Davis
Department of Physics, University of Leicester

Professor I. M. Ward
Department of Physics, University of Leeds

D. I. BOWER

Senior Lecturer
Department of Physics, University of Leeds

W.F. MADDAMS

Senior Visiting Fellow
Department of Chemistry, University of Southampton

The vibrational spectroscopy of polymers

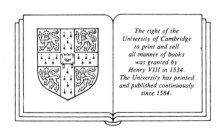

The right of the
University of Cambridge
to print and sell
all manner of books
was granted by
Henry VIII in 1534.
The University has printed
and published continuously
since 1584.

CAMBRIDGE UNIVERSITY PRESS

Cambridge
New York   New Rochelle   Melbourne   Sydney

Published by the Press Syndicate of the University of Cambridge
The Pitt Building, Trumpington Street, Cambridge CB2 1RP
32 East 57th Street, New York, NY 10022, USA
10 Stamford Road, Oakleigh, Melbourne 3166, Australia

First published 1989

Printed in Great Britain at the University Press, Cambridge

British Library cataloguing in publication data
Bower, D. I.
The vibrational spectroscopy of polymers.
1. Polymers. Vibration spectroscopy
I. Title   II. Maddams, W. F.
547.7′0448

Library of Congress cataloguing in publication data
Bower, D. I.
The vibrational spectroscopy of polymers/D. I. Bower and
W. F. Maddams.
   p.   cm. – (Cambridge solid state science series)
Bibliography: p.
Includes index.
ISBN 0 521 24633 4
1. Polymers and polymerization – Analysis. 2. Vibrational spectra.
I. Maddams, W. F. II. Title. III. Series.
QD381.8.B69 1989
547.8′43 – dc19   88-2956

ISBN 0 521 24633 4

MC

Contents

Preface

The aim of this book is to present a coherent introductory account of the theory of vibrational spectroscopy and of its application to the study of synthetic polymers. The level of presentation is intended to be suitable for the research student who has previously obtained a degree in either physics or chemistry and who is embarking on research in this area. Such a student would, we hope, read the book in its entirety and then be equipped with sufficient background knowledge and understanding to tackle the specialised literature of the subject with some confidence. We hope that in addition the book will fulfil a similar function for any research worker new to the subject and that parts of it may also be found useful by undergraduate students studying either vibrational spectroscopy or polymer science. To make the book accessible to those new to the study of polymers we have given a brief introduction to the subject in chapter 1. This chapter also contains a brief description of experimental methods in vibrational spectroscopy which is intended to give the reader the minimum amount of information required to follow the rest of the book and to feel that his feet are firmly on the ground. Although we regard the book primarily as one to be read rather than simply referred to, we have provided detailed indexes, since the nature of the subject is such that the spectroscopist needs to master a large number of ideas and facts and easy reference to these is vital if that mastery is to be quickly obtained.

Infrared spectroscopy has been used to characterise synthetic polymers for the whole of the period over which the majority of them have been known, and it is still the more popular of the two vibrational spectroscopic techniques discussed in this book. Raman spectroscopy was very difficult to apply to any material before the advent of the laser in 1960, but since that time the ease of use of the laser as a source has stimulated the development of better Raman spectrophotometers and Raman spectroscopy may now be undertaken quite as routinely as infrared. Its lesser popularity springs largely from its greater expense, but we feel that there is also an element of suspicion in the minds of some that despite the extra expense Raman spectroscopy does not offer any more than infrared for many investigations, and that in any case it is more difficult to understand. In fact, the two techniques are complementary in a variety of ways and there can be no excuse for neglecting either unless it be the one of expense. In this book the two methods are treated completely on a par and while we admit that Raman spectroscopy is slightly more difficult to understand than infrared we hope to dispel any fears on this ground.

One of the problems of all spectroscopic techniques is that they involve large amounts of specialised notation and usually attract a corresponding amount of jargon. It is one of the aims of this book to present both of these aspects of Raman and infrared spectroscopy, particularly as applied to polymers, in as simple a way as possible and to give many examples of their use. In order to leave room for a large amount of illustrative material, the theoretical sections do not go into more depth than is necessary for a beginner's understanding. For fuller details the reader is referred to the books listed in the Further Reading sections at the ends of chapters 2, 3 and 4. One piece of jargon that we have tried to avoid entirely is the use of the word 'band' to describe the broadened spectral lines that are observed in the infrared and Raman spectra of polymers. To us that word implies a set of discrete lines very close together and is best avoided since that is not what is implied by its use in the literature of the vibrational spectroscopy of polymers. We have preferred to use words such as 'peak' or 'absorption' and we hope that this idiosyncrasy on our part will not be found confusing.

We owe a great debt to many books, reviews and papers on vibrational spectroscopy in general and on the vibrational spectroscopy of polymers in particular. The books and reviews that we have found most useful are listed in the Further Reading sections at the ends of the chapters. We did feel, however, that in a textbook of this nature the pages should not be peppered with references to individual papers. Our aim has been primarily to illustrate the basic principles of the subject by means of examples where the interpretation of the spectra is now generally accepted as well established, at least in outline. Such acceptance is usually the outcome of the work of a large number of people. In those places where we have relied extensively on the work of particular authors their names will usually occur in the text, in acknowledgments for the use of tables or diagrams or in the list of references for chapters 5 and 6 given at the end of the book. This list is not meant to be comprehensive but to include references to classic papers and to papers that will lead the reader to the literature of a particular topic. We thank all authors whose work we have made use of in preparing the book and especially those who have given us permission to reproduce or adapt their material. We hope that all will forgive us for any inadvertent misinterpretation or oversimplification of their work.

We are very grateful to a number of people who have read drafts of various parts of the book, which has benefitted enormously from their comments. In particular, we should like to thank Professor N. Sheppard for reading and commenting in great detail on drafts of chapters 1 to 4: without these comments a large number of errors and obscurities would not have been avoided. Others who read various sections and made valuable suggestions were Professors J. S. Dugdale and G. J. Morgan, Drs G. C. East, L. J. Fina, P. Gans, J. B. Gill, S. J. Spells, A. P. Unwin and J. M. Woodhouse. Professor A. C. T. North rewrote our original version of subsection 6.6.4 and the present version is substantially his. Our greatest debt is to Professor I. M. Ward, both for the original suggestion to write the book and for reading at least two complete drafts. The present structure of the book and the content of a number of sections, particularly in chapters 5 and 6, were influenced greatly by his suggestions. To all these people who contributed to making the book better than it would otherwise have been we offer our thanks, while accepting that, in spite of their efforts, errors and obscurities will remain. We accept full responsibility for these and would be pleased to be informed about them.

Our thanks are due to Dr S. Parker of BP Research Centre, Sunbury, for providing the original spectra for figures 5.1, 5.3, 5.9, 5.14, 5.15, 5.17, 5.18, 5.19 and to Mr P. Spiby for providing the spectrum for figure 4.5. We also thank Mrs M. Edmundson and Mrs G. Garbett for typing parts of the early drafts and the Cambridge University Press for their patience during the long preparation time of this book.

August 1988   D. I. Bower
                                              W. F. Maddams

1 Introduction

1.1 Vibrational spectroscopy

1.1.1 Molecular vibrations

Molecules consist of atoms bound together by what are usually called *chemical bonds*. The nature of these bonds will be discussed more fully in chapter 2 and it is only necessary to note here that the bonds and the angles between them are not rigid. To a first approximation the force required to make a small change in the length of a bond, or a small change in the angle between two bonds, is proportional to the change produced; similarly, the torque required to twist one part of a molecule through a small angle with respect to the rest about a bond is approximately proportional to the angle of twist. The molecule thus consists of a set of coupled harmonic oscillators and if it is disturbed from its equilibrium state it will vibrate in such a way that the motion can be considered to be a superposition of a number of simple harmonic vibrations. In each of these so-called *normal modes* every atom in the molecule vibrates with the same frequency, and in the simplest molecules all atoms pass through their respective positions of zero displacement simultaneously.

There are three principle methods by which the vibrations may be studied: infrared and Raman spectroscopies and inelastic neutron scattering. The first two methods are available in very many laboratories, since the equipment required is relatively small and cheap. Neutron scattering is less readily available, since the technique requires a neutron source, which is usually a nuclear reactor, and relatively specialized and expensive equipment to analyse the energies of the neutrons scattered from the sample. In this book we shall not consider neutron scattering in any detail, although it will be mentioned occasionally. The main purpose of the book will be to explain what kinds of information infrared and Raman spectroscopies can provide about polymers and how the information may be obtained. Much of chapters 2 and 3 and parts of chapter 4 are, however, relevant to the study of the vibrations of molecules of any kind.

1.1.2 Infrared and Raman spectroscopy

A vibrating molecule may interact in two distinctly different ways with electromagnetic radiation of appropriate frequency. If the radiation has the same frequency as one of the normal modes of vibration, and this usually means that it will be in the infrared region of the electromagnetic spectrum, it may be possible for the molecule to *absorb* the radiation. The energy absorbed will later be lost by the molecule either by re-radiation or, more usually, by transfer to other molecules of the material in the form of heat energy. An *infrared absorption spectrum* of a material is obtained simply by allowing infrared radiation to pass through the sample and determining what fraction is absorbed at each frequency within some particular range. The frequency at which any peak in the absorption spectrum appears is equal to the frequency of one of the normal modes of vibration of the molecules of the sample.

The second way in which electromagnetic radiation may interact with a molecule is by being *scattered*, with or without a change of frequency. If light is allowed to fall on a sample which is homogeneous on a scale large compared with the wavelength of light both types of scattering will in general take place. The scattering without change of frequency may be thought of as scattering from the equilibrium states of the molecules and is called *Rayleigh scattering*. The scattering with change of frequency is called *Raman scattering* and the change in frequency is equal to the frequency of one of the normal modes of vibration of the molecules. The strongest scattering is at frequencies lower than that of the incident light and this is called the *Stokes Raman scattering*.

In general, some but not all of the modes of vibration of a particular type of molecule can be observed by means of infrared spectroscopy and these are said to be *infrared-active* modes. Similarly, some but not all modes are *Raman active*. Which modes are active for which process depends on the symmetry of the molecule. Chapters 2 and 3 are largely devoted to a discussion of molecular symmetry and how it affects the nature of the normal modes of molecules, polymer chains and polymer crystals. Chapter 4 considers the *selection rules* which determine whether each mode is infrared or Raman active and how observed spectral features can be *assigned* to normal modes.

1.2   Fundamentals of polymers

In this book the term *polymer* is used to mean a particular class of macromolecules which consist, at least to a first approximation, of a set of regularly repeated chemical units of the same type, or possibly of a very limited number of different types (usually only two), joined end to end to form a chain molecule. If there is only one type of chemical unit the corresponding polymer is a *homopolymer* and otherwise it is a *copolymer*. The synthetic polymers will form the main examples in this book although much of what is said applies equally to certain classes of biological macromolecules, or *biopolymers*. In this section we consider briefly the main types of chemical structural repeat units present in synthetic polymers, together with the kinds of structural regularities and irregularities that may occur. Further details of the structures of individual polymers will be given in later sections of the book. It should be noted that the term *monomer* or *monomer unit* is often used to mean either the chemical repeat unit or the small molecule which polymerizes to give the polymer. These are not always the same in atomic composition, as will be clear from what follows, and the chemical bonding must of course be different even when they are.

1.2.1   *Addition polymers*

The sequential addition of monomer units to a growing chain is a process which is easy to visualize and is the mechanism for the production of an important class of polymers. For the most common forms of this process to occur, the monomer must contain a double (or triple) bond. The process of addition polymerization occurs in three stages. In the *initiation* step an activated species, such as a free radical from an initiator added to the system, attacks and opens the double bond of a molecule of the monomer, producing a new activated species. In the *propagation* step this activated species adds on a monomer unit which becomes the new site of activation and adds on another monomer unit in turn. Although this process may continue until thousands of monomer units have been added sequentially, it always terminates when the chain is still of finite length. The *termination* process normally occurs by one of a variety of specific chain terminating reactions, which lead to a corresponding variety of end groups.The propagation process is normally very much more probable than the termination process, so that macromolecules containing thousands or tens of thousands of repeat units are formed.

The simplest type of additon reaction is the formation of polyethylene from ethylene monomer:

$$\text{+\!CH}_2\text{+}_n\text{CH}_2\!-\!\text{CH}_2^{\bullet} + \text{CH}_2\!=\!\text{CH}_2 \rightarrow \text{+\!CH}_2\text{+}_{n+2}\text{CH}_2\!-\!\text{CH}_2^{\bullet}$$

Polyethylene is a special example of a generic class that includes many of the industrially important macromolecules, the *vinyl* and *vinylidene* polymers. The chemical repeat unit of a vinylidene polymer is $-CH_2-CXY-$, where X and Y represent single atoms or chemical groups. For a vinyl polymer Y is H and for polyethylene both X and Y are H. If X is $-CH_3$, Cl, $-CN$, $-\!\langle\bigcirc\rangle$ or $-O.CO.CH_3$ and Y is H, the well-known materials polypropylene, poly(vinyl chloride) (PVC), polyacrylonitrile, polystyrene or poly(vinyl acetate), respectively, are obtained. When Y is not H, X and Y may be the same type of atom or group, as with poly(vinylidene chloride) (X and Y are Cl), or they may differ, as in poly(methyl methacrylate)(X is $-CH_3$, Y is $-COOCH_3$) and poly(α-methyl styrene) (X is $-CH_3$, Y is $-\langle\bigcirc\rangle$). When the substituents are small, polymerization of a tetra-substituted monomer is possible, to produce a polymer such as polytetrafluoroethylene (PTFE), $+CF_2-CF_2+_n$, but if large substituents are present on both carbon atoms of the double bond there is usually steric hindrance to polymerization.

Polydienes are a second important group within the class of addition polymers. The monomers have two double bonds and one of these is retained in the polymeric structure, to give one double bond per chemical repeat unit of the chain. This bond may be in the backbone of the chain or in a side group. If it is always in a side group the polymer is of the vinyl or vinylidene type. The two most important examples of polydienes are polybutadiene, containing 1,4-linked units of type $-CH_2-CH=CH-CH_2-$ or 1,2-linked units of type $-CH_2-CH(CH=CH_2)-$, and polyisoprene, containing corresponding units of type $-CH_2-C(CH_3)=CH-CH_2-$ or $-CH_2-C(CH_3)(CH=CH_2)-$. Polymers containing both types of unit are not uncommon, but special conditions may lead to polymers consisting largely of one type. Acetylene, $CH\equiv CH$, polymerizes by an analogous reaction in which the triple bond is converted into a double bond to give the chemical repeat unit $+CH=CH+$.

Ring-opening polymerizations, such as those in which cyclic ethers polymerize to give polyethers, may also be considered to be addition polymerizations:

$$n\overline{CH_2-(CH_2)_{m-1}-O} \rightarrow \{-(CH_2)_m O\}_n$$

The simplest type of polyether, polyoxymethylene, is obtained by the similar polymerization of formaldehyde in the presence of water:

$$nCH_2=O \rightarrow +CH_2-O+_n$$

1.2.2   Step growth polymers

Step growth polymers are obtained by the repeated process of joining together smaller molecules, which are usually of two different kinds at the beginning of the polymerization process. For the production of linear (unbranched) chains it is necessary and sufficient that the number of reactive groups on each of the initial 'building brick' molecules is two and that the molecule formed by the joining together of two of these molecules also retains two appropriate reactive groups. There is usually no specific initiation step so that any appropriate pair of molecules present anywhere in the reaction volume can join together. Many short chains are thus produced initially and the length of the chains increases both by the addition of monomer to either end of any chain and by the joining together of chains.

Condensation polymers are an important class of step growth polymers formed by the common condensation reactions of organic chemistry. These involve the elimination of a small molecule, often water, when two molecules join, as in *amidation*:

$$RNH_2 + HOOCR' \rightarrow RNHCOR' + H_2O$$

which produces the *amide linkage*

$$-\underset{\underset{H}{|}}{N}-\underset{\underset{O}{\|}}{C}-$$

and *esterification*:

$$RCOOH + HOR' \rightarrow RCOOR' + H_2O$$

which produces the *ester linkage*

$$-\underset{\underset{O}{\|}}{C}-O-$$

An important example of a reaction employed in step growth polymerization which does not involve the elimination of a small molecule is the reaction of an isocyanate and an alcohol

$$RNCO + HOR' \rightarrow RNHCOOR'$$

which produces the *urethane linkage*

$$-\underset{\underset{H}{|}}{N}-\underset{\underset{O}{\|}}{C}-O-$$

In all these reactions R and R' may be any chemical group.

The amidation reaction is the basis for the production of the *polyamides* or *Nylons*. For example, Nylon 6,6, which has the structural repeat unit $-HN(CH_2)_6NHCO(CH_2)_4CO-$, is made by the condensation of hexamethylene diamine, $H_2N(CH_2)_6NH_2$, and adipic acid, $HOOC(CH_2)_4COOH$, and Nylon 6,10 results from the comparable reaction between hexamethylene diamine and sebacic acid, $HOOC(CH_2)_8COOH$. In the labelling of these Nylons the first number is the number of carbon atoms in the amine residue and the second the number of carbon atoms in the acid residue. Two Nylons of somewhat simpler structure, Nylon 6 and Nylon 11, are obtained, respectively, from the ring-opening polymerization of the cyclic compound ε-caprolactam:

$$n\overline{OC(CH_2)_5N}H \rightarrow {+OC(CH_2)_5NH+}_n$$

and from the self-condensation of ω-amino-undecanoic acid:

$$nHOOC(CH_2)_{10}NH_2 \rightarrow {+OC(CH_2)_{10}NH+}_n + nH_2O$$

The most important *polyester* is poly(ethylene terephthalate), ${+(CH_2)_2OOC-\bigcirc-COO+}_n$, which is made by the condensation of ethylene glycol, $HO(CH_2)_2OH$, and terephthalic acid, $HOOC-\bigcirc-COOH$, or dimethyl terephthalate, $CH_3OOC-\bigcirc-COOCH_3$. There is also a large group of unsaturated polyesters that are structurally very complex because they are made by multi-component condensation reactions, e.g. a mixture of ethylene glycol and propylene glycol, $CH_3CH(OH)CH_2OH$, with maleic and phthalic anhydrides (see fig. 1.1).

One of the most complex types of step growth reactions is that between a glycol, HOROH, and a di-isocyanate, $O=C=NR'N=C=O$, to produce a polyurethane, which contains the structural unit $-ORO.CO.NHR'NH.CO-$. Several subsidiary reactions can also take place and although all of the possible reaction products are unlikely to be present simultaneously, polyurethanes usually have complex structures. Thermoplastic polyurethanes are copolymers which usually incorporate sequences of polyester or polyether segments and they are considered in more detail in subsections 5.3.5 and 6.6.3.

Fig. 1.1. (*a*) maleic anhydride; (*b*) phthalic anhydride.

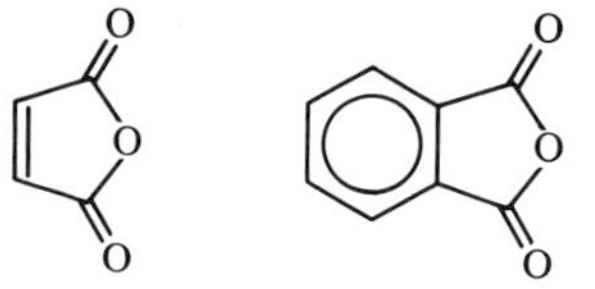

(*a*)   (*b*)

Formaldehyde, $CH_2{=}O$, provides a very reactive building block for step growth reactions. For example, in polycondensation reactions with phenol, ⟨O⟩—OH, or its homologues with more than one OH group, it yields the *phenolic resins* and with urea, $CO(NH_2)_2$, or melamine (see fig. 1.2a) it yields the *amino resins*. The products of such condensation reactions depend on the conditions employed but they are usually highly cross-linked. Acid conditions lead to the formation of methylene bridged polymers of the type shown in fig. 1.2b and c, whereas alkaline conditions give structures containing the methylol group, $—CH_2OH$, which may condense further to give structures containing ether bridges, of the form $R—O—R'$ (fig. 1.2d).

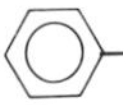

In general, step growth polymers are likely to produce rather complex spectra because of the size of the building blocks, even when only one type of unit is present, as for example in poly(ethylene terephthalate). When more than one type of reaction occurs to an appreciable extent, as with phenolic resins and polyurethanes, the additional complexity resulting from the two- or multi-component mixture will lead to still more marked difficulties of interpretation.

1.2.3 *Regular chains and defect structures*

A perfectly regular polymer chain would consist of the repetition of identical structural units at equal intervals and with the same orientation along the direction of the chain: sets of corresponding

Fig. 1.2. (a) melamine; (b), (c) and (d) various bridging structures in phenolic resins.

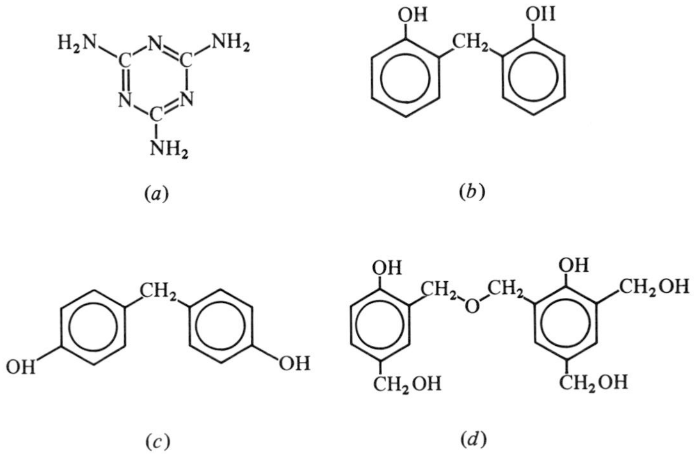

(a)

(b)

(c)

(d)

points in all structural units would lie on parallel straight lines. The term *defect structure* will be used to denote a chain-terminating group, a chain branch or a stereochemical imperfection in an otherwise regularly repeating polymer chain, i.e. a region in which either the internal bonding of a structural unit or its orientation with respect to the adjacent unit on one or both sides causes the chain to depart from regularity. This definition excludes the alternative types of chemical structural units that are frequently formed during condensation reactions. Polymers containing the latter structures may be regarded as copolymers. All real polymer chains contain defect structures, even those in the most perfect of polymer single crystals (see section 1.2.4).

Irregularities of addition, leading to defect structures, can occur during addition polymerization for several reasons. In polymers of the vinyl or vinylidene class, sequence isomerism may occur. The two carbon atoms of $-CH_2-CHX-$ or $-CH_2-CXY-$ bear different substituents. The units usually add *head-to-tail*, to give the regular structure $+CH_2-CHX+_n$ or $+CH_2-CXY+_n$, but they may add *head-to-head* or *tail-to-tail*, to give units of type $+CH_2-CHX-CHX-CH_2+$ or $+CHX-CH_2-CH_2-CHX+$, respectively, for a vinyl polymer. Most substituents are sufficiently bulky to cause some degree of steric hindrance in the head-to-head arrangement and this structure is usually of sufficiently higher energy that it is formed in very low concentration. The van der Waals radius of the fluorine atom is comparatively small and an appreciable degree of head-to-head addition occurs during the polymerization of vinyl fluoride and vinylidene fluoride, with a corresponding amount of tail-to-tail addition, necessarily.

The most important type of structural irregularity in polymers of the vinyl or vinylidene type occurs because the four bonds to each carbon atom are arranged approximately tetrahedrally in space, and this means that for vinyl polymers of the type $+CH_2-CHX+_n$ two regular types of spatial disposition of the substituent X with respect to the polymer backbone of carbon atoms are possible. The simplest types of structure that may be envisaged are those in which the carbon atoms form a planar zig-zag structure. If we look edge-on to the plane of this zig-zag, three distinct types of placing of the X atoms are possible (fig. 1.3). In the *isotactic* chain all the X atoms or groups are on the same side of the plane of the zig-zag, whereas in the *syndiotactic* chain they are alternately on opposite sides. These two structures are said to be *stereoregular*. In the *atactic* chain the X atoms or groups are randomly placed on the two sides of the plane. None of these three types of arrangement can be converted into any of the others without breaking and reforming chemical bonds and they are said to differ in *configuration*, or to be different *configurational isomers*. The term *tacticity* is used to refer to the type or degree of configurational regularity of vinyl chains. If the configuration is largely of one regular type, either syndiotactic or isotactic, but contains a small proportion of the other type, this latter may be regarded as a defect structure.

It is possible for a vinyl or vinylidene chain to take up the regular planar zig-zag structure referred to above, whatever its configurational structure, provided that the groups X (or X and Y) are small enough. Rotations about the C—C bond may occur, however, to give non-planar *conformational isomers* (or *conformers*). These structures may be of lower or higher energy depending on the size of the groups X (or X and Y). In the remainder of this book the terms *configurational* and *conformational* isomerism will be used strictly in accordance with the definitions just given, but the reader is warned that although this distinction is now made by most authors, passages will often be found in the literature in which the two types of isomerism are not clearly distinguished. The important distinction is that a new configurational

Fig. 1.3. Planar zig-zag conformations of vinyl polymers: (*a*) carbon backbone; (*b*) regular isotactic configuration; (*c*) regular syndiotactic configuration; (*d*) random atactic configuration.

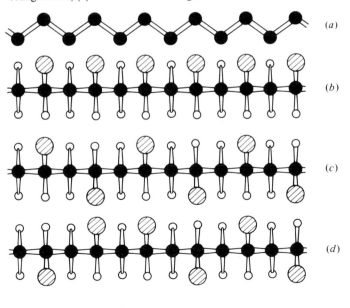

○ , H;　● , C;　⊘ , Cl or other group

isomer can only be formed by breaking and reforming one or more bonds, whereas a change of conformational isomer can be achieved simply by the rotation of parts of the molecule around single bonds.

In many vinyl polymers the X atom or group is rather large and the planar conformation is a high energy one for the isotactic configuration, because of the small mutual separation of those atoms or groups in that structure, and a different conformational isomer may result. Isotactic sequences in PVC are a good example of this. The equilibrium proportions of such conformers present at a particular temperature are determined by the differences between their Gibbs free energies. In isotactic vinyl polymers with bulky side groups a helical structure often occurs. This may be imagined to be formed from the planar zig-zag by rotating the part of the molecule on one side of a backbone C—C bond through 120° around the bond with respect to the rest of the molecule and repeating the operation for alternate C—C bonds, the rotation being always in the same direction.

The helical structure just considered illustrates two important conformational features which can occur in many polymers. In the planar zig-zag structure every backbone C—C bond may be considered to be the central member of a group of three C—C bonds which are coplanar, and in which the outer two bonds are on opposite sides of the projected line of the central bond (see fig. 1.4). The central bond of any such group is said to be a *trans* bond. In the helical structure alternate bonds are trans, but those about which the rotation has taken place, and which are parallel to the axis of the helix, are said to be *gauche* even though the rotation may be to the right. Left and right gauche bonds correspond to anticlockwise and clockwise rotations, respectively, of the third bond by 120° from the plane formed by the first two of the group of three, and a helix may be a regular right-handed or left-handed structure or it may have kinks where the handedness changes.

Polydienes which retain the double bond in the main polymer chain (1,4-polydienes) show a different type of isomerism from that of the vinyl polymers. This isomerism is called *cis/trans* isomerism and is illustrated in fig. 1.5. It must be stressed that unlike trans/gauche isomerism this is configurational rather than conformational isomerism, since rotation cannot take place about the double bond, and the properties of the all-cis and all-trans forms of the same polymer are often very different. An important example is polyisoprene, which occurs naturally in both forms; the cis form is natural rubber and the trans form is gutta percha, which is hard and inelastic at room temperature. In certain polymer structures three single bonds take up a cis or nearly cis conformation and a cis/trans isomerism that is conformational rather than configurational can also occur in poly(ethylene terephthalate) and similar compounds, as illustrated in fig. 1.6.

Chain branches are commonly encountered defect structures even for the simplest polymer, polyethylene, and with free radical

Fig. 1.4. Trans/gauche isomerism: (*a*) trans bond viewed normally to the plane of the three bonds required to define it; (*b*) view of (*a*) from the right-hand side, looking almost end-on to the trans bond; (*c*) and (*d*) nearly end-on views of left- and right-handed gauche bonds, respectively.

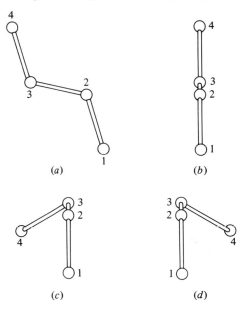

(*a*)   (*b*)

(*c*)   (*d*)

Fig. 1.5. Cis/trans isomerism: (*a*) cis 1,4-polydiene; (*b*) trans 1,4-polydiene. In the cis form the 1 and 4 carbon atoms which link the unit to the polymer chain are on the same side of the projected line of the double bond, whereas in the trans form they are on opposite sides of it.

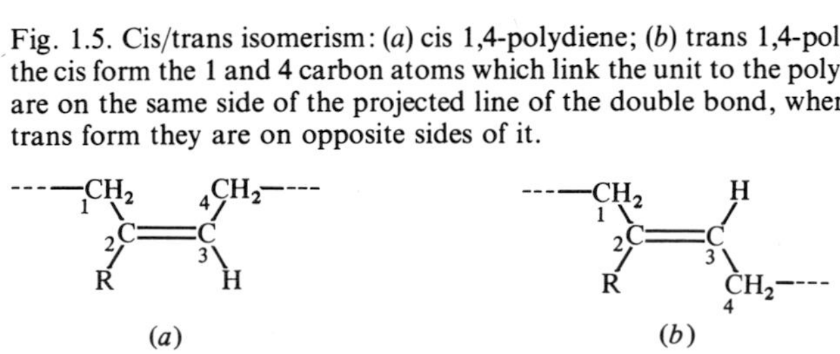

(*a*)   (*b*)

polymerizations branching is inevitable to some degree, because of *chain transfer*. A growing polymer chain collides with a second chain and there is transfer of a hydrogen atom. Termination of the growing chain occurs but the second chain involved in the collision now has an odd electron somewhere along its length. This acts as a site for polymerization, resulting in branch formation. Alternatively, the process of *back-biting* may occur. The growing chain bends back and abstracts a hydrogen atom situated on a carbon atom several removed from the chain end, again leading to termination of the original chain and branch formation. This process is favoured by a low monomer concentration and is therefore of more frequent occurrence during the later stages of radical polymerization.

1.2.4 *Polymer morphology and crystallization*

So far we have been concerned with the structure of single polymer chains at the detailed level of the individual monomer unit, its structure and its linking to adjacent monomers. If we now stand back, as it were, from the polymer chain, so that its structure at this level becomes blurred, we can regard the chain as being something like a flexible string, the degree of flexibility depending strongly on the particular type of polymer. If we consider the shapes, or *macroconformations*, taken up by these flexible strings in the polymer solid and how they are arranged relative to each other we are dealing with *polymer morphology*.

A great deal of evidence shows that the morphology of many types of polymer can be described, at least to a first approximation, in terms of

Fig. 1.6. Cis/trans isomerism of poly(ethylene terephthalate): (*a*) the trans conformation, and (*b*) the cis conformation of the terephthaloyl residue.

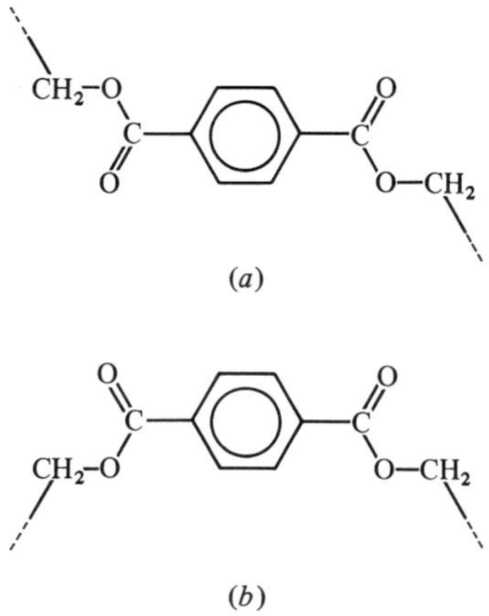

(*a*)

(*b*)

two simple types of structure or modifications of them. These structures are *randomly coiled chains*, which occur in the non-crystalline regions of polymers, and *lamellar chain-folded crystallites* or *lamellae*. In randomly coiled chains the various possible microconformational states of the types considered in the last section occur at random with overall concentrations determined by the Gibbs free energy differences between them. The non-crystalline regions of a polymer are often called the *amorphous* regions, but in oriented polymers and in some non-crystalline regions of partially crystalline polymers the chains do not have truly random coil form and the term should therefore be avoided or used with caution.

Polymer glasses, such as poly(methyl methacrylate) (PMMA, Perspex, Plexiglass), consist entirely of non-crystalline material which, in samples that have not been oriented, is present in the form of interpenetrating randomly coiled chains, and this is the reason for their high degree of transparency. PMMA is non-crystalline because it is an atactic vinylidene polymer with large side groups which cause it to be conformationally highly irreglar and so unable to crystallize. Polyethylene chains, on the other hand, are usually highly regular and can therefore take up the low-energy planar zig-zag form and crystallize under suitable conditions. In material cooled slowly from the molten state the crystallites usually aggregate into complicated structures called *spherulites*. The diameters of the spherulites are often comparable to that of the wavelength of light, which they therefore scatter strongly, so that the polymer has a milky appearance. Between the spherulites there is non-crystalline material.

If polyethylene is allowed to crystallize from dilute solution small, almost flat, pyramid-shaped single crystals form which may be a few micrometres across and about 100–150 Å thick. Electron diffraction shows that the chain axes in these lamellar crystals are normal to the mean plane of the crystal and the only possible conclusion is that the chains fold back and forth within it, as shown schematically in fig. 1.7. A great deal of evidence shows that the crystallites in material produced from the melt are also lamellar in nature and of a similar general type to those grown from solution, though less perfect, and that other polymers also form chain-folded lamellar crystals from both solution and melt. Much effort has been devoted to elucidating the precise nature of the folds and to determining whether the chain segments on either side of a particular fold lie adjacent to each other within the crystallite; the evidence that vibrational spectroscopy can bring to bear on these questions and other aspects of polymer crystals and crystallinity is discussed in chapter 6.

1.3   The infrared and Raman spectra of polymers

When the vibrations of small molecules are considered in chapter 2 we shall see that a non-linear molecule consisting of N atoms has $3N - 6$ normal modes of vibration, which may be classified as belonging to various symmetry species. A poymer molecule may contain tens of thousands of atoms and may thus have tens to hundreds of thousands of normal modes. The infrared or Raman spectrum of a polymer might thus be expected to be impossibly complicated. Fig. 1.8 illustrates that this is not so. The basic reason is that the molecule of a homopolymer consists of a large number of chemically identical units, each of which usually contains only tens of atoms or fewer. This leads to a considerable reduction in the complexity of the infrared or Raman spectrum.

Fig. 1.7. Schematic diagram of chain folding in a solution-grown single crystal of polyethylene.

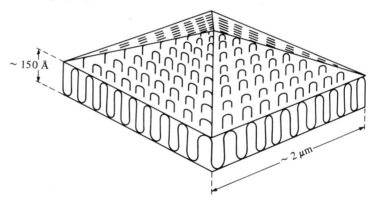

Fig. 1.8. The infrared spectrum of a sample of 1,2-polybutadiene. (Reproduced by permission from J. Haslam, H. A Willis and D. C. Squirrel, *Identification and Analysis of Plastics*.)

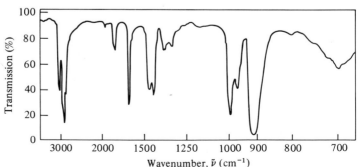

One can understood this simply by realizing that the repeat unit of a polymer has typical dimensions of order 10^{-9} m, whereas the wavelength of the light involved in Raman scattering is of order 5×10^{-7} m and the wavelength of the infrared radiation that is absorbed by the vibrating molecules is usually greater than 2×10^{-6} m. There are thus many repeat units within one wavelength and the interaction of the polymer molecule with the radiation will therefore depend on the sum of the interactions over many repeat units. If the polymer chain is a perfectly regular one in a crystalline region of a polymer, only a small fraction of the possible number of normal modes of vibration are such that all the repeat units in the chain vibrate in phase. The other modes are essentially standing-wave modes in which the vibrations of all repeat units in the chain are similar but the phases alternate along the chain. In these modes the interactions of the individual repeat units with the radiation cancel out over the length of the chain and the modes are thus inactive in both Raman and infrared spectroscopy. The infrared- and Raman-active modes belong to the group of modes in which all repeat units vibrate in phase. A further reduction in the complexity of the spectra often occurs because, as will be discussed in chapter 4, it is possible for even some of these modes to be inactive in either the Raman or infrared spectrum, or both.

For an irregular chain in a non-crystalline region of a polymer all normal modes are potentially both Raman and infrared active because there is no molecular symmetry. The chemical repeat units are, however, no longer also physical repeat units, because of the geometrical irregularity, and the coupling forces between repeat units are also no longer all the same. It is therefore often a useful approximation to regard each chemical repeat unit as a separate molecule in its own local environment. This is particularly so when the vibration concerned involves mainly the motion of side-group atoms and little motion of the atoms of the chain backbone. We then expect to see broad infrared absorption or Raman scattering peaks due to *group vibrations*, the broadening arising because the groups of atoms of the same kind in different repeat units have slightly different frequencies of vibration as a consequence of the different physical environments of the units to which they belong. The widths of the broad peaks are, however, usually small compared with the separations between peaks that are due to quite different types of vibration of the repeat unit.

We see, therefore, that the Raman or infrared spectrum of a polymer will usually contain a number of peaks which is of order $3n$ or less, where n is the number of atoms in the repeat unit, rather than $3N$, where N is the number of atoms in the whole molecule. Since the spectrum of a polymer is to a first approximation that of its repeat unit, it follows immediately that the spectrum is an aid to qualitative analysis, that is, to finding out what kinds of repeat units are present in a sample of an unknown polymer or polymer blend. Further, since the strength of the absorption or scattering will depend on the concentration or number of scatterers, the spectrum can also be used for quantitative analysis of polymer blends or copolymers. If residual monomer or additives are present or if the polymer has degraded and contains degradation products these will usually have their own characteristic spectra which, if strong enough, can be distinguished and used for qualitative or quantitative analysis. Chapter 5 is largely concerned with these types of characterization or analysis of polymers and with the necessary background understanding of how the chemical structure of a polymer affects its spectrum.

In addition to providing information about the chemical structures of a polymer, vibrational spectroscopy can also give very useful information about the physical structure, which also influences strongly the physical properties of the polymer. Any two regions of the polymer which differ in the way the repeat units are arranged may show detectable differences in their spectra. Sometimes these differences are small and give rise merely to broadening of peaks, which can itself provide evidence that one sample is more disordered than another, but sometimes peaks are split and sometimes totally new peaks appear as a result of some treatment, for instance when a polymer is allowed to crystallize or is subjected to physical deformation. These features signify either a change in the interactions between repeat units or the production of new conformations within repeat units, and these kinds of effects may be studied both qualitatively and quantitatively in the spectra.

Some polymers, while consisting of identical chemical repeat units, may have these units bonded together in different ways, as discussed in subsection 1.2.3, and this will also affect the spectra, which can thus be used to study details of structure such as tacticity and the presence of head-to-head structures in vinyl polymers or chain branching in a variety of polymers. A related type of study is that of the distribution of the different units along the chain in copolymers.

Finally, a quite important influence on the physical properties of many polymers in actual use is the fact that their molecules have been preferentially oriented at some stage in processing, either deliberately, as in textile fibres and in sheet material and tapes used for packaging, or incidentally, as in moulding or pressing processes. If spectra are obtained using polarized radiation, information can be obtained about the degree of molecular orientation, which leads to a better understanding of the properties of the materials.

The way in which vibrational spectroscopy can give useful information about all these physical details of polymer microstructure is the subject of chapter 6.

1.4 The electromagnetic spectrum – symbols and units

It has already been indicated that the regions of the electro-magnetic spectrum with which we are involved when studying the vibrational spectra of polymers are the visible and infrared regions. We now wish to be rather more specific and to introduce some of the units usually used in discussing spectra. Any simple harmonic wave may be referred to by its wavelength, its frequency or its wavenumber; the symbols used for these quantities in this book are the standard ones, namely, λ, ν and $\tilde{\nu}$, respectively.

The frequencies of the normal vibrational modes of small molecules generally lie in the region 6×10^{12}–1.2×10^{14} Hz or 6–120 THz; overtone and combination vibrations can extend the range up to about 250 THz. There are some very low frequency torsional vibrations of small molecules or polymers and in crystals there are various low frequency modes which are either essentially *lattice modes*, in which whole chains move with respect to each other, or *acoustic modes*, in which crystals vibrate on a macroscopic scale. These extend the low frequency end of the range down to about 0.3 THz. As already explained, the actual frequencies of infrared absorption coincide with the frequencies of the vibrational modes, so that the range of frequencies involved in infrared spectroscopy is roughly 0.3–250 THz, although few single instruments will cover the whole range.

Historically, spectroscopists have tended to specify the wavelength rather than the frequency of the radiation, because this is the quantity most easily determined directly in the visible and infrared regions, by using diffraction gratings of known ruling spacing. Since the velocity of light is 3×10^8 m s^{-1}, the wavelength range involved in infrared spectroscopy is about 1–1000 μm. The region above about 50 μm is sometimes called the *far infrared* and the region below the *near infrared*.

Older infrared instruments are often calibrated to read directly in wavelength. More recently it has become almost universal to work in reciprocal wavelength units, or wavenumber units, which specify the number of waves per metre or, more usually, per cm. This is because in theoretical treatments frequencies are calculated, or in quantum-mechanical calculations energy levels, and the frequencies or differences between energy levels are proportional to the wavenumbers of the corresponding radiation. In terms of wavenumbers the range of the infrared spectrum, including overtones, is roughly 10–8000 cm^{-1} or 10^3–8×10^5 m^{-1}. Most infrared spectroscopists and almost all polymer spectroscopists work in cm^{-1}. Absorption peaks in polymers usually have a width between about 5 and 30 cm^{-1} and for many purposes it is thus sufficient to specify the position of an absorption peak to the nearest whole cm^{-1} and to work with a spectral resolution of this order.

In the Raman spectrum the difference in frequency between the incident light and the Raman-scattered light is equal to the frequency of the vibration. It is therefore customary to specify only this difference in frequency, or more usually the corresponding wavenumber difference which is proportional to it, and these differences span exactly the same ranges as the infrared frequencies or wavenumbers. The wavelength or frequency of the incident light used is not critical unless one is interested in either observing or specifically avoiding the *resonance Raman effect*, which is dealt with briefly in chapter 5. It is very convenient to use a visible laser beam, and since the efficiency of the very weak Raman scattering process is inversely proportional to the fourth power of the wavelength it is usual to use light in the green or blue regions of the spectrum except for compounds which absorb strongly in these regions, for which light in the yellow or red regions of the spectrum may be used. If light of wavelength 500 nm is used the Stokes Raman spectrum of the vibrational fundamentals extends up to about 600 nm.

In the quantum-mechanical theory of the vibrations of molecules the classical frequency of vibration, v, is replaced by an energy difference $\Delta E = hv$ between vibrational energy levels and these energy differences are sometimes quoted in electron volts (eV) by dividing hv by e, the electronic charge. It is also often convenient to consider at what temperature the thermal energy kT would be equal to the energy difference between two levels, or how many kilojoules per mole would give each molecule (or repeat unit in a polymer) the corresponding energy. Fig. 1.9 shows the relationships between these and other units referred to in this section.

1.5   Spectrometers and experimental methods in infrared spectroscopy

1.5.1   *Introduction*

The spectacular increase in the use of infrared spectroscopy, both as a structural diagnostic tool and for analytical purposes, during the past 40 years owes much to instrumental developments. There has been an evolution from delicate, temperamental spectrometers, which gave satisfactory results only in the hands of skilled and patient operators, to the present-day robust, reliable instruments, which provide very satisfactory spectra on an almost routine basis. Nevertheless, an understanding of the design, operation and limitations of such instruments, and of the types of sample required in using them, is necessary if they are to be used to best advantage. In this and the following subsections a brief account of these topics will be given, but the reader is referred to the book edited by Willis *et al.* specified in section 1.8 for a fuller account of experimental techniques in general.

When infrared radiation passes through a sample of any substance the intensity is reduced by the same factor for each equal increase in distance travelled. Thus, if the reduction in intensity is due only to absorption, rather than to scattering and absorption:

$$I = I_0 e^{-ax \ln 10} = I_0 \cdot 10^{-ax} \tag{1.1}$$

Fig. 1.9. Nomogram for the interconversion of units. The dashed line shows, as an example, that radiation of wavelength $7\,\mu m$ has wavenumber $1.43 \times 10^3\,cm^{-1}$ and that its quanta have energy 2.84×10^{-20} J or 0.178 eV. The temperature at which kT is equal to this energy is 2.06×10^3 K and the energy required to give one mole of a substance this amount of energy per molecule is 1.71×10^4 J. Only one decade is shown. For other decades, multiply or divide by the appropriate powers of 10. (Adapted from W. H. J. Childs, *Physical Constants*, Chapman & Hall, 1972.)

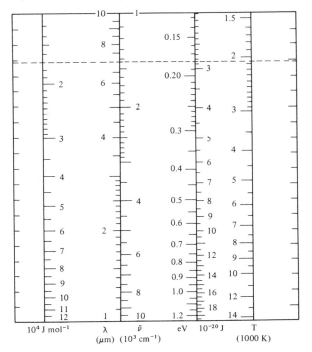

where x is the distance travelled from a reference surface within the substance, I_0 is the intensity at the reference surface and a is called the *absorption coefficient*. The quantity I/I_0 is the *transmittance*, T, of the thickness x and a quantity A given by

$$A = \log_{10}(I_0/I) \qquad (1.2)$$

is called the *absorbance* of the thickness x of the substance. The infrared absorption spectrum of a substance is ideally a plot of the transmittance or the absorbance of a suitable thickness against the wavelength or wavenumber of the infrared radiation. In practice the losses of intensity by reflection at the surfaces of a sample can often be neglected and then, from equations (1.1) and (1.2),

$$A = ax = \log_{10}(I_0/I) \qquad (1.3)$$

with x equal to the thickness of the sample, I_0 equal to the intensity of the radiation incident on the sample surface and I equal to the intensity of the radiation emerging from the other side of the sample. For most of this book reflection losses will not be specifically considered but it should be noted that they must be taken into account for certain kinds of quantitative investigation.

If the sample is very thick the transmittance will approach zero and the absorbance will tend to infinity for all wavelengths, and even for thinner samples there may be wavelength regions where $A > 3$, the practical limit for most spectrometers. In order to avoid too high an absorbance for a solid polymer it is usually necessary to work with samples of thickness between about 30 and 300 μm. Methods of producing such thin film samples, and other forms of sample which provide similar absorbances, will be discussed in subsection 1.5.5. For some polymers it is very difficult to produce samples with suitably low absorbance and for some investigations, such as those on surface coatings, it is not even possible to obtain the appropriate material free from a substrate. For such samples the method of *attenuated total reflection* (ATR) spectroscopy, described in subsection 1.5.6, may be useful because in this method the radiation passes only through a limited thickness, usually only a few μm, of the sample.

Until recently the conventional way to obtain an infrared spectrum of a substance, whether by transmission or ATR, has been to use a spectrometer consisting of (i) a source of radiation with a continuous spectrum over a wide range of infrared wavelengths, (ii) a means of dispersing the radiation into its constituent wavelengths, (iii) an arrangement for allowing the radiation to pass through the sample or be reflected from its surface, (iv) a means of measuring intensities using a detector which responds over the whole range of wavelengths of interest and (v) a method of displaying, and possibly performing calculations on, the spectrum. A more recent type of instrument replaces (i) and (ii) by a tunable diode laser which produces radiation of an extremely small range of frequencies, the centre of which may be varied in a systematic way to scan the spectrum. The principle use of such instruments is in high resolution infrared spectroscopy and they will not be further discussed here.

There are two distinct types of infrared spectrometer in common use for studying polymers: dispersive instruments and Fourier transform instruments. In the dispersive instruments the radiation is physically split up into its constituent wavelengths by a monochromator either before, or more usually after, it passes through the sample and the different wavelengths are processed in sequence. In the Fourier transform instrument radiation of all wavelengths passes through the sample to the detector simultaneously. The detector measures the total transmitted intensity as a function of the displacement of one of the mirrors in a double beam interferometer, usually of the Michelson type, and the separation of the different wavelengths is subsequently done mathematically using a dedicated computer. The principles of operation of each type of instrument are briefly explained in the next two subsections and the sources and detectors used for both types are described in subsection 1.5.4.

All modern spectrometers are capable of producing their output as an absorbance or transmittance spectrum plotted on a chart recorder, but many spectrometers are now equipped with computers which permit other forms of display and various kinds of mathematical processing of the data. Section 1.7 describes briefly some of the techniques employed in processing polymer spectra using either dedicated or non-dedicated computers.

1.5.2 *The dispersive infrared spectrometer*

A block diagram of a dispersive instrument is given in fig. 1.10*a*. The radiation from the source is split into two equal beams which travel along different optical paths: the sample and reference beams. The two beams are subsequently directed alternately to the detector via the monochromator, which scans slowly through the spectral region of interest by rotating the dispersive element. The beam splitting or recombination is usually achieved by a rotating mirror/chopper. Since each beam falls on the detector for half the time, two AC signals 180° out of phase with each other are generated in the detector. If there is absorption by the sample at a particular wavelength the intensity of the

Fig. 1.10. The dispersive infrared spectrophotometer. (*a*) Block diagram of a dispersive infrared spectrophotometer. (*b*) The optical diagram of a double-beam ratio-recording infrared spectrophotometer (reproduced by permission of Perkin Elmer Ltd). The rotating chopper/reflector C_1 generates the sample and reference beams from the source beam in consecutive quarter revolutions and absorbs the beam from the source in the remaining half revolution. This chopping sequence, together with the action of a second chopper C_2, which is synchronized with C_1, makes it possible to correct for radiation emitted, rather than transmitted, by the materials in the sample and reference beams. The instrument has four diffraction gratings, which are used for different regions of the spectrum, and these are brought into use automatically by rotation of the tables A or B or of the mirror pair assembly. The detector, D, is a thermocouple.

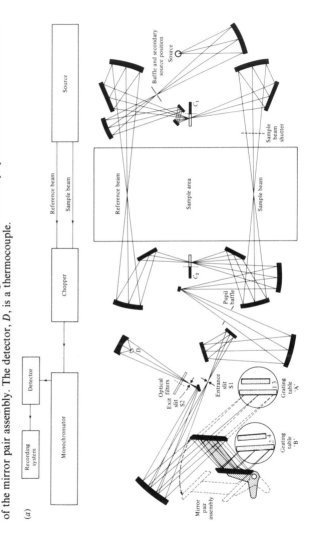

radiation in the sample beam will be lower than that in the reference beam and the corresponding electrical signal from the detector will be proportionately smaller. The transmittance and absorbance of the sample may be deduced from the magnitudes of the two signals, as explained below. This double-beam optical system has two major advantages. It largely eliminates the effects of absorption by atmospheric water vapour and carbon dioxide, which occurs in both beams and can give strong absorptions capable of causing serious interference at several wavelengths in single-beam instruments. Also, because of the rapid alternation between the two beams it is possible to eliminate the effects of thermal drift, which gives only a DC signal.

Two methods are in common use for deriving the transmittance, and hence the absorbance, from the electrical signals. The first is the principle of optical null balance. The total AC electrical signal from the detector, caused by the difference in energy between the two beams, is used to drive a mechanical 'comb' in or out of the reference beam until the two out of phase signals from the detector are equal in magnitude and the total AC signal is therefore zero. The shape of comb is chosen so that its transmittance is related linearly to its displacement and thus the position of the comb at this balance point is a direct measure of the absorption of the sample. Although widely used, this null-balance system has well-known disadvantages. When the sample absorption is high the signal from the detector tends to zero and the servo system driving the balancing comb becomes sluggish and unreliable. Even when the energy is not low the response time constant of the system is relatively long and distortion occurs if sharp absorption peaks are scanned rapidly. Finally, it is not easy to design and manufacture a comb that gives a linear response over the whole operating range, particularly for low transmissions. These problems do not occur with the method of ratio recording, in which a direct ratio of the electrical signals from the reference and sample beams is obtained using electronic circuitry. This requires high stability low noise circuits and the method was until recently found only in high performance instruments. Following the advent of solid state electronics it is now becoming the norm for spectrometers in the medium price range. The advantages for quantitative work are such that there has been a renaissance in this area of the application of infrared spectroscopy to polymers.

In early instruments the dispersive element of the monochromator was often a sodium chloride or potassium bromide prism. Such monochromators have the drawback that the dispersion falls off badly with decreasing wavelength and is nowadays considered to be inadequate in the region where C—H, O—H and N—H stretching modes occur. In recent years the ready availability of competitively priced replica diffraction gratings has led to their almost universal use, providing monochromators for which the dispersion is much less wavelength dependent. The resolution which can be achieved depends on the scanning speed and on the slit width required to provide sufficient energy to give a satisfactory signal-to-noise ratio at any wavelength (see subsection 1.5.4) and usually lies in the range $1-2\,cm^{-1}$, which is appreciably smaller than the widths of most infrared peaks for the liquid and solid phases. The optical diagram of a modern instrument is shown in fig. 1.10b.

1.5.3   *The Fourier transform spectrometer*

The Fourier transform infrared (FTIR) spectrometer, which is usually based on the Michelson interferometer, as shown schematically in fig. 1.11, provides a completely different approach to the recording of infrared spectra. Its advantages and disadvantages may be appreciated from a brief consideration of the principles involved.

Referring to fig. 1.11, consider first the effect of movement of the mirror M_1 if monochromatic radiation of wavelength λ passes through the interferometer. When the path difference between the two beams is an odd integral multiple of $\lambda/2$ destructive interference will occur for radiation transmitted in the direction of the detector and the measured intensity of the recombined beams will be zero. When, however, the path difference is $n\lambda$, where n is an integer, the two beams constructively interfere, to give maximum measured intensity. The intensity measured by the detector varies cosinusoidally with the displacement of the mirror, as shown in fig. 1.12a. If the radiation passing through the interferometer and reaching the detector consists of two wavelengths of equal intensity, the intensity of the combined beam at the detector is a more complex but still symmetrical function of the displacement of the mirror, as shown in fig. 1.12b. When the two wavelengths are of unequal intensity owing, for example, to the presence of material at the sample position which absorbs at one of them, the pattern is further modified (fig. 1.12c).

For monochromatic illumination of wavenumber $\tilde{v}(=1/\lambda)$ and path difference x, the measured intensity, $I(\tilde{v})$, is related to the incident intensity, $I_0(\tilde{v})$, from the source by the equation

$$I(x) = \tfrac{1}{2}I_0(\tilde{v})T(\tilde{v})(1 + \cos 2\pi x\tilde{v}) \tag{1.4}$$

where $T(\tilde{v})$ is the transmittance of the sample at $\tilde{v}$. The path difference x is equal to twice the mirror displacement, measured from the position where the two paths are equal, and the factor $\tfrac{1}{2}$ arises because, on

Fig. 1.11. Schematic diagram of a Fourier transform infrared spectrophotometer based on a Michelson interferometer.

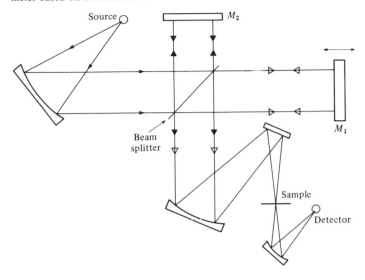

Fig. 1.12. Output from a Michelson interferometer, shown on the right, for different spectral distributions of radiation, shown on the left: (a) monochromatic radiation; (b) two wavelengths of equal intensity; (c) two wavelengths of different intensities; (d) seven wavelengths of different intensities and spacings. In infrared spectroscopy the spectrum is essentially continuous except for 'missing' regions where absorption has taken place, and this leads to a very large peak in $I(x)$ for $x = 0$. To obtain resolution w wavenumbers the output must be scanned up to a path difference $1/w$.

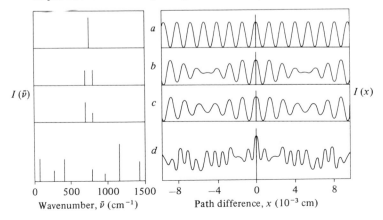

average, half the radiation is reflected back to the source by the beam splitter. When polychromatic radiation passes through the interferometer, as it does when the interferometer is used for recording infrared spectra, the intensity measured for any value of x is the integral of the above expression over the range of wavenumbers involved. The intensity expressed as a function of x is then, apart from a constant added term $\frac{1}{2} \int I_0(\tilde{v})T(\tilde{v})d\tilde{v}$, proportional to the Fourier transform, $\int I_0(\tilde{v})T(\tilde{v}) \cos 2\pi x\tilde{v} \, d\tilde{v}$, of the spectrum $I_0(\tilde{v})T(\tilde{v})$ and this is called the *interferogram*. The detector signal and corresponding mirror displacement are fed to an on-line computer which calculates the inverse Fourier transform to provide a spectrum in the conventional form. The effect of a double-beam instrument is obtained by recording an additional interferogram corresponding to $I_0(\tilde{v})$ after either removing the sample or rotating a mirror or mirrors to provide a new path from the interferometer to the detector which does not pass through the sample but is otherwise equivalent to the original path. In both methods the transmittance or absorbance of the sample is obtained by calculation from the stored spectra.

Fig. 1.12a shows the output of the interferometer as a function of path difference for a monochromatic input. It is equally logical to display the output amplitude or intensity as a function of time and the periodicity may then be defined in terms of a modulation frequency, which is clearly a function of the wavelength of the monochromatic input radiation and the velocity of the mirror movement. For example, with a mirror velocity of 0.2 cm s^{-1} the modulation frequency for an input signal at 400 cm^{-1} is 160 Hz and for 4000 cm^{-1}, the short-wavelength limit of most instruments, it is 1600 Hz. This calls for a detector with a fast response and it was not until the advent of pyroelectric detectors, which have this property, that interferometry in the mid-infrared region became a practical proposition. Detector response speed is less critical in the far infrared region and interferometric spectrometers became well established for this type of work about two decades ago.

In view of the relative complexity of the data processing stage, compared with dispersive spectrometers, it is pertinent to question the advantages of the FTIR approach. They stem from two features of the optical system. As discussed above, radiation of the whole range of wavelengths being examined passes through the instrument at all times, so that the intensity falling on the detector is much greater than for a dispersive instrument with the same optical aperture; this is known as the *multiplex* or *Fellgett advantage*. It is also possible to show that a higher optical aperture can be used for a given resolution than with a dispersive instrument, again increasing the intensity falling on the detector; this is called the *throughput* or *Jacquinot advantage*. The combined effect is to give a very much higher signal-to-noise ratio for a given time taken to record a spectrum. As a consequence, and also because spectral scanning merely involves the linear displacement of a mirror through a distance $1/(2w)$, where w is the spectral resolution in wavenumbers, it is possible to record spectra of similar quality much more rapidly than with dispersive instruments, sometimes in seconds or less rather than minutes. The higher signal-to-noise ratio also permits the study of more highly absorbing samples and so provides a greater range of options for specimen thickness than is provided by dispersive instruments, particularly those of the null-balance type. This is particularly useful in polymer spectroscopy.

Since FTIR instruments are necessarily equipped with a computer they are usually also provided with a wide range of data processing facilities. The value of the information they can provide, given these advantages, often justifies the somewhat increased cost of an FTIR spectrometer over that of a medium performance dispersive instrument, and the technique is coming into increasing use. It should be noted, however, that dispersive instruments are increasingly being equipped with dedicated computer systems and, since ratio recording instruments can deal with absorbances up to about 3, it may be that the rapid scanning ability of Fourier transform instruments, which makes them useful for dynamic studies, will turn out to be their major advantage in the future.

1.5.4 *Sources, detectors and polarizers*

The source of the infrared radiation is a heated rod, usually of a ceramic material. The temperature of operation is about 1200°C and this represents a compromise between total radiative output, the proportion of energy at longer wavelengths and the life of the source. The output approximates to the spectrum of a black body with its maximum in the near infrared region. The relative energy in the interval 400–600 cm^{-1}, close to the long-wavelength limit of some instruments, is some two orders of magnitude lower than at the emission maximum, and many instruments now scan down to 250–200 cm^{-1}. The spectrometer energy throughput is held approximately constant in dispersive instruments by increasing the monochromator slit width in scanning from short to long wavelengths.

The important properties of a detector are its sensitivity and its speed of response. High sensitivity is essential because the energy falling on the detector of a dispersive instrument is in the microwatt range, and the speed of response must be such as to cope with the chopping frequency of a dispersive instrument, typically about 15 Hz, or with the much higher modulation frequencies involved in Fourier transform instruments, discussed in subsection 1.5.3. Until recently, thermal detectors have been dominant because they can respond to all wavelengths and, of these, the thermocouple has been used most frequently, with the sensitive but mechanically not very robust Golay pneumatic cell falling into disfavour. There is now a move towards pyroelectric detectors, typified by doped triglycine sulphate. These are somewhat less sensitive then thermocouples but have a considerably faster response and this makes them particularly suitable for use in FTIR spectrometers.

For a number of investigations on oriented samples it is necessary to study the absorption of polarized infrared radiation and many instruments incorporate a polarizer or polarizers to permit such studies. The most important type of polarizer in use at present is the *wire grid* polarizer, which consists of a set of parallel, equally spaced strips of metal on a substrate transparent to the infrared radiation. If the separation of the strips is less than about a quarter of a wavelength a high degree of polarization is produced in infrared radiation which passes through the grid, and the electric vector of the transmitted beam is normal to the strips. Since spectrometers of any type have different transmission factors for two principle polarization directions at right angles, the polarization direction of the polarizer should be set at 45° to these directions if it, rather than the sample, is to be rotated when making polarization measurements on oriented samples. The polarizer is sometimes placed in the common beam of a double-beam instrument, but for the most accurate work it should be placed next to the sample, with a compensating absorber in the reference beam if necessary.

1.5.5   *The preparation of samples for transmission spectroscopy*

In most instances transmission measurements provide the simplest and most direct approach for the recording of infrared spectra. As explained in subsection 1.5.1, the requirement here is to obtain the optimum sample thickness or optical density. Transmission spectra of low molecular weight organic compounds are obtained by four widely applicable sampling techniques. Liquids may be examined as thin films, obtained by squeezing a drop between a pair of infrared transmitting plates, or as solutions in solvents such as carbon tetrachloride, cyclohexane, carbon disulphide and chloroform, which are reasonably transparent over the range 400–4000 cm^{-1} and have good solvating properties. Solids may be examined in solution or, as powders, by mulling them with liquid paraffin (Nujol) or other suitable liquids or incorporating them in potassium bromide (or other alkali halide) disks; these techniques considerably reduce the scattering. Potassium bromide disks are made by mixing about 1% of the sample, as a finely divided powder, with potassium bromide and cold sintering under pressure to give an optically clear material.

These four sampling methods, using thin films, solutions, mulls or potassium bromide disks, are applicable to polymers but more care, experience and sometimes ingenuity are required to obtain satisfactory results. The two principle problems are the relative intractability of many polymeric materials and the wide intensity range of the various absorptions encountered. The majority of polymer solvents have a number of strong infrared absorption peaks and examination in solution is therefore not usually a practical proposition. Nevertheless, the approach is useful in some circumstances, particularly for quantitative work, and should not be overlooked. Most polymers are studied in the solid state, either as thin films or powders.

If thin films are used, thicknesses differing by an order of magnitude may be required for different polymers or spectral regions. For a saturated hydrocarbon polymer, such as polyethylene, 300 μm is usually satisfactory whereas with materials containing oxygenated groups, e.g. poly(methyl methacrylate), a reduction to 30 μm, or less, is necessary. The latter represents the limit obtainable by hot pressing, which is a convenient and relatively rapid method for preparing films of thermoplastic materials. The polymer is pressed between polished stainless steel plates, using a temperature at which plastic flow occurs readily, in a small hydraulic press which generates a ram pressure of about 2 kN cm^{-2}. For films with thicknesses greater than about 0.1 mm feeler gauge blades may be used as spacers, but for thinner films the sample thickness is adjusted by altering the amount of material used, or the pressing temperature.

Thin films of strongly absorbing material are often best prepared by casting from solution. A good solvent for the polymer is required and it must not be so volatile that poor quality films are produced by too rapid evaporation or of such low volatility that it is difficult to remove, so that solvent peaks occur in the spectrum. Films may be cast on a glass plate and subsequently stripped off, or cast directly on a sodium chloride or potassium bromide plate. Aqueous latices may be cast on silver chloride or KRS5 (thallium bromoiodide) plates. Solvent may be removed from the film by the use of an infrared lamp and vacuum oven or by washing with a suitable liquid with which the solvent is miscible; extraneous peaks may, however, still occur with solvents such as tetrahydrofuran which form relatively involatile oxidation products on storage.

It may also be possible to prepare thin films by the use of a microtome. Rigid and tough polymers are best suited to this approach and some control and ease of cutting may be obtained by warming or cooling the specimen, as appropriate. Swelling with a solvent prior to sectioning may also prove beneficial. In general, however, use of the microtome is best reserved for the limited number of samples for which other methods have failed.

Polymer samples are not usually available in powder form and are often difficult to comminute. Nevertheless, it may be necessary to examine materials which are insoluble, and which will not press, as powders. Low temperature grinding is the best method of producing such powders, particularly for rubbers, and suitable equipment is now available commercially. The other principal source of powder samples is material recovered from solution following separational procedures such as gel permeation chromatography. In this case, the small amount of material available precludes hot pressing into a film. Powders are examined as mulls or potassium bromide disks, but the former technique suffers from the disadvantage that if liquid paraffin is used for mulling it gives peaks which interfere badly with the spectra of hydrocarbon polymers and although quantitative work on powder samples is not particularly easy, pressed disks have considerable advantages over mulls for this type of work.

1.5.6   *Reflection spectroscopy*

Although diffuse reflectance spectra may be recorded, as yet they are little used and attenuated total reflection spectroscopy (ATR) dominates the reflectance approach. The recognition of the optical principles involved dates back to Newton, who noted that if light propagating in one optical medium meets the boundary with a less dense medium at a suitable angle of incidence, it will transverse a small air gap 'above the ten hundred thousandth part of an inch' before being reflected back into the first material without penetrating the second. There have subsequently been many elegant demonstrations of the effect, which may be described in terms of a standing wave normal to the surface in the denser medium and an evanescent, non-propagating field in the rarer medium, whose electric field amplitude decays exponentially with distance from the surface. In practice, the effective depth of penetration is usually a few μm, and depends on the angle of incidence and the difference between the refractive indices of the two media.

Despite the antiquity of the principle involved, the technique has only been in use for about two decades for the recording of infrared spectra. The sample under examination is the medium of lower refractive index and a hybrid type of reflection/absorption spectrum, commonly known as an ATR spectrum, is obtained. Although, as predicted by theory, the intensities of the peaks at shorter wavelengths are reduced relative to those of peaks at longer wavelengths, by comparison with conventional transmission spectra, the ATR spectra of polymers prove very suitable for certain characterizational purposes.

The medium of higher refractive index must, of course, be transparent to infrared radiation and three materials, silver chloride, KRS5 and germanium are in common use. As the difference between the refractive indices of the sample and the other medium increases the ATR spectrum approximates more closely to the absorption spectrum, but the peak intensities decrease. A compromise is necessary and the choice may depend upon the polymer being examined. In view of the small depth of penetration, the greatest practical problem is to achieve good optical contact between the sample and the material of higher refractive index, which is flat and polished. Hence, the sample must be reasonably smooth and planar, and have some flexibility, so that it may be squeezed to promote good contact. Predictably, elastomers often give the best results. Attachments are available for both dispersive and FTIR instruments which permit ATR rather than transmission spectra to be recorded. The absorption intensities are often low and this has led to the widespread use of multireflection units, which are reasonably effective. ATR spectroscopy is an essentially qualitative technique but it has proved of considerable value and its largest area of application in the polymer field is probably for the examination of surface coatings.

1.6 Spectrometers and experimental methods in Raman spectroscopy

1.6.1 *Introduction*

Although infrared and Raman spectra both originate from the vibrational modes of molecules, the frequencies of the modes are observed, in essence, directly and indirectly, respectively. As implied in subsection 1.1.2, the two methods are also to some extent overlapping and to some extent complementary in the information that they provide. Although the Raman effect was discovered in 1928 and the first Raman spectrum of a polymer (polystyrene) was obtained in 1932, Raman spectroscopy has only become widely used since laser sources became available in the early 1960s, and even now it is not as widely used as infrared spectroscopy because of the high capital outlay required, which is almost an order of magnitude greater than that for a medium performance infrared spectrometer. Its simplicity and versatility with respect to sampling are, however, leading to a steadily increased use, particularly for the examination of polymers. The measurements are made in the visible spectrum, which permits the use of glass containers for liquid samples. In addition, water is both transparent to the exciting radiation and weak in Raman scattering, so that aqueous solutions are readily amenable to examination, in marked contrast to infrared spectroscopy, and this is a particular advantage for studies of biopolymers and some synthetic polymers.

The use of a laser as source confers a further advantage because the small diameter of the incident beam, which can be reduced to about 10 μm, makes it ideal for the examination of small samples, such as inclusions in polymers in sheet or rod form. Conversely, large samples may be accommodated, in contrast to infrared transmission spectroscopy, and this provides a potentially valuable approach to the characterization of orientation and strain in compression moulded sheets and extruded pipes. For these reasons Raman spectroscopy is increasingly gaining acceptance as a technique to be used in conjunction with infrared spectroscopy in the study of polymers. It is worth noting here an important difference between infrared and Raman spectroscopy which is relevant to quantitative studies. In infrared spectroscopy the absorbance of a sample is directly proportional to the concentration of the absorbing species. In Raman scattering the total intensity scattered in all directions from unit volume would be proportional to the concentration, but it is usual to make only relative intensity measurements, using an internal standard of some kind, for instance a solvent peak for a solution or a peak due to one component of a mixture.

In principle, all that is needed for obtaining a Raman spectrum is a source of high intensity monochromatic light, which is used to irradiate the sample, a lens system to collect scattered radiation and a spectrometer to disperse it into its component wavelengths and record the intensity as a function of wavelength or wavenumber. A block diagram of such a system is shown in fig. 1.13. The main difficulties arise because the efficiency of Raman scattering is very low, the scattered energy being typically about 10^{-8} to 10^{-6} of that of the incident energy and about 10^{-3} of that of the elastic Rayleigh scattering, and because polymer samples often produce a background scattering (see, e.g., fig. 5.3), which is usually called *fluorescence*, over a wide range of wavelengths. The true origin of this background is often not fluorescence in the generally accepted technical meaning and is often not understood.

The next subsection describes sources, sample illumination, the collection of scattered light and the use of polarized radiation; the subsequent two subsections discuss spectrometers and the preparation of samples, respectively.

1.6.2   *Sources and optical systems*

The source must provide one or more discrete wavelengths at high intensity. Until the advent of the laser, the 435.8 Å and 546.16 nm lines from a high pressure mercury arc were in common use. Even with power dissipations of a few kilowatts, which posed cooling problems, the line intensities were such that photographic recording with exposures of several hours was necessary. The appearance of laser sources in the 1960s transformed Raman spectroscopy, primarily because they provide monochromatic radiation of high intensity but also because they give a wider choice of exciting wavelengths. This is particularly true of the argon ion laser which has 10 lines of useful intensity (hundreds of milliwatts to watts) over the range 454.5 to 528.7 nm. The krypton ion laser provides lines at 520.8, 530.9, 568.2 and 647.1 nm and the third and fourth of these are particularly useful for the examination of coloured samples because absorption, and therefore heating, is lessened. Lasers of these types usually incorporate a tuning device which permits the selection of the wavelength at which the laser will oscillate, but it is necessary to employ either a filter which transmits only a narrow band of wavelengths or a subsidiary monochromator to eliminate the non-lasing lines emitted by the laser discharge, since the Rayleigh or other elastic scattering at these wavelengths can be comparable in intensity to the Raman scattering excited by the lasing line.

It is possible to illuminate the sample directly with the laser beam after it has passed through the filter or monochromator, but a lens or lens

Fig. 1.13. Block diagram of a minimal system for Raman spectroscopy.

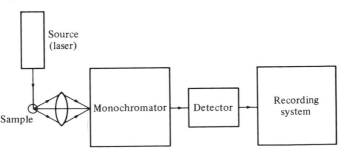

system may be used to focus the beam into a very small sample or region of a sample if required. The laser radiation is almost always linearly polarized and this makes studies using polarized light easy (see chapters 4 and 6). For these studies it is usual to pass the beam from the laser through a suitably oriented half-wave plate so that its polarization direction can be rotated through 90° by rotation of the half-wave plate through 45°. It is not easy to get exact equality of intensity for the two polarization directions and in accurate work it is necessary to allow for this. The scattered radiation is usually collected in a direction at 90° to that of the incident radiation, although 180° scattering is also used, particularly for studies of oriented samples or crystals. The lens system which collects the light images the sample on the entrance slit of the monochromator. Because the scattering is weak, high aperture systems are often used, up to f/1.0, and multipass systems are also sometimes used to increase the observed scattering, but systems involving either high apertures or multiple reflections are best avoided in polarization studies. For such studies an analyser, usually of high quality polaroid sheet, is included either in the collection optics or between them and the entrance slit of the monochromator. Since the fraction of the energy entering the monochromator which is transmitted to the exit slit, for any given wavelength, usually depends strongly on the polarization of the light it is necessary to place a polarization scrambler between the analyser and the monochromator. This may be a quartz wedge or a quarter-wave plate. Perfect scrambling is difficult to achieve and for accurate work it is necessary to allow for the residual differential polarization sensitivity of the system.

1.6.3   *Monochromators and detectors*

The problem posed by the disproportionate intensities of the elastically and inelastically scattered radiation calls for a monochromator which offers high rejection of light at the laser wavelength, particularly when low frequency modes are being examined. An example is the longitudinal acoustic mode of polyethylene (chapter 6), which frequently appears in the interval $10-20$ cm^{-1}. For an exciting wavelength of 488.0 nm the scattered radiation lies between 488.2 and 488.4 nm, on the 'wing' of the much stronger elastic scattering. The high intensity observed in the wings of the exciting line, and the general background scattering observed throughout the spectrum, is due to imperfections in the diffraction gratings and mirrors which scatter the light within the monochromator. The only way that sufficiently high rejection can be obtained so that low frequencies can be studied and the noise at higher frequencies can be reduced to an acceptable level is by passing the scattered radiation through two or more monochromators in series, all set to transmit the same wavelength. In this way the transmission function of the monochromator is raised to the second or higher power so that the function is narrowed and its wings are drastically reduced relative to the peak transmission. Until recently only triple monochromators gave the required rejection for studies below about $50\,cm^{-1}$ on polymers, though double spectrometers were used for studies at higher wavenumbers. The availability of holographic gratings and the redesign of monochromators to reject most of the scattering from the mirrors has led to cheaper, simpler monochromators with the required performance. The optical diagram of a double monochromator instrument is shown in fig. 1.14.

Until recently, monochromators have scanned through the spectrum by rotating the diffraction gratings to bring the various wavelengths sequentially onto the exit slit from which the light is directed to the detector. The third instrumental factor which has transformed the recording of Raman spectra is the advent of new and more sensitive

Fig. 1.14. Semi-schematic optical diagram of a double monochromator Raman system. The lens L_1 focusses the laser beam onto the sample and L_2 images the illuminated sample region onto the entrance slit of the spectrometer. M_1 and M_4 are concave mirrors which serve to collimate the beams falling on the diffraction gratings and to focus the diffracted beams onto the intermediate or exit slits. M_2 and M_3 are plane mirrors. The pre-monochromator may be adjusted to permit the use of different laser wavelengths.

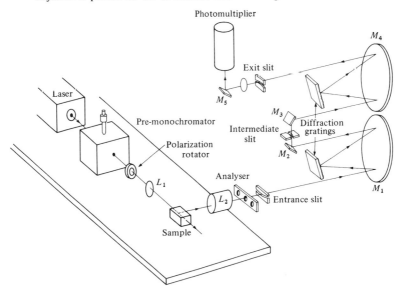

detectors, typified by the gallium arsenide photomultiplier. The overall improvement is such that the average time to record a spectrum over the range 50–2000 cm^{-1} is about 20 minutes. This is still very long but recently it has become possible to record Raman spectra in times as short as those permitted by FTIR spectroscopy. This has been achieved by the use of multidetector arrays. Typically, 1024 diode detectors are located at the focal surface of the monochromator in place of the exit slit and detector of the more conventional instrument. Each of these detectors records the signal within an interval of 1 cm^{-1} simultaneously with the others so that the scanning involved in the conventional instrument is eliminated and it is possible, for instance, to obtain a satisfactory spectrum from a polyethylene pellet in a fraction of a second. In addition to speeding up the recording of Raman spectra for general work and providing the means for accumulating spectra from weak scatterers in a shorter time than was possible before, this new approach makes it possible to undertake dynamic measurements, such as those on polymers being drawn at strain rates as high as 100 % per minute, to complement comparable studies by FTIR spectroscopy. Other novel forms of Raman spectrometer have also been introduced with which microscopic images of samples can be obtained in the Raman scattered light at a particular frequency. This permits the identification of small inclusions and makes possible other specialized studies.

1.6.4   *Sample preparation and mounting*

It has already been pointed out that one of the advantages that Raman spectroscopy has over infrared spectroscopy is that it may be used for a very wide variety of sizes and forms of sample and it is true that a Raman spectrum of some sort can usually be obtained from any sample of polymer in powder, chip or sheet form by placing it in the laser beam in almost any way. If, however, the best possible spectrum is to be obtained some simple precautions and perhaps preparative treatment are necessary.

If the sample is available only in the form of a thin sheet and if 90° collection optics are to be used it is best to place the sample so that its plane is parallel or nearly parallel to the laser beam, since in this way it is possible to image a long length of illuminated sample on the spectrometer slit and there is no danger of reflecting a large fraction of the incident laser beam into the spectrometer, with the possibility of overloading the detector, producing a spectrum with a rather poor signal-to-noise ratio or producing grating ghosts of significant intensity. Opaque samples and surface coatings may be studied in a similar way, but the spectrum of the substrate will of course be strongly observed in the latter type of study. If thick transparent samples are available it is preferable to allow the laser beam to traverse the sample parallel to the surface but just inside it and the points of entry and exit should preferably not be imaged on the spectrometer slit so as again to minimize the amount of scattered light at the exciting wavelength which enters the spectrometer slit. If the surfaces of the sample can be polished, including that at which the Raman scattered light leaves, it is useful to do so. In careful polarization studies such polishing of the sample face is important and it is also necessary to check that neither the exciting beam nor the Raman light passes through a large enough thickness of sample for significant polarization scrambling to take place. Some scrambling always takes place because of inhomogeneities in the sample and this is more serious at optical wavelengths than at infrared wavelengths. If a sample shows the background fluorescence referred to in subsection 1.6.1 it is often possible to diminish the level of this background by 'burning out', which involves leaving the sample in the laser beam for periods of as much as a few hours before recording a spectrum.

Samples in powder form or in the form of thin fibres are rather difficult to examine without scattering large amounts of the exciting radiation into the spectrometer. Fibres may be examined by illuminating them end-on so that they behave like waveguides but it is difficult to get reliable polarization data from them unless they are illuminated normal to their axes and immersed in a liquid of similar refractive index to cut down multiple reflections which tend to scramble the polarization. If a liquid is used its spectrum will also be recorded and will need to be subtracted. It is usually almost impossible to obtain reliable polarization data from powder samples, although immersion of the powder in a liquid of similar refractive index may help and will also cut down the scattered light of the exciting wavelength at the expense of complicating the observed spectrum. Dry powders may be examined by packing them into the open end of a glass capillary tube with a pointed tool which produces a conical depression in the powder. If the capillary is placed parallel to the laser beam, which is focussed into this depression, good illumination of the powder is obtained without too much scatter of the exciting light into the spectrometer, but no polarization data can be obtained.

Solutions or melts can be studied in glass containers, preferably with optical quality faces for the best spectra, particularly for polarization studies, and although the Raman scattering from most glasses is rather weak, care should be taken not to image the points of entry and exit of the beam into the sample on the spectrometer slit.

1.7   Mathematical techniques for processing spectra

The infrared or Raman spectra of polymers usually consist of peaks which vary in width from about 5 to 30 cm^{-1}. The broadening arises from a variety of causes and a study of the shapes and widths of the peaks and of how they depend on temperature and on the way in which the polymer was prepared could in principle give a great deal of information about the microstructure of the polymer, but with one or two exceptions such studies have not been undertaken. The finite width of the peaks is usually regarded as a nuisance because it generally leads to the overlapping of peaks due to different modes of vibration and makes the interpretation of the spectra more difficult. Various ways of dealing with this problem have been devised and a very brief account will be given here of the principles of some of the more widely used mathematical techniques.

A simple method of dealing with the problem of overlapping peaks when determining absorbances is to use the *pseudo-baseline* method, which perhaps barely justifies the description of a mathematical technique. The method is illustrated in fig. 1.15 for a hypothetical infrared spectrum. The pseudo-baseline *AB* is chosen for determining the absorbances at peaks 1 and 2, which are taken as A_1 and A_2, respectively. Similarly, the pseudo-baseline *CD* is chosen for determining the absorbance at peak 3, which is taken as A_3. The method is based on an attempt to allow for the fact that the absorbance at the position of peak 2, for instance, includes not only the absorbance due to peak 2 but contributions from the overlapping 'tails' of peaks 1 and 3 and of even more distant peaks. The choice of pseudo-baselines is clearly rather arbitrary and the method cannot give fully quantitative results,

Fig. 1.15. The use of pseudo-baselines. See text for explanation.

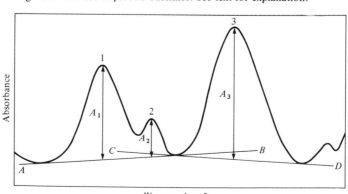

Wavenumber, $\bar{\nu}$

particularly if there is a great deal of peak overlap. Nevertheless, if used in a systematic way on suitable spectra it can sometimes give useful semiquantitative results. A similar technique can be used for estimating intensities from a Raman spectrum. The method has been largely superseded by the more quantitative methods which will now be discussed.

Another simple method of processing spectra is simply to scale one spectrum and subtract it from another. This *subtraction* method may be used, for instance, in finding the spectrum of one component of a two-component mixture by subtracting the known spectrum of the other component from the spectrum of the mixture. The known spectrum is repeatedly scaled and subtracted until a suitable peak, which occurs only in its spectrum, disappears from the resulting difference spectrum. A similar method may be used to deduce the changes introduced into a spectrum by the crystallization of a polymer by subtracting from it the spectrum of the totally amorphous polymer. This simple method assumes that the spectrum of the mixture is simply the weighted sum of the spectra of the individual components and this may not be so. In addition, the method does not help in separating the overlapping peaks due to a single component and other methods have therefore been devised for separating peaks. These methods may be broadly divided into two types: *curve-fitting* methods and *peak-narrowing* methods.

It has been shown empirically that many isolated infrared absorbance and Raman scattering peaks conform fairly closely to a Lorentzian line shape (see fig. 1.16), of the form

$$L(\tilde{v}) = \frac{I_0(w/2)^2}{(\tilde{v} - \tilde{v}_0)^2 + (w/2)^2} \tag{1.5}$$

where $\tilde{v}_0$ is the wavenumber at the line centre, w is the width of the line at half the peak intensity, $L(\tilde{v})$ represents the infrared absorbance or the intensity of Raman scattering at wavenumber $\tilde{v}$ and I_0 is the peak value of $L(\tilde{v})$. If a complex region of overlapping peaks is assumed to be composed of a known number of Lorentzian peaks with unknown centres, half-widths and heights the region can be computer fitted to find the values of these quantities. Such fits can also be made for other assumed line shapes, such as the Gaussian form, $G(\tilde{v}) = I_0 \exp[-4(\tilde{v} - \tilde{v}_0)^2 \ln 2/w^2]$, or combinations of Gaussian and Lorentzian shapes. Since real peaks may deviate somewhat from the form assumed and real spectra are noisy, the fit obtained may not be unique, particularly if there is any doubt about the number of component peaks. For this reason peak-narrowing methods are often used, either alone or in order to help determine the number of peaks and their positions before applying curve fitting.

Two principal peak-narrowing techniques have been used. In the first, derivatives of the spectrum are calculated, usually the second or fourth, $d^2I/d\tilde{v}$ or $d^4I/d\tilde{v}^4$. Fig. 1.16 shows a Lorentzian peak and its second and fourth derivatives, which are clearly narrower than the original peak. The difficulty of this method is that the subsidiary structure produced in the derivative spectra can lead to spurious peaks when the derivatives due to two or more overlapping peaks are added together. The second technique attempts a real narrowing of the line by a process called *deconvolution*.

Convolution is essentially the smearing of one line shape by another, such as happens when a narrow spectral line is studied using a spectrometer with slits of such a width that the instrumental line width is not negligible compared to the width of the spectral line. The true spectral line shape may be considered to be equivalent to a large set of very narrow adjacent lines, each with intensity appropriate to a particular position in the true line. The observed spectrum then consists of the sum of a set of overlapping instrumental functions scaled to different heights, one for each of the fictitious very narrow lines, and the observed spectrum is said to be a convolution of the true line shape with the instrumental line shape. If the instrumental line shape is known it is possible by various mathematical means to deconvolute it from the observed line shape to obtain the true shape of the spectral line.

Fig. 1.16. A Lorentzian function and its derivatives: (*a*) The Lorentzian function $I = L(\tilde{v})$; (*b*) $D_2 = -d^2I/d\tilde{v}^2$; (*c*) $D_4 = d^4I/d\tilde{v}^4$. The derivatives are arbitrarily normalized to the same peak height as the Lorentzian.

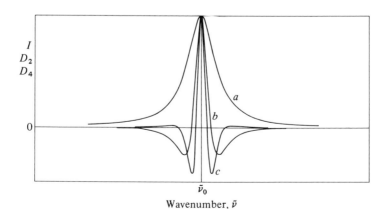

I
D_2
D_4

0

a

b

c

$\tilde{v}_0$

Wavenumber, $\tilde{v}$

It may be shown that if two Lorentzians are convoluted the result is a third Lorentzian with a width equal to the sum of those of the other two. Thus a Lorentzian line can be narrowed by deconvolution with a Lorentzian of suitably chosen width without introducing any spurious structure and if a whole overlapping complex of Lorentzians is deconvoluted with a Lorentzian improved resolution will result. In practice, this method is not as easy to apply as it sounds because the deconvoluting Lorentzian must be narrower than any of the peaks in the line complex or spurious peaks may be introduced. Furthermore, noise in the original spectrum is greatly magnified by the deconvolution process. Nevertheless the method is beginning to come into use, particularly as better signal-to-noise ratios are available with modern instrumentation.

1.8   Further reading

In this chapter we have attempted to give only sufficient general background information on polymers to enable the complete beginner in the study of polymers to follow the subsequent chapters. Such a complete beginner may find the book *Introduction to Polymer Science* by L. R. G. Treloar in the Wykeham Science Series (Wykeham Publications, London, 1970) useful in giving further details at about the same level. Another introductory book at a somewhat more advanced level is *Introduction to Polymers* by R. J. Young, Chapman and Hall, London, 1981 (reprinted with additional material 1983, 1986).

The sections on experimental methods are intended to provide an introduction to these topics which should enable the beginner in vibrational spectroscopy not only to understand how the types of spectra discussed in the book may be obtained, but also to have some idea of the difficulties involved and of the possible sources of error. A very useful book on the instrumental aspects of infrared and Raman spectroscopy, which gives many more details and which has chapters specifically about polymers, is *Laboratory Methods in Vibrational Spectroscopy* 3rd Edn edited by H. A. Willis, J. H. van der Maas and R. G. J. Miller, John Wiley and Sons, 1987.

2   Symmetry and normal modes of vibration

2.1   Interatomic forces and molecular vibrations

Before considering the vibrations of polymers we shall consider those of isolated small molecules. The vibrational motion of such a molecule is independent of its overall translational motion and we shall neglect the effect of overall rotational motion of the molecule on its vibrational motion, since it can be shown that the two types of motion are independent to a good approximation. For polymers, particularly in the solid state, there is no significant rotational or translational motion of the molecules as a whole, but we shall see later, in chapter 3, that there are vibrational modes of crystalline polymers which are related to these types of motion.

When talking about the vibrations of a molecule it is convenient to consider the molecule to be a collection of *atoms* held together by *chemical bonds*. The atoms are imagined to be point masses and the bonds rather like springs, so that the whole assembly can vibrate as the masses move to and fro and the springs stretch and compress. This is the simple classical model that will be adopted in this chapter and it is worth a brief justification because at first sight it appears, when the true natures of the atoms and of the bonds are considered, to be rather far from the truth.

2.1.1   *Interatomic forces*

A more correct picture of a simple diatomic molecule is that it consists of two atomic nuclei, each of which is closely surrounded by a number of *core electrons* and a number of *bonding electrons*. The core electrons interact mainly with the particular nucleus which they surround and with each other, whereas the bonding electrons interact strongly with both nuclei and their surrounding core electrons, as well as with each other. The bonding electrons are thus shared between the two cores (nucleus + core electrons). When two atoms form a molecule the bonding electrons thus interact with two cores rather than with a single one. The consequent lowering of their potential energy is an important contribution to the binding energy of the molecule. The other major contribution is a lowering of the kinetic energy of the bonding electrons. This is due to the quantum-mechanical effect that the kinetic energy of a system of particles can be lower when they move in a larger region of space.

When the atoms come together to form the molecule the motions of the core electrons are perturbed, though considerably less so than those of the bonding electrons, so that the picture given is oversimplified. The molecule is in fact a single assemblage of particles all interacting with each other, and the quantum-mechanical wave-function which describes it is a function of the coordinates of the two nuclei and of all the electrons. When the energy of the system changes it can only change by an amount which corresponds to the difference between two energy levels of the complete system. Fortunately, because of the large difference between the masses of the atomic nuclei and those of the electrons which surround them and bind them together, the motions of the electrons are much faster than those of the nuclei. It is thus possible, when considering the quantum-mechanical description of the electronic states, to regard the nuclei of the molecule as being effectively at rest. The electronic-state wave-functions depend, however, on the instantaneous positions of the nuclei, and so do the corresponding energies of these states. The force between the nuclei which causes them to tend to return to their equilibrium separation when disturbed arises from the dependence of both the electronic energy and the direct coulomb interaction energy of the nuclei on their separation. The motions of the nuclei can thus be discussed in terms of an effective interaction force between them which depends on their separation, and the fact that this force is due in part to the rearrangement of the electrons and their motions can be forgotten, provided that the electrons do not make a discontinuous change of state. Such a discontinuous change of state of the electrons, an *electronic transition*, corresponds to a large change in the energy of the molecule and usually gives rise to the emission or absorption of radiation in the ultraviolet or visible region of the spectrum. Such transitions will not be considered in this book, except when discussing the *resonance Raman effect* in chapter 5.

The motion of the interacting nuclei is most correctly described in terms of quantum mechanics, as has already been indicated. We shall see, however, that the effective mutual potential energy is approximately of the form which gives rise in a classical system to simple harmonic motion. The quantum-mechanical treatment of a simple harmonic oscillator shows that the frequency of the radiation emitted or absorbed when a transition between two energy levels of such a system takes place is exactly the same as the frequency of vibration of a classical system with the same potential energy function. We shall therefore discuss the vibrations of molecules using classical dynamics, and because we think of atoms joined together by bonds in the simplified picture we shall generally refer to the positions or motions of atoms rather than nuclei.

2.1.2   *Normal modes of vibration*

We are concerned only with the degrees of freedom of a molecule that are associated with the translational motions of its constituent atoms, and then only with those motions that do not correspond to translation or rotation of the whole molecule but to *genuine vibrational modes*. A diatomic molecule has only one vibrational degree of freedom, associated with changes in the separation of the atoms. More generally, a molecule consisting of N atoms has $3N - 6$ vibrational degrees of freedom. The six degrees of freedom subtracted from the total of $3N$ translational degrees of freedom of N atoms are the degrees of freedom of the undistorted molecule, i.e. the translations and rotations of the whole molecule parallel to or about three mutually perpendicular axes. For a linear molecule there is no rotational degree of freedom around the axis of the molecule associated with translations of the atoms; such a degree of freedom can only be associated with the electronic states or with internal states of the nuclei, and therefore the number of vibrational degrees of freedom for such a molecule is $3N - 5$.

Let the masses of the atoms in a diatomic molecule be m_1 and m_2 and let their separation be r. If it is assumed that the potential energy $V(r)$ of the molecule is given by

$$V(r) = V_0 + \tfrac{1}{2}k(r - r_0)^2 \tag{2.1}$$

where V_0 and k are constants and r_0 is the equilibrium separation of the atoms, it is easy to show that when the molecule is disturbed from equilibrium it will vibrate with simple harmonic motion of frequency v, where

$$v = \frac{1}{2\pi} \sqrt{\frac{k(m_1 + m_2)}{m_1 m_2}} \tag{2.2}$$

The quantity k is called a *force constant* and the restoring force acting on either atom when $r \neq r_0$ is $k(r - r_0)$. Equation (2.1) does not, however, represent the exact form of the potential energy function for a real molecule. $V(r)$ cannot contain a term linear in $(r - r_0)$, since r_0 would not then be the equilibrium separation, but it can, and generally will, contain terms in powers of $(r - r_0)$ higher than two. Neglect of these terms is called the *harmonic approximation*.

In a similar way one can make the harmonic approximation for a polyatomic molecule by writing the potential energy as

$$V = V_0 + \tfrac{1}{2} \sum k_{ai,bj}(x_{ai} - x^0_{ai})(x_{bj} - x^0_{bj}) \qquad (2.3)$$

where x_{ai} for $i = 1, 2$ or 3 represents the x, y or z coordinate, respectively, of the position of atom a with respect to axes $Oxyz$, x^0_{ai} is the equilibrium value of x_{ai}, the $k_{ai,bj}$ are force constants and the summation is over all a, b, i and j. The force on atom a in the direction Ox_i is obtained by differentiating V with respect to $-x_{ai}$ and since it can be shown that $k_{ai,bj} = k_{bj,ai}$ for all a, i, b and j, the equation of motion for x_{ai} is

$$m_a\left(\frac{\mathrm{d}^2 x_{ai}}{\mathrm{d}t^2}\right) = -\sum_{bj} k_{ai,bj}(x_{bj} - x^0_{bj}) \qquad (2.4)$$

where m_a is the mass of atom a.

If equation (2.3) holds, so that equation (2.4) holds for all a and i, it can be shown that after any disturbance of the atoms from their equilibrium positions the molecule will always vibrate in such a way that the displacements of the individual atoms can be expressed as a weighted sum of their displacements in a particular set of $3N - 6$ (or $3N - 5$) simple harmonic vibrations called the *normal modes* of vibration of the molecule. It is possible to choose these normal modes of vibration so that in any one of them all the nuclei move in straight (or almost straight) lines with simple harmonic motions of the same frequency and all the atoms pass through their equilibrium positions simultaneously. Fig. 2.1 shows the normal modes of CO_2.

Note that two normal modes (or more, in general) may, because of the symmetry of the molecule, be constrained to have the same frequency and such modes are said to be *degenerate*. The frequencies of the normal modes depend on the values of the force constants, on the masses of the atoms and on the geometry of the molecule, and so do the directions of vibration of the different atoms in each normal mode, the ratios of their amplitudes of vibration and the phase relationships between their motions. A vector in a $3N$-dimensional space with components proportional to the simultaneous values of $x_{ai} - x^0_{ai}$ for all values of a and i is called the *eigenvector* of the normal mode.

In a non-degenerate mode all the atoms necessarily vibrate in such a way that they pass through their equilibrium positions simultaneously. This follows simply from the fact that in any system where no energy is dissipated, the equations of motion, such as equation (2.4), are independent of the sign of t, where t represents time, and any valid solution of the equations must therefore give rise to a second valid solution if $-t$ is written instead of t. Thus if the coordinate x_{ai} is given by

$$(x_{ai} - x_{ai}^0) = X_{ai} \cos(2\pi v t + \delta_{ai}) \tag{2.5}$$

in a mode of frequency v, where δ_{ai} is the phase with respect to the motion of a chosen atomic coordinate, there must be another mode of the same frequency in which x_{ai} is given by

$$(x_{ai} - x_{ai}^0) = X_{ai} \cos(-2\pi v t + \delta_{ai}) = X_{ai} \cos(2\pi v t - \delta_{ai}) \tag{2.6}$$

Thus the mode is only non-degenerate if $\delta_{ai} = 0$ or π for all a and i.

For degenerate modes it is possible to choose eigenvectors in which the atoms do not move in straight lines but it is also always possible to choose eigenvectors in which they do. As an illustration of this, consider the degenerate modes (c) and (d) of CO_2 illustrated in fig. 2.1. If the displacements of corresponding atoms in these two modes are added together a linear, circular or elliptical motion results for each atom, depending on the phase relationship between the vibrations of the original two modes. This motion persists with the original frequency and is thus a perfectly good choice of one eigenvector for the degenerate pair of modes. In describing any eigenvector in which the motions of

Fig. 2.1. Normal modes of vibration of CO_2: (a) symmetric stretch; (b) antisymmetric stretch; (c) and (d) bend. The arrows show the relative directions and magnitudes of the simultaneous displacements of the atoms in one phase of the motion, but the actual displacements are exaggerated. In (d) the $+$ and $-$ signs indicate displacements up and down, respectively, with respect to the plane of the diagram. The displacements in (d) are obtained by rotating the molecule as shown in (c) by $90°$ around the line through the three atoms. In the opposite phase of the vibration all the displacement directions are reversed for any mode. The modes illustrated in (c) and (d) form a degenerate pair and the total number of modes is $3N - 5 = 4$, for $N = 3$.

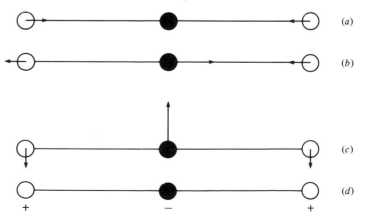

some or all atoms are not linear the relative phases of the displacements $x_{ai} - x_{ai}^0$ must be specified, in addition to their relative magnitudes.

We shall consider in chapter 4 what information is required if the frequencies and eigenvectors of a molecule are to be calculated in detail and how the calculation may be done. We shall also discuss briefly some effects due to the departure of the true potential energy function from the form assumed in the harmonic approximation. In the remainder of the present chapter we shall consider how the symmetry of the molecule influences the possible forms, or eigenvectors, of the normal modes of vibration and how the normal modes may be grouped into various *symmetry species*.

2.2 Symmetry and small molecules

2.2.1 *Point symmetry operations and symmetry elements*

We first consider what is meant by saying that a molecule has some form of *symmetry*. Consider the molecule of carbon tetrachloride, CCl_4. The chlorine atoms, atoms 2, 3, 4 and 5 in fig. 2.2, are arranged tetrahedrally around the central carbon atom, 1. If we rotate the molecule through 120° around the line joining atoms 1 and 2 so that atom 4 replaces atom 5, atom 5 replaces atom 3 and atom 3 replaces atom 4 we shall not be able to distinguish the new state from the old because all chlorine atoms are identical and the labels 3, 4 and 5 are arbitrary. (We assume that all the chlorine atoms are of the same isotope.) Such an operation, which (*a*) displaces every point in the molecule according to the same rule, (*b*) leaves the separation between any two points unchanged and (*c*) leaves the molecule unchanged when the labels on the atoms are removed, is called a *symmetry operation*.

Two symmetry operations are considered to be identical if they lead to the same final position for any general initial point within the molecule. Thus there are only two different symmetry operations of rotation around the line joining atoms 1 and 2. They may be regarded either as rotations through plus and minus 120°, or as rotations through 120° and 240° in the same direction, or in many other equivalent ways. The direction 1–2 is called a *three-fold rotation axis* and the operation of rotation through 120° about such an axis (or any equivalent rotation) is given the symbol C_3. More generally the operation of rotation through $2\pi/n$ is given the symbol C_n.

There are several other types of symmetry operation, some of which may also be illustrated by referring to fig. 2.2. If the molecule is imagined to be reflected in a mirror coincident with the plane through atoms 1, 2 and 4, the atoms 3 and 5 are interchanged and atoms 1, 2 and 4 remain in their original places. There are six such *mirror planes* in the molecule, any one of which interchanges two out of the four chlorine atoms. The operation of *reflection* in a mirror plane is given the symbol σ. A slightly less obvious symmetry operation is that of rotation through the angle $2\pi/n$ followed by reflection in the plane normal to the axis of rotation. Such an operation of *rotation–reflection* is given the symbol S_n and is called an *improper rotation*. The dashed lines in fig. 2.2 coincide with S_4 axes. The S_4 operation corresponding to the axis labelled a leaves atom 1 undisplaced and interchanges the other atoms as follows; $2 \rightarrow 4$, $3 \rightarrow 5$, $4 \rightarrow 3$, $5 \rightarrow 2$ or $2 \rightarrow 5$, $3 \rightarrow 4$, $4 \rightarrow 2$, $5 \rightarrow 3$, depending on the direction of rotation. There are two other S_4 axes and thus a total of six operations of type S_4. Since the same operation S_4 performed twice is equivalent to a simple rotation through π, any S_4 axis is also a C_2 axis, and there are thus a total of three C_2 symmetry operations for CCl_4. (Note that a two-fold rotation axis corresponds to only one C_2 operation, since rotation by π in either direction around the axis leads to the same interchanges of all points and is thus the same symmetry operation.)

Fig. 2.2. The CCl_4 molecule. Atom 1 is a carbon atom and atoms 2, 3, 4 and 5 are chlorine atoms. The dashed lines coincide with S_4 axes.

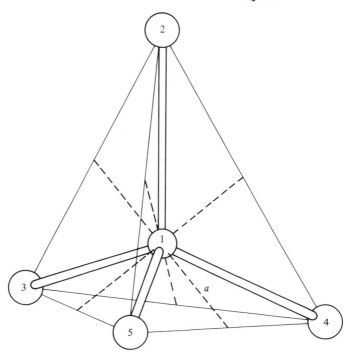

In general the rotation or rotation–reflection axis of highest n is called the *principal axis* and is often imagined to be *vertical*. (For CCl_4, as for many other molecules, there is no unique principle axis.) The operation of reflection in a plane which contains the principal axis is given the symbol σ_v (v for *vertical*) and the operation of reflection in a plane perpendicular to the principal axis is given the symbol σ_h (h for *horizontal*). A two-fold axis perpendicular to the principal axis is called a *dihedral axis* and the corresponding operation is given the symbol C_2'. A mirror plane which contains the principal axis and bisects the angle between two C_2' axes or two σ_v planes is called a *dihedral plane* or a *diagonal plane* and the corresponding symmetry operation is given the symbol σ_d.

Two further symmetry operations must be mentioned. The first is the *identity* operation, E, the effect of which is to leave every point of the molecule in its original position. E is thus a symmetry operation for all molecules and the reason for its introduction will be explained later. For many molecules there is a point called the *centre of inversion*. For every atom in such a molecule there is another atom of the same kind situated on the other side of the centre of inversion at the same distance from it and the two atoms and the centre of inversion lie on a straight line. The operation of *inversion*, given the symbol i, consists of interchanging all pairs of points in the molecule which are situated in this way with respect to each other.

Sometimes, for a particular type of molecule, two different symmetry operations can produce the same interchange of atoms. Consider, for example, a diatomic molecule consisting of two identical atoms. Reflection in the mirror plane which bisects normally the line joining the two atoms interchanges the two atoms and so does rotation through $180°$ about any one of the infinite number of two-fold axes which pass through the centre of the molecule and lie in this mirror plane. The two operations are, nevertheless, different because any point which does not lie either on the line joining the atoms or on the two-fold axis is moved to two different points by the two operations. Similarly, rotations through $180°$ about any two of the infinite number of two-fold axes are different operations.

The operations so far defined are called *point symmetry operations* because there is one point of the molecule which is undisplaced by any of the operations. This point corresponds to the position of atom 1 for CCl_4, though it does not necessarily correspond to the position of an atom in other molecules. All the possible point symmetry operations are listed in table 2.1. If any symmetry operation P is applied m times, the resulting operation, written P^m, is also a symmetry operation which may

Table 2.1. *The point symmetry operations*

Symmetry element		Corresponding operations	Definition of symmetry operation	Order of element	Relationship between operations
Symbol	Name				
—	—	E	The identity	1	$E^k = E$
			Proper operations		
C_n	n-fold rotation axis (Assumed to lie in the z direction, except in cubic and icosahedral point groups)	$C_n^1, C_n^2, \ldots, C_n^k, \ldots, C_n^n$	Rotation through an angle $2\pi k/n$ about the axis (when $k = 1$, the superscript is usually omitted, i.e. $C_n^1 = C_n$)	n	$C_n^n = E$ $C_n^{n+k} = C_n^k$ $(C_n^k)^{-1} = C_n^{-k} = C_n^{n-k}$ Factoring: $C_n^k = C_{n/k}^1$ if n/k is integral (operations are always written in their simplest form)
Special cases					
C_2', C_2''	2-fold rotation axis perpendicular to the principal C_n axis	C_2', C_2''	Rotation through π about the axis	2	
$C_n(x)$ (etc.)	n-fold rotation axis in the x direction (etc.)				
C_∞	Infinite-fold rotation axis in the z direction (present in all linear molecules)	C_∞^ϕ	Rotation through an arbitrary angle ϕ about the axis	∞	$(C_\infty^\phi)^{-1} = C_\infty^{-\phi}$

Improper operations

Symbol		Operations	Meaning	Number	Relations
S_n	n-fold rotation–reflection axis (or 'alternating axis'. Assumed to lie in the z direction except in the cubic and icosahedral point groups)	$S_n^1, S_n^2, \ldots, S_n^k, \ldots, S_n^n$ $S_n: S_n^2, \ldots, S_n^k, \ldots, S_n^n$ $\ldots, S_n^{n+k}, \ldots, S_n^{2n}$	Rotation about the axis through $2\pi k/n$, combined with reflection k times in a plane normal to the axis (i.e. if k is odd, net result is one reflection; if k even, no reflection and the operation becomes proper)	n (n even) $2n$ (n odd)	$S_n^n = E$ (n even) $S_n^{2n} = E$ (n odd) $S_n^k = C_n^k$ (k even) $S_n^{n+k} = C_n^k$ (n and k odd)

Special cases

Symbol		Operations	Meaning	Number	Relations
$i\,(=S_2)$	Inversion centre	$i\,(=S_2^1)$	Inversion of all points through the origin of coordinates	2	$i^2 = E$ $i = S_{2n}^n$ (n odd)
$\sigma\,(=S_1)$	Mirror plane	$\sigma\,(=S_1^1)$	Reflection of all points in a plane	2	$\sigma^2 = E$
σ_v	A mirror plane containing the principal axis (v = 'vertical')	σ_v	Reflection in the plane specified		
σ_h	A mirror plane normal to the principal axis (h = 'horizontal')	σ_h	Reflection in the plane specified		$\sigma_h = S_n^n$ (n odd)
σ_d	A dihedral mirror plane containing the principal axis and bisecting the angles between C_2' axes	σ_d	Reflection in the plane specified		—

Reproduced by permission from J. A. Salthouse and M. J. Ware, *Point Group Character Tables and Related Data*, Cambridge University Press, 1972.

or may not be identical to a symmetry operation with a simpler description. Note also that all point symmetry operations may be considered to be proper or improper rotations, since $\sigma = S_1$ and $i = S_2$.

It is useful to distinguish between an operation such as C_n and the n-fold axis which a molecule possesses if C_n is a symmetry operation. The n-fold axis is called a *symmetry element* of the molecule, and similarly a mirror plane or a centre of inversion is a symmetry element. The *order* of a symmetry element is the number of times that the corresponding symmetry operation must be applied in order to return every point in the molecule to its original position. Thus the order of an n-fold rotation axis is n and the order of a mirror plane or centre of inversion is two. A knowledge of the symmetry elements of a molecule enables a classification of the various normal modes of vibration of the molecule to be made. Before discussing the formal aspects of this classification, which requires the introduction of some ideas of group theory, we show in a simple way how the symmetry of a molecule restricts the forms of its normal modes.

Consider a molecule undergoing one of its normal modes of vibration with frequency v. Imagine now that we apply one of the symmetry operations of the (static) molecule to the vibrating molecule at an instant during its vibration when the atoms are displaced from their equilibrium positions. The potential energy V of the molecule remains unchanged by the symmetry operation whatever the particular instant chosen in the vibration, because the distances between all pairs of points in the molecule are unchanged by the operation. The force on each atom must therefore be the same in magnitude as before the symmetry operation and be directed towards the new equilibrium position of the atom, as it was towards the old equilibrium position before the symmetry operation was performed. The molecule thus vibrates with the same frequency after the symmetry operation is performed as before. Let us now assume that the normal mode under consideration is non-degenerate, so that the new vibration must be indistinguishable from the old, except possibly by its phase. For a non-degenerate mode all the atoms move in straight lines and pass through their equilibrium positions simultaneously so that they also attain their extreme displacements simultaneously. These extreme displacements for the two vibrations would be interchanged by the symmetry operation, and the phases of the two vibrations can thus differ only by 0 or π; in other words, the atomic displacement vectors for a non-degenerate mode must be such that the symmetry operation multiplies each of them by $+1$ or -1, i.e. the mode must be symmetric or antisymmetric with respect to every symmetry operator.

We see immediately that the symmetry of the molecule restricts the nature of the possible normal modes. It may also be shown that the vibrational displacements of a pair of degenerate modes for a molecule with a unique n-fold axis $(n > 2)$ are either both symmetric or both antisymmetric with respect to the operations i, σ_h and a C_2 axis parallel to the n-fold axis if these are present and that it is always possible to choose the eigenvectors of such a pair so that one is symmetric and one antisymmetric with respect to any C_2' axis or σ_v plane present. The importance of these restrictions on the nature of the normal modes is that many different molecules may possess the same symmetry elements, so that the forms of their normal modes will have something in common: they will belong to the same possible types of *symmetry species*, as explained in more detail in subsection 2.3.3.

It is worth noting that the assertion that the potential energy of the molecule depends only on the separations of its constituent particles or, in our approximation, the separation of its nuclei, is a statement about the nature of space. It implies that space is isotropic and homogeneous. If the potential energy depends on something other than the separations of the nuclei we usually say that that there is an external *field* present, rather than saying that space is not homogeneous and isotropic. We shall deal in the next chapter with the effects of the electric fields in crystals on the vibrational modes of the molecules.

2.2.2 *Point groups*

A *group*, in the mathematical sense, consists of a set of *members*, or *elements*, which obey certain rules. A group which contains n members is said to be a group of *order n*. Let us call the members of a group $A, B, C, \dots$ etc. Then these members must obey the following rules:

(1)   Any two members of the group may be combined according to some *rule of combination*. Such a combination of two typical members of the group, P and Q, is usually called *multiplication* and the combination is written PQ or QP. The nature of the members and the rule of combination may be such that PQ is not always equal to QP. If PQ is equal to QP, P and Q are said to *commute*.

(2)   Any combination such as PQ (or QP) is also a member of the group.

(3)   The *associative law* of multiplication holds, i.e. $(PQ)R = P(QR)$, where P, Q and R are any members of the group.

Table 2.2. *The point groups, excluding the cubic and icosahedral groups*

Type		Symbol	Generating operations	Symmetry elements	Order	Number of classes and irreducible representations	Comments
Non-axial		C_1	E	None	1	1	No symmetry
		C_s	σ	σ_h	2	2	$C_s = C_{1h} = C_{1v} = S_1$
		C_i	i	i	2	2	$C_i = S_2$
Axial: Cyclic	R	C_n	C_n	C_n	n	n	$n = 2, 3, 4, \ldots$
		S_{2n}	S_{2n}	C_n, S_{2n}	$2n$	$2n$	$S_6 = C_{3i}$
		C_{nh}	C_n, σ_h	C_n, σ_h, S_n	$2n$	$2n$	
		C_{nv}	C_n, σ_v	$C_n, n\sigma_v$	$2n$	$2n$	
Dihedral	R	D_n	C_n, C_2'	C_n, nC_2'	$2n$	$\begin{cases}\frac{1}{2}(n+3) \\ \frac{1}{2}(n+6)\end{cases}$ if n odd / if n even	$D_2 = V$
		D_{nh}	C_n, C_2', σ_h	$C_n, nC_2', S_n, \sigma_h, n\sigma_v$	$4n$	$\begin{cases}n+3 \\ n+6\end{cases}$ if n odd / if n even	$D_{2h} = V_h$
		D_{nd}	C_n, C_2', σ_d	$C_n, nC_2', n\sigma_d, S_{2n}$	$4n$	$n+3$	$D_{2d} = V_d$ $D_{nd} = S_{2nv}$
Linear		$C_{\infty v}$	C_ϕ, σ_v	$C_\infty, \infty\sigma_v$	∞	∞	
		$D_{\infty h}$	C_ϕ, C_2', σ_h	$C_\infty, \infty\sigma_v, S_\infty, \infty C_2'$	∞	∞	

Reproduced by permission from J. A. Salthouse and M. J. Ware, *Point Group Character Tables and Related Data*, Cambridge University Press, 1972.

(4)   One member of the group is the *identity*, usually given the
        symbol E. E has the property that $PE = EP = P$, where P is any
        member of the group.
(5)   For each member of the group, such as P, there is an *inverse*,
        written P^{-1}, which is also a member of the group and which has
        the property $PP^{-1} = P^{-1}P = E$. (Note that P may sometimes
        be its own inverse, so that $P^2 = E$.)

It is important to realize that groups may be formed with members of
many different types and suitable combination rules.

If we define multiplication, i.e. the rule of combination, as successive
operation, then the point symmetry operations for a molecule form a
group, called its *point group*. It is obvious from the definition of the
symmetry operations of a molecule that they form a set which satisfies all
the above rules, provided that the identity operation is included but not
otherwise. We need, however, to define multiplication slightly more
precisely: if P and Q are symmetry operations then by PQ we shall mean
the operation Q followed by the operation P.

A list of point groups is given in table 2.2, excluding the cubic and
icosahedral groups which have no direct relevance for polymers and
which will not be considered in any detail in this book. All the
operations of any particular group can be generated as powers or
products of at most three *generating operations* which may usually be
chosen in many different ways. A possible set of generating operations
for each group is shown in table 2.2.

Any group has a *multiplication table* which expresses the results of
multiplying all possible ordered pairs of members of the group. To
illustrate this we shall consider the point group C_{3v} to which the
ammonia molecule, fig. 2.3, belongs. (This group is simpler than that to
which carbon tetrachloride belongs, which is one of the cubic groups,
T_d.)

There is a C_3 axis which passes through atom 1 and is normal to the
plane of atoms 2, 3 and 4. We define the direction of the operation C_3 so
that it takes atom 2 to atom 3 and atom 3 to atom 4. The operation C_3^2
then takes atom 2 to atom 4 and atom 3 to atom 2. The other symmetry
elements are the σ_v mirror planes $\sigma(2)$, $\sigma(3)$ and $\sigma(4)$, which contain the
C_3 axis and the positions originally occupied by atoms 2, 3 and 4,
respectively. (The symmetry elements are to be considered fixed in
space.) The multiplication table is shown as table 2.3. Each operation
shown is that obtained by multiplying the operation at the left of the row
containing it by the operation at the top of the column containing it so
that the latter operation is the right-hand factor in the multiplication, i.e.

the operation performed first. Note that some pairs of operations, such as $\sigma(2)$ and $\sigma(3)$, do not commute.

Symmetry operations are only one of an infinite number of different kinds of elements that may be defined in such a way as to permit the existence of groups of such elements. Any two groups are said to be *isomorphous* if their elements can be placed in one to one correspondence so that the multiplication tables for the two groups have the same form, in the following sense. Replace each element in the multiplication table for the first group by the corresponding element in the second group; if the multiplication table is now correct for the second group then the two multiplication tables have the same form and the two groups are isomorphous. It is not necessary that the two groups should consist of elements of the same type. The consideration of isomorphous groups and *homomorphous* groups, with a similar many to one correspondence, is of great importance for the following sections.

Table 2.3. *The multiplication table for the group* C_{3v}

C_{3v}	E	C_3	C_3^2	$\sigma(2)$	$\sigma(3)$	$\sigma(4)$
E	E	C_3	C_3^2	$\sigma(2)$	$\sigma(3)$	$\sigma(4)$
C_3	C_3	C_3^2	E	$\sigma(4)$	$\sigma(2)$	$\sigma(3)$
C_3^2	C_3^2	E	C_3	$\sigma(3)$	$\sigma(4)$	$\sigma(2)$
$\sigma(2)$	$\sigma(2)$	$\sigma(3)$	$\sigma(4)$	E	C_3	C_3^2
$\sigma(3)$	$\sigma(3)$	$\sigma(4)$	$\sigma(2)$	C_3^2	E	C_3
$\sigma(4)$	$\sigma(4)$	$\sigma(2)$	$\sigma(3)$	C_3	C_3^2	E

Fig. 2.3. The NH_3 molecule. Atom 1 is a nitrogen atom and atoms 2, 3 and 4 are hydrogen atoms.

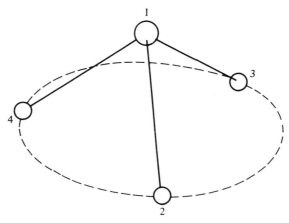

2.3. **Group representations**

2.3.1 *Representation of groups by matrices*

Any set of elements which combine under a specified rule to form a group which is isomorphous or homomorphous with a particular group is said to form a *representation* of that group. The most useful representations of point groups for our purposes are those in which the elements are *square matrices* and the rule of combination is *matrix multiplication*. A set of square matrices, each of $n \times n$ elements, which form a representation of a group is called an *n-dimensional representation*. It is possible to find one or more representations of any point group for all possible values of n. In particular, if $n = 1$ we can find the trivial representation of any point group in which each matrix is simply the number 1. This is called the *identity representation*. For most, but not all, point groups there exists at least one other one-dimensional representation, in which some members of the group are represented by $+1$ and others by -1. We have already said that the atomic displacements of a non-degenerate normal mode of vibration are multiplied by ± 1 when any symmetry operation is applied to the molecule and it should now be clear that this set of multipliers, one for each symmetry operation, forms a one-dimensional representation of the point group of the molecule. Let us now consider a two-fold degenerate mode.

Suppose that we could directly observe a molecule undergoing its normal modes of vibration and that we found two different modes which had the same frequency v. This means that there are at least two possible eigenvectors corresponding to the frequency v. Let us now imagine applying each of the symmetry operations of the group to either of these eigenvectors or, more precisely, to the molecule with its atoms displaced as specified by the eigenvector. It was shown in subsection 2.2.1 that this can only lead to a vibration of the same frequency. If we represent the set of displacements for the two eigenvectors by the symbols S and T there are three possibilities: (*a*) each member of the set of displacements S transforms into $\pm$ itself under each of the symmetry operations; (*b*) for each of the symmetry operations the set S transforms into a linear combination of S and T; (*c*) the set S transforms in some other way. A set, W, of displacements of the atoms is a linear combination of the sets S and T if, for every atom a in the molecule, the displacement vector w_a of the atom in the set W is given by

$$w_a = \alpha s_a + \beta t_a \tag{2.7}$$

where s_a and t_a are the displacement vectors of the atom in the sets S and T, respectively, and α and β are the same for every atom.

If the eigenvector transforms as in (*a*) it is clear that the two modes are not related to each other by symmetry and for some change of force constants, masses or geometry which do not change the symmetry of the molecule the corresponding frequencies will be different. Such modes are said to be *accidentally degenerate*. If the eigenvector transforms as in (*c*) there is at least one more mode of frequency *v* for which the eigenvector is not a linear combination of the eigenvectors of the first two degenerate modes considered. If these are not accidentally degenerate the mode is thus at least three-fold degenerate. For eigenvectors which transform as in (*b*) we have true *symmetry-determined two-fold degeneracy*, and this is the only type of symmetry-determined degeneracy which can occur for the point groups listed in table 2.2.

All the sets of displacements generated by operations of the group on either of the pair of original eigenvectors of a symmetry-determined two-fold degenerate mode are given by

$$W = \alpha S + \beta T \tag{2.8}$$

where W, S and T each represent $3N$-dimensional vectors whose components are the N sets of components of the vectors w_a, s_a and t_a. Because the equations of motion, equation (2.4), are linear in the displacements, any W given by any values of α and β must also be a possible eigenvector with the same frequency. The possible eigenvectors are therefore said to 'span' a two-dimensional sub-space of the $3N$-dimensional space spanned by any general set of displacement vectors w_a. The $3N$-dimensional vector specifies what will be called the *generalized position* of the molecule, or its 'position'. Note that S and T cannot be parallel or the mode is not degenerate because S and T would then simply correspond to vibrations of different amplitude and possibly different phase.

In a two-dimensonal space we can choose any two non-parallel vectors and refer all other vectors in that space to this *basis set*. It is convenient, but not necessary, to choose *basis vectors* which are orthogonal and of unit length. If S_0 and T_0 are two such orthonormal basis vectors then

$$S_0 \cdot T_0 = 0, \qquad S_0 \cdot S_0 = 1, \qquad T_0 \cdot T_0 = 1 \tag{2.9}$$

It is now possible to represent any vector in the two-dimensional subspace by a two-component column matrix **W** and to represent the effect of any operation P of the symmetry group on this vector as the premultiplication of the matrix **W** by a 2×2 matrix **P**. From now on italic boldface characters will represent vectors and roman boldface characters will represent matrices.

If $\tilde{W}$ represents the transpose of W, the square of the length of the vector W is $\tilde{W}W$ and is equal to the sum of the squares of the lengths of the real three-dimensional vectors w_a. Since each of these lengths remains constant for any operation of the group, so does the length of W. The operation P and the corresponding matrix P thus produce a proper or improper rotation in the two-dimensional subspace and the set of matrices P for all the operations of the group forms a two-dimensional representation of the group. Thus a doubly degenerate mode corresponds to a two-dimensional representation of the molecular point group. In general, the eigenvectors of an n-fold degenerate mode span an n-dimensional subspace and correspond to an n-dimensional representation of the group. Let us consider a specific two-dimensional example.

The square-pyramidal molecule has two degenerate pairs of modes and possible orthogonal eigenvectors of one pair are shown in fig. 2.4. (Atoms 2, 3, 4 and 5 are assumed to be identical.) The symmetry operations of this molecule, which belongs to the point group C_{4v}, are E, $\pm C_4(z)$, $C_2(z)$, $\sigma_v(xz)$, $\sigma_v(yz)$, $\sigma_d(1)$ and $\sigma_d(2)$, where $\sigma_d(1)$ contains the z axis and the bisector of angle yOx and $\sigma_d(2)$ contains the z axis and is perpendicular to $\sigma_d(1)$. Choosing the two eigenvectors shown in figs. 2.4a and 2.4b as the basis for the two-dimensional representation we can write them, by definition, as $S_0 = \begin{pmatrix} 1 \\ 0 \end{pmatrix}$ and $T_0 = \begin{pmatrix} 0 \\ 1 \end{pmatrix}$, respectively. Note that strictly we cannot equate a vector and a matrix, and equations such as the last two will only be used when specifying the components of a vector with respect to a given basis or set of axes. We shall similarly often refer to column matrices, or the corresponding row matrices, as vectors when we mean that the matrix is the representation of the vector with respect to a given basis or set of axes.

Fig. 2.4. A pair of degenerate modes of the square-pyramidal molecules.

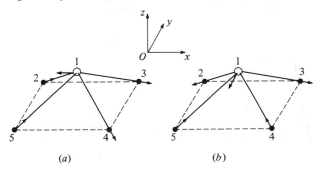

(a)   (b)

If we consider the operations of the group in turn we find that the 2×2 matrices corresponding to them are as follows:

$$E \to \begin{pmatrix} 1 & 0 \\ 0 & 1 \end{pmatrix} \qquad \pm C_4(z) \to \pm \begin{pmatrix} 0 & -1 \\ 1 & 0 \end{pmatrix} \qquad C_2(z) \to \begin{pmatrix} -1 & 0 \\ 0 & -1 \end{pmatrix}$$

$$\sigma_v(xz) \to \begin{pmatrix} 1 & 0 \\ 0 & -1 \end{pmatrix} \qquad \sigma_v(yz) \to \begin{pmatrix} -1 & 0 \\ 0 & 1 \end{pmatrix} \qquad \sigma_d(1) \to \begin{pmatrix} 0 & 1 \\ 1 & 0 \end{pmatrix}$$

$$\sigma_d(2) \to \begin{pmatrix} 0 & -1 \\ -1 & 0 \end{pmatrix}$$

$$(2.10)$$

For example, the statement that the operation C_4 converts the eigenvector S_0 into the eigenvector T_0 corresponds to the equation

$$\begin{pmatrix} 0 & -1 \\ 1 & 0 \end{pmatrix}\begin{pmatrix} 1 \\ 0 \end{pmatrix} = \begin{pmatrix} 0 \\ 1 \end{pmatrix} \tag{2.11}$$

and the statement that the operation $\sigma_v(xz)$ followed by the operation C_4 is equivalent to the operation $\sigma_d(1)$ corresponds to the equations

$$C_4\sigma_v(xz) = \sigma_d(1) \tag{2.12a}$$

and

$$\begin{pmatrix} 0 & -1 \\ 1 & 0 \end{pmatrix}\begin{pmatrix} 1 & 0 \\ 0 & -1 \end{pmatrix} = \begin{pmatrix} 0 & 1 \\ 1 & 0 \end{pmatrix} \tag{2.12b}$$

The reader is encouraged first to construct the multiplication table for the group C_{4v} using the definitions of the operations and then to check that the matrices (2.10) do indeed form a representation.

2.3.2   *Reducible, irreducible and equivalent representations*

A particularly simple choice of basis was made for the representation we have just given, but any two different linear combinations of the two eigenvectors shown in figs. 2.4*a* and 2.4*b* could be chosen as a basis, say $X = \begin{pmatrix} s_1 \\ t_1 \end{pmatrix}$ and $Y = \begin{pmatrix} s_2 \\ t_2 \end{pmatrix}$ with respect to the old basis, provided that they do not represent parallel vectors. By definition, these two combinations would then become vectors $\begin{pmatrix} 1 \\ 0 \end{pmatrix}$ and $\begin{pmatrix} 0 \\ 1 \end{pmatrix}$ with respect to the new basis. Each operation P, which is represented by a 2×2 matrix $\mathbf{P}$ in the old representation, will be represented by a new 2×2 matrix $\mathbf{P}'$ in the new representation. It is important to consider whether any basis could be chosen which would produce a set of representation matrices every one of which had the diagonal form

$$\mathbf{P}' = \begin{pmatrix} p_1 & 0 \\ 0 & p_2 \end{pmatrix} \qquad (2.13)$$

If this were possible, then for any two operations P and Q it would be true that

$$\mathbf{P}'\mathbf{Q}' = \begin{pmatrix} p_1 & 0 \\ 0 & p_2 \end{pmatrix}\begin{pmatrix} q_1 & 0 \\ 0 & q_2 \end{pmatrix} = \begin{pmatrix} p_1 q_1 & 0 \\ 0 & p_2 q_2 \end{pmatrix} \qquad (2.14)$$

and thus each of the sets of numbers p_1, q_1 etc., and p_2, q_2 etc., would separately form a one-dimensional representation of the group, i.e. the two-dimensional representation would be *reducible* to one-dimensional representations. It would also follow, however, that neither of the new eigenvectors $\begin{pmatrix} 1 \\ 0 \end{pmatrix}$ and $\begin{pmatrix} 0 \\ 1 \end{pmatrix}$ would be converted into a linear combination of itself and the other by any of the symmetry operations of the group. This is contrary to the assumption that such a pair corresponds to a doubly degenerate mode. Thus the two-dimensional representation corresponding to a doubly degenerate mode is *irreducible*; it cannot be reduced to two one-dimensional representations by choosing a new basis. More generally, the n-dimensional representation corresponding to an n-fold degenerate mode is irreducible in the sense that it cannot be reduced to two or more representations of dimension less than n.

Whether a representation of a group is reducible or not, the new matrices $\mathbf{P}'$ form an alternative representation to the matrices $\mathbf{P}$. Let the original eigenvectors S_0 and T_0 take the form $S_0' = \begin{pmatrix} S_{0x} \\ S_{0y} \end{pmatrix}$ and $T_0' = \begin{pmatrix} T_{0x} \\ T_{0y} \end{pmatrix}$ with respect to the new basis. Then any vector $\mathbf{w} = \begin{pmatrix} s \\ t \end{pmatrix}$ with respect to the old basis takes the form $\mathbf{w}' = \begin{pmatrix} x \\ y \end{pmatrix}$ with respect to the new basis, where

$$\begin{pmatrix} x \\ y \end{pmatrix} = \mathbf{A}\begin{pmatrix} s \\ t \end{pmatrix} \quad \text{and} \quad \begin{pmatrix} s \\ t \end{pmatrix} = \mathbf{A}^{-1}\begin{pmatrix} x \\ y \end{pmatrix} \quad \text{with } \mathbf{A} = \begin{pmatrix} S_{0x} & T_{0x} \\ S_{0y} & T_{0y} \end{pmatrix} \qquad (2.15)$$

Let $\begin{pmatrix} s_i \\ t_i \end{pmatrix}$ represent a particular initial vector with respect to the old basis and let $\begin{pmatrix} s_f \\ t_f \end{pmatrix}$ represent (still with respect to the old basis) the final vector into which it is transformed by multiplying by the matrix $\mathbf{P}$ which represents the operation P. Then

$$\mathbf{P}\begin{pmatrix} s_i \\ t_i \end{pmatrix} = \begin{pmatrix} s_f \\ t_f \end{pmatrix} \tag{2.16}$$

so that

$$\mathbf{PA}^{-1}\begin{pmatrix} x_i \\ y_i \end{pmatrix} = \begin{pmatrix} s_f \\ t_f \end{pmatrix} \tag{2.17}$$

and

$$\mathbf{APA}^{-1}\begin{pmatrix} x_i \\ y_i \end{pmatrix} = \mathbf{A}\begin{pmatrix} s_f \\ t_f \end{pmatrix} = \begin{pmatrix} x_f \\ y_f \end{pmatrix} \tag{2.18}$$

In obtaining equation (2.17) and the right-hand part of equation (2.18), equation (2.15) has been applied, first with the vectors $\begin{pmatrix} x \\ y \end{pmatrix}$ and $\begin{pmatrix} s \\ t \end{pmatrix}$ representing the initial vector with respect to the two bases and secondly with these vectors representing the final vector with respect to the two bases. A comparison of equations (2.16) and (2.18) shows that the matrix $\mathbf{P}'$ in the new representation corresponding to the matrix $\mathbf{P}$ in the old representation is equal to $\mathbf{APA}^{-1}$. Such a transformation of each of a set of matrices $\mathbf{P}$ using the same matrix $\mathbf{A}$ is called *similarity transformation*. The transformed matrices $\mathbf{P}'$ clearly obey the same multiplication table as $\mathbf{P}$ and therefore form a representation of the group. Representations related by a similarity transformation are said to be *equivalent representations*.

So far we have tacitly assumed that all the representation matrices are real. This is not a necessary assumption and the use of complex matrices leads to a reduction of the two-dimensional representations for the doubly degenerate modes of groups C_n for $n \geqslant 3$, S_n for n even and $\geqslant 4$, and C_{nh} for $n \geqslant 3$. These are all the groups with three-fold or higher-order axes which have no mirror planes containing the principal axis and no dihedral axes. Consider a molecule which has an n-fold axis and belongs to one of these symmetry groups. There must be $n - 1$ equivalent atoms into which any given starting atom is transformed by $n - 1$ successive operations of C_n or S_n. We choose a set of reference axes for the displacement of each of the n atoms with its origin at the equilibrium position of the atom and we assume that the sets are oriented so that they may be brought into coincidence by the operation C_n or S_n. If, for a given mode of vibration, any component x_r of the displacement of the atom r of these n equivalent atoms is given by $x_r = x_{r0} \cos(\omega t + \delta_r)$ then, since the n atoms are equivalent, it must be possible to find solutions of the equations of motion such that x_{r0} and $\delta_{r+1} - \delta_r$ are the same for all values of r. The latter requirement can be met if and only if $\delta_r = \delta_0 \pm 2m\pi r/n$, where m is an integer. Hence the component of the displacement of atom r is given by

$$x_r = x_{r0} \cos(\omega t + \delta_0 \pm 2m\pi r/n) \tag{2.19}$$

Equally valid solutions are

$$x_r = x_{r0} \exp[i(\omega t + \delta_0 \pm 2m\pi r/n)] \tag{2.20}$$

We see immediately that for these solutions the operation C_n or S_n simply multiplies each component of the displacement of each of the atoms, and hence the eigenvector, by $\exp(\pm 2\pi i m/n)$. For the groups under consideration the other operations of the group are multiples of C_n or S_n, the identity and possibly σ_h. For groups C_{nh}, which contain σ_h, the displacements must be such that the operation σ_h multiplies the eigenvector by -1 or $+1$. Otherwise the eigenvectors given by equation (2.19) would not be the only two corresponding to the frequency ω. Thus the two-dimensional representation for the degenerate mode is reduced to two one-dimensional representations, the elements of which are $\exp(+2\pi i m/n)$, $\exp(-2\pi i m/n)$ and, for groups C_{nh}, $+1$ or -1. The two corresponding vibrational modes may be considered to be waves travelling round the principal axis in opposite directions. Excluding phase differences of 0 and π, which correspond to non-degenerate modes, the number of values of m which give distinguishable phase differences between consecutive equivalent atoms is $\frac{1}{2}(n-1)$ for odd n and $\frac{1}{2}n - 1$ for even n. For the groups C_n and S_n these are the numbers of different pairs of one-dimensional representations which correspond to degenerate pairs of modes. For the groups C_{nh} there are twice as many pairs, since for any phase relationship there can be modes symmetric or antisymmetric with respect to σ_h. The groups C_n have important applications to polymers.

It is easily seen that the operations σ_v, σ_d or C_2' would interchange the solutions with $+2\pi i m r/n$ and $-2\pi i m r/n$ and thus the two-dimensional representations are not reducible for groups such as D_n which contain any of these operations. The pairs of eigenvectors which correspond to waves travelling in opposite directions round the principal axis do, however, form legitimate sets of basis vectors for the corresponding two-dimensional representations.

2.3.3 *Symmetry species and character tables*

There is a very important property associated with each of the matrices which constitute a group representation. This property is called

Table 2.4. *The character table for the point group* C_{4v}

C_{4v}	E	$2C_4$	C_2	$2\sigma_v$	$2\sigma_d$		
A_1	$+1$	$+1$	$+1$	$+1$	$+1$	z	$x^2 + y^2, z^2$
A_2	$+1$	$+1$	$+1$	-1	-1	R_z	
B_1	$+1$	-1	$+1$	$+1$	-1		$x^2 - y^2$
B_2	$+1$	-1	$+1$	-1	$+1$		xy
E	$+2$	0	-2	0	0	$(x, y)(R_x, R_y)$	(xz, yz)

its *character*, χ, and is simply the sum of the diagonal elements of the matrix. If we consider the two-dimensional representation explicitly given in subsection 2.3.1 for the group C_{4v}, the characters of the matrices corresponding to the various operations are:

E	$C_4(z)$	$C_2(z)$	$\sigma_v(xz)$	$\sigma_v(yz)$	$\sigma_d(1)$	$\sigma_d(2)$
2	0	-2	0	0	0	0

For a one-dimensional representation of any point group the characters are simply $+1$, -1 or $\exp(\pm 2\pi im/n)$. It is easy to show, using the properties of matrices, that if the basis of any representation of dimension greater than one is changed to give an equivalent representation the character of the matrix which represents any operator remains unchanged, even though the elements of the matrix do generally change. It is obvious that any two non-equivalent one-dimensional representations have different sets of characters and it may be shown that this is true for representations of higher dimension. Non-equivalent irreducible representations may therefore be distinguished by the sets of characters which correspond to the various operations of the group. Two normal modes, or two sets of degenerate normal modes, which correspond to the same irreducible representation of the group are said to belong to the same *symmetry species*. The possible symmetry species for all point groups have been evaluated and the corresponding characters are listed in *character tables*. As an example, the character table for the group C_{4v} is shown in table 2.4.

Several features of such tables need explanation. The symbols in the extreme left-hand column label the different symmetry species according to rules given in the next subsection. It is not necessary to list the character for a species for every individual operation of a group, because the operations fall into *classes* and it is easy to show that all operations in the same class have the same character for any representation. Imagine for a moment that the symmetry element which corresponds to a particular operation P were fixed with respect to the molecule rather than being fixed in space. If this symmetry element could then be superimposed upon that of another operation Q, which is of exactly the same type as P, by performing some other operation R of the group then P and Q are said to be in the same class. As an example, the operations of reflection in the three σ_v mirror planes of the NH_3 molecule, discussed in subsection 2.2.2, are in the same class because any two of the mirror planes would be interchanged by reflection in the third if they were considered fixed with respect to the molecule.

Character tables usually have several columns at the right-hand side which may be used to deduce the spectral activity of normal modes of vibration belonging to each symmetry species; two such columns are shown in table 2.4. The explanation of the use of these columns will be deferred until subsection 2.4.3 and chapter 3 and we now explain the notation for the various symmetry species.

2.3.4 *Notation for symmetry species*

The symbols A and B are used to denote one-dimensional representations, or non-degenerate species, and the symbol E is used to indicate two-fold degenerate species. (This standard notation is unfortunate, since it clashes with the equally standard notation for the identity operation.) Species of higher degeneracy do not occur for polymers. The particular A species which has character 1 for all operations, corresponding to the identity representation, is called the *totally symmetric species*.

Axial point groups can contain only one symmetry axis (C_n or S_n) of order higher than two, which is called the *principal axis*. If an axial group does not contain the operation i or σ_h, a non-degenerate symmetry species which is symmetric ($\chi = 1$) with respect to the operation C_n or S_n is given the symbol A and a species which is antisymmetric ($\chi = -1$) is given the symbol B. If there is more than one A or B species, subscripts 1 and 2 are used to distinguish between species which are symmetric and antisymmetric, respectively, with respect to a C_2' axis normal to the principal axis or with respect to a σ_v plane if there is no C_2' axis. (This does not uniquely differentiate between different B species for the point groups C_{nv}, D_n and D_{nh} and specific conventions must then be adopted.) Integer subscripts, m, are used on the symbol E to distinguish different irreducible two-dimensional representations or symmetry species where necessary. The value of m is $(n/2\pi) \cos^{-1}(\frac{1}{2}\chi_R)$, where χ_R is the character of the representation for the rotation C_n or S_n (or the sum of the two complex characters for reducible two-dimensional representations). In the modes of symmetry species E_m the phases of the motions of corresponding atoms (atoms related by C_n or S_n) differ by $2m\pi/n$ (see subsection 2.3.2). Species with degeneracy higher than two do not occur for axial point groups.

If an axial point group, which we shall call G, contains the operation $i\,(=S_2)$, its operations may be divided into two sets. One set consists of the identity and the real rotations of G; these together form a point group which we shall call G'. The other set consists of the operations of G' multiplied by i. G is said to be the *direct product* of the groups G' and C_i, since its elements are all the possible products of an element of G and an element of C_i (C_i contains only E and i). Such a group G has two symmetry species for each species of the corresponding group G', and each pair is given the same label as the corresponding species of G' together with an additional subscript g (*gerade*) or u (*ungerade*) according to whether the species is symmetric or antisymmetric, respectively, with respect to i.

Similarly, if an axial point group G contains σ_h but not i it is the direct product of the groups G' and C_s, where G' again contains the identity and the real rotations of the group and C_s contains E and σ_h. The group G again has two symmetry species for each species of the corresponding group G', and each pair is given the same label as the corresponding species of G' but single and double dashes are used to distinguish between species which are symmetric and antisymmetric, respectively, with respect to σ_h.

The groups D_2 and D_{2h}, which are important for polymers, each contain three C_2 axes which are in different classes. The group D_2 has only one A species but three different B species. If axes $Oxyz$ are chosen in the molecule, B_1, B_2 and B_3 species modes are defined to be symmetric with respect to the $C_2(z)$, $C_2(y)$ and $C_2(x)$ axes, respectively. The group D_{2h} is the direct product of D_2 and C_s and the rules for subscripts u and g are as for the simpler axial groups. Unfortunately there is not always agreement about the orientation of the set of axes $Oxyz$ for a particular molecule and great care must therefore be taken when specifying symmetry species for molecules belonging to these groups.

The groups C_i and C_s each have only two symmetry species, which are labelled A_g or A_u and A' and A'', respectively, according to their symmetry with respect to i or σ_h. The cubic and icosahedral groups will not be considered, since they are not directly relevant to a study of polymers.

2.4   The number of normal modes of each symmetry species

We turn now to a consideration of how the character table for the point group of a molecule may be used to calculate the number of normal modes of vibration of the molecule which belong to each symmetry species. The importance of being able to do this is that the symmetry species of a normal mode determines its spectral activity, i.e. whether it is infrared or Raman active and, if it is, how the observed intensity depends on the polarization of the radiation involved. We consider first a $3N$-dimensional representation of the point group, where N is, as usual, the total number of atoms in the molecule.

2.4.1 *A reducible 3N-dimensional representation of the point group*

Imagine a set of coordinate axes $Ox_a y_a z_a$ to be defined with its origin at the equilibrium position of atom a and let such a set be defined for each of the atoms of the molecule. It is convenient for the moment to think of these sets of axes as all being rectangular and all having the same orientation in space, although this is not necessary for the argument and it will later be convenient to define the axes in a different way. Imagine now a general set of displacements of the atoms from their equilibrium positions. This set of displacements can be expressed in terms of a set of $3N$ vectors, each of which is a unit vector parallel to one of the $3N$ axes of the N sets of axes of type Ox_a, Oy_a or Oz_a. This set of vectors forms a basis for a representation of the group in terms of $3N \times 3N$ matrices. Equivalently, as already indicated in subsection 2.3.1, we can say that the 'position' of the molecule can be expressed in terms of $3N$ coordinates. Consider the effect of any one of the symmetry operations, say P, on the set of displacements. In performing the operation we leave the sets of axes where they are in space and move all points in the molecule according to the rule appropriate to P. Since individual atoms of the same type are indistinguishable, the effect of the operation P is to produce a different 'position' of the molecule, i.e. a different set of $3N$ coordinates, each of which depends linearly on the original set of $3N$ coordinates. The operation P can thus be represented by a $3N \times 3N$ matrix which, when it multiplies the $3N \times 1$ column matrix of original coordinates, produces the $3N \times 1$ column matrix of new coordinates. Let us consider a simple example.

Fig. 2.5a illustrates a bent triatomic molecule, in which the two outermost atoms, a and c, are identical (e.g. the water molecule). The three sets of axes are shown located at the equilibrium positions of the atoms. The Oz axes have been chosen parallel to the C_2 axis and the Ox axes are directed vertically out of the page. Fig. 2.5b illustrates a general 'position', or set of displacements from equilibrium, and fig. 2.5c illustrates the new 'position' after the operation C_2 is performed. If the column vector **I** which represents the initial position of the molecule before the operation C_2 is performed is given by

$$\tilde{\mathbf{I}} = (x_a \quad y_a \quad z_a \quad x_b \quad y_b \quad z_b \quad x_c \quad y_c \quad z_c) \qquad (2.21)$$

it is easy to see that the new column vector $\mathbf{F}$ which represents the final position is given by

$$\tilde{\mathbf{F}} = (-x_c \quad -y_c \quad z_c \quad -x_b \quad -y_b \quad z_b \quad -x_a \quad -y_a \quad z_a) \ (2.22)$$

Since, by definition,

$$\mathbf{F} = \mathbf{C}_2\mathbf{I} \qquad (2.23)$$

where $\mathbf{C}_2$ is the 9×9 matrix which represents the operation C_2, we have, by inspection,

$$
\mathbf{C}_2 =
\begin{pmatrix}
\cdot & \cdot & \cdot & \cdot & \cdot & \cdot & -1 & \cdot & \cdot \\
\cdot & \cdot & \cdot & \cdot & \cdot & \cdot & \cdot & -1 & \cdot \\
\cdot & \cdot & \cdot & \cdot & \cdot & \cdot & \cdot & \cdot & 1 \\
\cdot & \cdot & \cdot & -1 & \cdot & \cdot & \cdot & \cdot & \cdot \\
\cdot & \cdot & \cdot & \cdot & -1 & \cdot & \cdot & \cdot & \cdot \\
\cdot & \cdot & \cdot & \cdot & \cdot & 1 & \cdot & \cdot & \cdot \\
-1 & \cdot & \cdot & \cdot & \cdot & \cdot & \cdot & \cdot & \cdot \\
\cdot & -1 & \cdot & \cdot & \cdot & \cdot & \cdot & \cdot & \cdot \\
\cdot & \cdot & 1 & \cdot & \cdot & \cdot & \cdot & \cdot & \cdot
\end{pmatrix}
\qquad (2.24)
$$

where the dots represent zero elements. In a similar way we could write out the 9×9 matrices for the other symmetry operations and these matrices would then form a nine-dimensional representation of the group.

Fig. 2.5. The bent triatomic molecule. Atoms a and c are assumed to be identical. The undistorted molecule is shown at (a) with the sets of reference axes used for the displacements of the atoms. A general set of displacements of the atoms is shown at (b) and the result of performing the operation C_2 is shown at (c). The C_2 axis coincides with Oz_b. The dashed lines in (b) and (c) show the undistorted molecule.

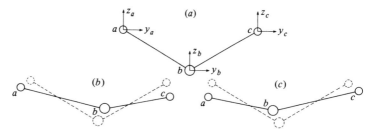

2.4.2   *Reduction of the 3N-dimensional representation by the use of normal coordinates*

The 'position' of any molecule can be expanded as a linear combination of the $3N$ normal-mode eigenvectors, where the three translations and three rotations are regarded as normal modes of zero frequency. This is because the $3N$ eigenvectors are linearly independent. It may be shown (see subsection 4.4.3) that if distances are measured along the axes Ox_a etc. in units of length proportional to $m_a^{-\frac{1}{2}}$, where m_a is the mass of atom a, the eigenvectors are also orthogonal in the sense that the scalar product of the $3N$-dimensional vectors for any two normal modes is zero, provided that the vectors for degenerate modes are suitably chosen.

Suppose that the molecule is vibrating in the normal mode k and that at a particular instant during the vibration its 'position' can be described by the $3N$-dimensional displacement vector $^k\mathbf{d}$, where

$$^k\mathbf{d} = (^kx_a \quad ^ky_a \quad ^kz_a \quad ^kx_b \quad \ldots \quad ^kz_{3N}) \tag{2.25}$$

where the ^kx_a etc. are assumed to be the mass-weighted displacements just defined. We can define a unit vector

$$\mathbf{D}_k = {}^k\mathbf{d}/l_k \tag{2.26a}$$

where

$$l_k^2 = \sum_a (^kx_a^2 + {}^ky_a^2 + {}^kz_a^2) \tag{2.26b}$$

If we do this for each normal mode we can express the vector $\mathbf{I}$ corresponding to any 'position' of the molecule as

$$\tilde{\mathbf{I}} - U_1\mathbf{D}_1 + U_2\mathbf{D}_2 + \ldots + U_{3N}\mathbf{D}_{3N} - \sum_{k=1}^{3N} U_k\mathbf{D}_k \tag{2.27}$$

i.e. as a weighted sum of orthogonal unit vectors whose 'directions' in the generalized mass-weighted position space are those of the eigenvectors of the normal modes. Thus by taking these unit vectors as new coordinates axes in the 'position' space the vector $\mathbf{I}$ can be expressed as

$$\tilde{\mathbf{I}}' = (U_1 \ U_2 \ldots U_i \ldots U_3^N) \tag{2.28}$$

where U_1, U_2 etc. are called the *normal coordinates* of the 'position'. The 'position' vector is now written $\mathbf{I}'$ in matrix form because although it is the same vector as that represented by $\mathbf{I}$ its coordinates are expressed in terms of a new set of axes which form the basis for a new representation of the group.

Assume now that the normal coordinates U_k are listed in equation (2.28) so that all those that correspond to non-degenerate modes occur first, all those that correspond to doubly degenerate modes occur next and so on. Since the eigenvector of a non-degenerate mode is simply multiplied by ± 1 by any operation and the eigenvector of a degenerate multiplet is in general converted into a linear superposition of all the eigenvectors of the multiplet, we see that in the new representation the matrix $\mathbf{P}'$ for the general operation P must take the diagonalized form

$$\mathbf{P}' = \begin{pmatrix} p_1 & & & & & & & & \\ & p_2 & & & & & & & \\ & & \cdot & & & & & & \\ & & & \mathbf{p}_i & & & & & \\ & & & & \mathbf{p}_{i+1} & & & & \\ & & & & & \cdot & & & \\ & & & & & & \mathbf{p}_j & & \\ & & & & & & & \cdot & \end{pmatrix} \tag{2.29}$$

where p_1, p_2 etc. represent $+1$ or -1; $\mathbf{p}_i$, $\mathbf{p}_{i+1}$ etc. represent 2×2 matrices; $\mathbf{p}_j$, $\mathbf{p}_{j+1}$ etc. represent 3×3 matrices and so on. Dashes represent zero elements. The original $3N$-dimensional representation has now been expressed in terms of irreducible representations.

2.4.3   *The number of normal modes of each symmetry species*

Although we have formally reduced the original $3N$-dimensional representation, it is not necessary to do so in practice when determining the number of normal modes which belong to each symmetry species. The important point is that the character of the matrix $\mathbf{P}'$ is the same as that of the original matrix $\mathbf{P}$, since changing the basis of a representation does not change the character of any of the matrices. It is clear that the character $\chi_{P'}$ of $\mathbf{P}'$ and hence the character χ_P of $\mathbf{P}$ is simply given by

$$\chi_{P'} = \sum_i n_i \chi_{Pi} = \chi_P \tag{2.30}$$

where n_i is the number of normal modes which belong to the ith irreducible representation, or symmetry species, and χ_{Pi} is the character of $\mathbf{P}$ in the ith irreducible representation, which is given in the character tables.

It may be shown by group theory that there are always exactly the same number of distinct irreducible representations of a group as there are classes in the group, so that if equation (2.30) is written for all the different classes of operator there will be the same number of equations as unknown quantities n_i. It may further be shown that these equations are independent, so that they can be solved for the values of n_i. The values of n_i can be obtained more directly and usually more simply by the use of the equation

$$n_i = \frac{1}{g} \sum_P h_P \chi_P \chi_{Pi} \tag{2.31}$$

where g is the order of the group and h_P is the number of operators in the class to which P belongs. This equation may be derived from equation (2.30) by the use of certain formal results of group theory. Before giving a simple example of the use of equation (2.31) we consider how the characters χ_P may be written down directly, even for complicated molecules, without deducing the complete forms of the corresponding matrices.

Examination of the form of $\mathbf{C}_2$, given in equation (2.24), and its derivation, shows that it may be divided into 3×3 submatrices and that only those submatrices that lie on the leading diagonal can possibly contribute to the character. Each of the sub-matrices on the leading diagonal describes how the final coordinates of an atom in a specified position depend on the initial coordinates of the atom in that position before the operation C_2 was performed. The submatrix contains non-zero elements, and is thus a potential contributor to the total character of the matrix which represents C_2, only if the atom originally in the specified position (assuming for the moment that individual atoms are identifiable) would be left in that position if the operation C_2 were applied to the molecule in its equilibrium 'position'. Similar remarks apply to the matrix which represents any operation for any molecule.

An atom can remain in its original position after the performance of an operation only if (a) the operation is C_n and the atom lies on the corresponding n-fold axis, or (b) the operation is σ and the atom lies on the corresponding mirror plane, or (c) the operation is S_n and the atom lies on the corresponding axis and reflection plane. Let us consider each of these in turn, remembering that we may choose the axes of coordinates situated at the position of any atom with any orientation we please, since we are only interested in the character of the submatrix and characters are unchanged by rotation of axes.

(a)   Choose the Oz axis parallel to the C_n axis. The operation C_n, i.e.

rotation through the angle $2\pi/n$, is then represented by the matrix

$$
C_n = \begin{pmatrix} \cos(2\pi/n) & \sin(2\pi/n) & 0 \\ -\sin(2\pi/n) & \cos(2\pi/n) & 0 \\ 0 & 0 & 1 \end{pmatrix} \tag{2.32}
$$

with character $1 + 2\cos(2\pi/n)$.

(b)   Choose Oz normal to the plane σ. The operation σ is then represented by

$$
\sigma = \begin{pmatrix} 1 & 0 & 0 \\ 0 & 1 & 0 \\ 0 & 0 & -1 \end{pmatrix} \tag{2.33}
$$

with character 1.

(c)   Choose Oz parallel to the S_n axis. The operation S_n is then represented by

$$
S_n = \begin{pmatrix} \cos(2\pi/n) & \sin(2\pi/n) & 0 \\ -\sin(2\pi/n) & \cos(2\pi/n) & 0 \\ 0 & 0 & -1 \end{pmatrix} \tag{2.34}
$$

with character $-1 + 2\cos(2\pi/n)$.

The above rules for the value of the character may be summarized as follows: the contribution $\Delta\chi_P$ made to χ_P by each atom which is not displaced during the operation P ($=S_n$ or C_n) is $\pm 1 + 2\cos(2\pi/n)$, where the plus sign applies for C_n and the minus sign for S_n. Note that $E = C_1$, so that $\Delta\chi_E = 3$, that $\sigma = S_1$, so that $\Delta\chi_\sigma = 1$ (as shown directly) and that $i = S_2$, so that $\Delta\chi_i = -3$.

Let us now apply these ideas and equation (2.31) to the simple bent triatomic molecule already considered. We choose a set of axes $Oxyz$ coincident with $Ox_b y_b z_b$ of fig. 2.5. The symmetry operations for the molecule are:

(i)   the identity, E;
(ii)   a C_2 axis coincident with Oz;
(iii)   a mirror plane $\sigma_v(zx)$ containing the C_2 axis and the normal to the plane of the molecule;
(iv)   a mirror plane $\sigma'_v(yz)$, the plane of the molecule.

The corresponding point group is C_{2v} and the character table for this group is given in table 2.5. There are four classes in the group, each of

Table 2.5. *The character table for the point group* C_{2v}

C_{2v}	E	$C_2(z)$	$\sigma_v(zx)$	$\sigma'_v(yz)$	
A_1	$+1$	$+1$	$+1$	$+1$	z
A_2	$+1$	$+1$	-1	-1	R_z
B_1	$+1$	-1	$+1$	-1	x, R_y
B_2	$+1$	-1	-1	$+1$	y, R_x

which contains only one operation. All three atoms remain unmoved for the identity operation and for $\sigma'_v(yz)$ but only atom b remains unmoved for $C_2(z)$ and $\sigma_v(zx)$. The corresponding total characters for the 9×9 representation are thus as follows:

class	E	C_2	$\sigma_v(zx)$	$\sigma'_v(yz)$
χ	9	-1	1	3

Equation (2.31) now gives

$$n_{A1} = \tfrac{1}{4}(1 \times 9 \times 1 \quad + \quad 1 \times -1 \times 1 \quad + \quad 1 \times 1 \times 1 \quad + \quad 1 \times 3 \times 1) = 3$$

$$n_{A2} = \tfrac{1}{4}(1 \times 9 \times 1 \quad + \quad 1 \times -1 \times 1 \quad + \quad 1 \times 1 \times -1 \quad + \quad 1 \times 3 \times -1) = 1$$

$$n_{B1} = \tfrac{1}{4}(1 \times 9 \times 1 \quad + \quad 1 \times -1 \times -1 \quad + \quad 1 \times 1 \times 1 \quad + \quad 1 \times 3 \times -1) = 2$$

$$n_{B2} = \tfrac{1}{4}(1 \times 9 \times 1 \quad + \quad 1 \times -1 \times -1 \quad + \quad 1 \times 1 \times -1 \quad + \quad 1 \times 3 \times 1) = 3$$

The original 9×9 representation is thus equivalent to:

$$3A_1 + A_2 + 2B_1 + 3B_2$$

The total number of irreducible representations is nine, corresponding to the $3N = 9$ degrees of freedom and the nine normal modes (including those of zero frequency) of a bent triatomic molecule.

We are interested in the modes with non-zero frequency, so that we wish to remove the representations which correspond to the three translations and three rotations of the molecule as a whole. This is easily done by considering the effect of each of the operations of the group on a new 'position' of the molecule which corresponds to a translated or rotated molecule. Consider, for example, translation parallel to Oy. The operations E, C_2, $\sigma_v(zx)$ and $\sigma'_v(yz)$ multiply all coordinates by 1, -1, -1 and 1, respectively. Thus translation parallel to Oy belongs to symmetry species B_2. Translations parallel to Oz and Ox can similarly be shown to belong to symmetry species A_1 and B_1, respectively, and rotations parallel to Ox, Oy and Oz to symmetry species B_2, B_1 and A_2, respectively. Note that character tables usually contain a column, such as that on the extreme right of table 2.5, in which the symmetry species of the translations (x, y, z) and rotations (R_x, R_y, R_z) are indicated. The symmetry species corresponding to the genuine vibrations are thus $2A_1 + B_2$. Sets of displacements which belong to these symmetry species are illustrated in fig. 2.6. The choice of axes Ox and Oy has been made according to the usual convention; the B_1 and B_2 species would be interchanged if x and y were interchanged.

2.5   Symmetry coordinates and internal coordinates

The precise form of the B_2 vibration, illustrated in fig. 2.6c, is completely determined by symmetry and the requirement that the angular and linear momenta of the molecule are zero. A non-degenerate genuine vibrational mode can have no angular momentum because the equations of motion are unchanged if all displacements and velocities are reversed, which would reverse the angular momentum if it were not zero and thus lead to a second mode of the same frequency. The only way that this restriction can be satisfied for B_2 symmetry is for the directions of motion of atoms a and c to be precisely along the bond directions, so that the lines of motion of all three atoms meet at a point. Fig. 2.6c thus represents the actual relative displacements of the atoms during the B_2 antisymmetric stretching vibration, i.e. it represents the eigenvector or normal coordinate vector for the mode, provided only that the magnitudes of the displacements shown are chosen correctly for the masses and the angle of the molecule so that there is no linear momentum at any instant.

Suppose that we draw a diagram like fig. 2.6c but with the displacement of atom b shown as zero (see fig. 2.7a). This still represents a set of displacements of the atoms corresponding to symmetry species B_2, but in a vibration corresponding to this set of displacements the molecule would clearly have, at any general instant, a net linear

Fig. 2.6. Symmetry modes of the bent triatomic molecule: (a) A_1 symmetric stretch; (b) A_1 bend; (c) B_2 antisymmetric stretch.

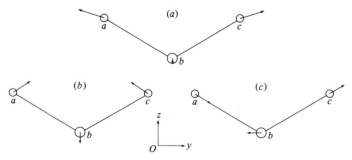

momentum parallel to Oy and a net angular momentum about an axis parallel to Ox and passing through the centre of gravity of the molecule. If, however, we add to this set of displacements those represented by figs. 2.7b and 2.7c, which correspond to pure rotation around an axis parallel to Ox and to pure translation parallel to Oy, respectively, we obtain the set represented by fig. 2.7d, which is the same as fig. 2.6c. The set of displacements (a) is thus equal to the sum of the sets (d), $-(b)$ and $-(c)$, i.e. it is a linear combination of three normal mode vectors (all of B_2 symmetry). The set of displacements (a) is called a *symmetry mode*. A normalized vector in the generalized position space of the molecule corresponding to this 'position' of the molecule would be a *symmetry mode vector* and if any general position of the molecule were expressed in terms of linearly independent symmetry mode vectors the corresponding set of components would be a set of *symmetry coordinates* of the 'position' in the same way that normal coordinates are the set of components of the 'position' vector with respect to the normal mode eigenvectors. In principle, any 'position' of the molecule which belongs to a definite symmetry species may be chosen as a symmetry mode, but it is usual to define symmetry coordinates in such a way that rotations and translations are excluded. The only symmetry mode of B_2 symmetry for the bent triatomic molecule is then the same as the only normal mode of that species, represented by figs. 2.6c and 2.7d.

For symmetry species A_1 there are, however, an infinite number of symmetry modes, but any such mode may be expressed as a linear combination of any two which differ in some way other than simply by their amplitude or phase, such as the stretching and bending modes shown in figs. 2.6a and 2.6b. These two have been arbitrarily chosen so that fig. 2.6a represents a set of displacements in which the angle between the two bonds does not change and fig. 2.6b represents a set in

Fig. 2.7. Symmetry modes for the B_2 species of the bent triatomic molecule: (a) antisymmetric stretch; (b) pure rotation; (c) pure translation; (d) the sum of the displacements in (a), (b) and (c). The vector diagrams alongside each atom in (d) show the addition of the appropriate vectors. See also text.

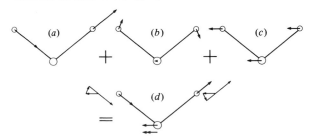

which the bond lengths do not change. For either set the precise directions and relative magnitudes of the displacements then depend only on the geometry and the relative masses of the atoms. The two true normal modes of A_1 symmetry will, in general, be combinations of the symmetry modes of figs. 2.6a and 2.6b and the particular form of the combination will depend on the geometry of the molecule, the masses of the atoms and the forces between them.

To illustrate this, consider first a hypothetical molecule in which there are significant restoring forces tending to keep the lengths of bonds *a–b* and *b–c* constant but only a very small force tending specifically to maintain the equilibrium separation of atoms *a* and *c* or the angle *abc*. The normal modes of A_1 symmetry are then almost precisely those shown in figs. 2.6a and 2.6b, with fig. 2.6b representing a mode of almost zero frequency, because of the small restoring force, and fig. 2.6a representing a mode with frequency dependent on the masses and the restoring force acting along the bonds. Now consider, in contrast, a hypothetical molecule in which the only significant force is one tending to keep atoms *a* and *c* at constant separation, with much smaller forces tending to keep atoms *a* and *b* or *b* and *c* at constant separation or to maintain the angle *abc* constant. The normal modes of A_1 symmetry are then as illustrated in fig. 2.8. Fig. 2.8a represents the simple stretching of the bond which must be considered to exist between atoms *a* and *c*, with a frequency dependent on the force constant for this bond and the mass of atom *a* (or *c*), and fig. 2.8b represents a mode of almost zero frequency in which the separation of atoms *a* and *c* remains constant.

In a real molecule the forces would not be as simple as either of these, but a reasonable approximation might be forces tending to keep the *a–b* and *b–c* separations constant, together with an independent angular force tending to keep the angle *abc* constant. For this set of forces it would be sensible to express any general position of the molecule in terms of three coordinates, r_{ab}, r_{bc} and δ, where r_{ab} and r_{bc} are the

Fig. 2.8. Normal modes of A_1 symmetry for the bent triatomic molecule for a special simple force field (see text).

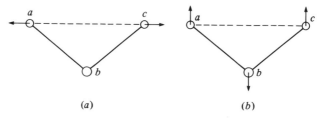

(a)   (b)

increases in length of the corresponding bonds and δ is the increase in the angle *abc*. Such a set of coordinates is called a set of *internal coordinates* because it is not possible to express any translational or rotatonal motion of the molecule as a whole (i.e. any *external motion*) in terms of them. A different set of internal coordinates would be $R_1 = r_{ab} + r_{bc}$, δ and $R_2 = r_{ab} - r_{bc}$, and these latter would be a set of *internal symmetry coordinates*, since the set of displacements of fig. 2.6a has $R_2 = \delta = 0$, the set of fig. 2.6b has $R_1 = R_2 = 0$ and the set of fig. 2.6c has $\delta = R_1 = 0$, so that each of the symmetry modes has only one non-zero internal symmetry coordinate.

The main reason for introducing symmetry coordinates is that if they are used as the basis of a representation of the group the calculation of the normal mode frequencies is made easier. The use of internal symmetry coordinates enables a further simplification in the calculation and, of course, eliminates the translations and rotations from the problem at the outset. We shall briefly discuss the use of symmetry and internal coordinates for the calculation of the frequencies of normal modes in chapter 4 but consider here a secondary use of symmetry coordinates.

It is often easy to write down the possible symmetry mode vectors by inspection, because each can readily be associated with a simple type of motion of certain atoms in the molecule, as in the illustrations just given. Under certain circumstances, i.e. for particular values of the masses and force constants, these symmetry modes approximate quite closely to true normal modes of the corresponding symmetry species. They may thus be used to describe the normal modes when the purpose is only to distinguish between two normal modes of the same symmetry species and not to describe them accurately. Let us consider a further example.

Fig. 2.9 shows three symmetry modes of A_g species for the ethylene molecule. The reader is encouraged to check that this molecule belongs to point group D_{2h} and that there are three genuine normal modes of A_g symmetry. Since the three symmetry modes illustrated are independent (only (*a*) involves displacements of carbon atoms and the displacements of the H atoms in (*b*) are at right angles to those in (*c*)), the true normal mode vectors of A_g symmetry must be linear combinations of these three. The masses of the hydrogen atoms are, however, very much lower than those of the carbon atoms and, since the restoring force for angle bending is very much smaller than that for bond stretching, the three normal modes are not very different from the symmetry modes shown. The true normal mode which involves mainly C—H stretching will, however, obviously involve a small amount of C=C stretching, since the corresponding symmetry mode could only be the true normal mode if either the masses of the carbon atoms or the restoring force for C=C stretching were infinitely large. Similar remarks apply to the other two modes shown. Descriptions such as 'C—H stretching mode' must therefore be treated with caution when a normal mode is implied, and this is true for all such simple descriptions when there is more than one mode of the same symmetry species.

2.6   Further reading

There are a number of excellent books which cover the topics of this chapter and some of those of chapter 4 in greater detail. Among those which may be found particularly useful are: *Infrared and Raman Spectra of Polyatomic Molecules* by G. Herzberg, Van Nostrand, NY, 1945, and *Introduction to the Theory of Molecular Vibrations* and *Vibrational Spectroscopy* by L. A. Woodward, Oxford University Press, 1972.

The first of these is the classic work on the subject and any vibrational spectroscopist will need to refer to it from time to time, but it is a very detailed book which, in addition to dealing fully with the theory of vibrational spectroscopy, considers specifically a large number of types of molecule and individual examples of these types. The book by Woodward aims 'to expound the essential features of vibrational theory' but 'at the same time to explain and develop the special mathematical

Fig. 2.9. Symmetry modes of A_g species for the ethylene molecule, $CH_2=CH_2$: (a) C=C stretch; (b) C—H stretch; (c) C—H bend.

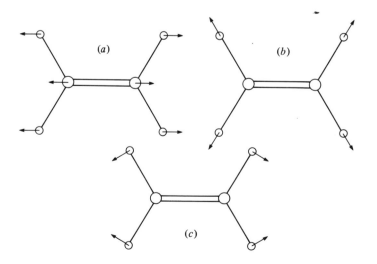

ideas and methods involved'. This includes a treatment of the relevant properties of matrices. A more general book on this latter topic which may be found useful is *Matrix Theory for Physicists* by J. Heading, Longmans, Green and Co., London, 1958.

The reader will by now have realized that access to a set of character tables is essential for the vibrational spectroscopist. A handy compendium of such tables and much other very useful material relating to molecular spectroscopy and group theory is *Point Group Character Tables and Related Data* by J. A. Salthouse and M. J. Ware, Cambridge University Press, 1972.

3 The vibrational modes of polymers

3.1 The vibrations of regular polymer chains

3.1.1 *Introduction*

It has been pointed out already, in section 1.3, that the spectra of polymers are a great deal simpler than might at first be expected, because of the repetitive nature of the structure of the polymer chain. In this section we wish to make the arguments presented earlier more precise by first considering the vibrations of a single perfectly regular polymer chain.

We define a *perfectly regular chain* to be one with which a straight line can be associated in such a way that all translations of the chain parallel to this line by integral multples of some basic distance are symmetry operations, i.e. they bring the chain into exact coincidence with itself when the indistinguishability of all atoms of the same type is taken into account. Such a polymer chain cannot exist in reality, since it would have to be of infinite length, but in polymer crystals there are considerable lengths of chain for which such translations may be considered to be symmetry operations to a first approximation. Even in the non-crystalline regions of solid polymers there can sometimes be appreciable lengths of such straight chains. Since the interactions between the atoms in the same chain are often, but not always, considerably greater than those between the atoms in adjacent chains, the isolated perfectly regular polymer chain is a good starting point for the discussion of the normal vibrations and vibrational spectra of polymers. The effects of interactions between chains in a crystal will be considered in the next section and the effects of departures from perfect regularity in the next chapter.

In the remainder of this chapter we shall restrict attention to the limited set of vibrational modes of regular polymer chains to which all infrared- or Raman-active modes belong. These are modes in which corresponding atoms in all translational repeat units of the chain vibrate in phase with each other. In anticipation of the next subsection we shall call these the *factor group modes*. As explained in section 1.3, all the other modes of a regular chain are definitely inactive for both infrared absorption and Raman scattering, because the interactions of the different repeat units with the radiation cancel out. Whether any one of the factor group modes is in fact active in infrared absorption or Raman scattering depends on its symmetry species. We consider now how to determine the symmetry species of the factor group modes and defer until the next chapter a discussion of the *selection rules* which determine the activities of the modes of various symmetry species. The results of the application of the selection rules will, however, sometimes be quoted in discussing the illustrative examples.

3.1.2 *Line group and factor group: polyethylene*

Consider the perfect polyethylene chain, a section of which is illustrated in fig. 3.1. There are six different ways in which the orientation of the reference axes $Oxyz$ may be chosen. (The handedness of the axes is not important here.) Different authors make different choices and this affects the designations of the B-type symmetry species, as pointed out in subsection 2.3.4. The choice of orientation in fig. 3.1 is made so that the comparison of the symmetry of the vinyl polymers with that of polyethylene in the next subsection is simplified. Later in this section the relationships between the species designations for all the possible orientations of the axes are tabulated (table 3.2). The *translational repeat unit* of any chain is the smallest unit from which the whole molecule can be constructed by repeating the unit at intervals of its length, a, along the direction of the molecule without rotating or changing it in any way. For polyethylene it is two methylene ($-CH_2-$) groups, whereas the *chemical repeat unit* is one methylene group. The distance a is called the *translational period* of the molecule.

The notation for the symmetry operations shown in fig. 3.1 is an extension of that already used for point groups. For instance, $C_2(z, k)$ and $C_2^+(z, k)$ are different C_2 axes parallel to Oz in repeat unit k. Symbols for symmetry elements with and without a superscript $+$ will be used in this section to indicate the two operations of the same type for a particular translational repeat unit. A similar notation will be used for the two carbon atoms of the repeat unit. The only other operations that require further explanation, since they are different from those involved in point groups, are $T(x)$, $\sigma_g(xy)$ and $\bar{C}_2(x)$. The symbol $T(x)$ stands for a *translation axis* and the corresponding operation is *translation* parallel to Ox through a distance a. The symbol $\sigma_g(xy)$ stands for a *glide plane* and the corresponding operation is reflection in the plane Oxy followed by translation parallel to Ox through $a/2$. The symbol $\bar{C}_n(x)$ stands for a *screw axis* and the corresponding operation is rotation through the angle $2\pi/n$ about Ox followed by translation parallel to Ox by a/n. If it is necessary to specify the translational part of $\sigma_g(xy)$ or $\bar{C}_n(x)$ the symbols $\sigma_g(xy, d)$ or $\bar{C}_n(x, d)$ will be used, where d is the distance through which the translation is made. Since there are an infinite number of symmetry

Fig. 3.1. Symmetry elements for the regular polyethylene chain: (a) view normal to the plane of the zig-zag backbone, with the positive Oy axis pointing away from the viewer; (b) view along the chain axis, with the positive Ox axis pointing towards the viewer. See text for explanation of notation.

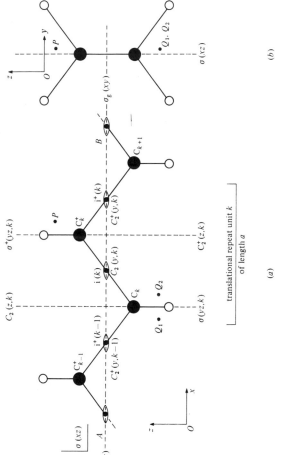

carbon atom

hydrogen atom

centre of inversion and C_2 axis parallel to Oy

operations, the corresponding group has an infinite number of members; it is called a *line group* because all the intersections of the symmetry elements lie on a line, *AB*. We shall come back to a consideration of the full line group in the next chapter, and we consider here only a finite group related to it which enables us to determine the symmetry species of the infrared- and Raman-active modes of the infinite regular chain.

Since the active modes are included among those in which the motions of corresponding atoms in all translational repeat units have identical displacements from their equilibrium positions, their symmetry species must be contained within a $3n \times 3n$ representation based on any general displacement of the n atoms of any particular translational repeat unit. In writing down the $3n \times 3n$ representation matrix corresponding to any operation we must bear in mind that if the real operation moves any point P within the repeat unit under consideration into the position of a point Q' within any other repeat unit, the point P must be considered to have moved into the position of the point Q in the original unit which may be reached from Q' by a translation $T(x)$. This is because the translation must have no effect on the generalized position of the polymer molecule as a whole for the restricted set of generalized positions under consideration, i.e. those in which corresponding atoms in all repeat units have identical displacements from their equilibrium positions. When an operator is thus redefined it will be placed between vertical bars, e.g. $|C_2(y)|$.

To see how this works in practice, choose the repeat unit k which consists of the carbon atoms C_k and C_k^+, together with their attached hydrogen atoms, as the repeat unit on which to base the $3n \times 3n$ representation (here $n = 6$) and consider the effect of the operation $|C_2^+(y, k)|$. If we operate with $C_2^+(y, k)$ the atom C_k moves to the position C_{k+1}^+, which must be considered as equivalent to moving to C_k^+, and the atom C_k^+ moves to C_{k+1}, which must be considered as equivalent to moving to C_k. The effect is similar for the attached hydrogen atoms and for any general point within the molecule, so that the effect of the operation $|C_2^+(y, k)|$ on the repeat unit is equivalent to the effect of the operation $C_2(y, k)$. In a similar way, it may be shown that the operations $|C_2(y, k')|$ and $|C_2^+(y, k')|$ for any value of k' are equivalent to $C_2(y, k)$, so that we can label any of this set of operations simply $|C_2(y)|$.

Consider, as a second example, the effect of $\sigma_g(xy)$. Atom C_k goes to C_k^+ and C_k^+ goes to C_{k+1}, which is equivalent to C_k. The effects of $|\sigma_g(xy)|$ on the atoms considered as points are exactly the same as the effects of $|C_2(y)|$, but the two operations are distinct because they do not have the same effect on any general point within the repeat unit, such as the point P, which goes to Q_1 under $|C_2(y)|$ and to Q_2 under $|\sigma_g(xy)|$.

The operations E, $|C_2(z)|$, $|C_2(y)|$, $|\bar{C}_2(x)|$, $|i|$, $|\sigma_g(xy)|$, $\sigma(xz)$ and $|\sigma(yz)|$ form a group which will be called the *factor group* of the line group. (This name is not strictly correct, but it is the one usually used by vibrational spectroscopists.) It should be noted, however, that it is not possible to form a group by taking any actual set of true symmetry operations of the line group which contains E, $\sigma(xz)$ and just one representative of each of the remaining types of operation of the line group. By considering the effects of the operations of the factor group of polyethylene on a general point within the repeat unit, it is easy to show that the group multiplication table has exactly the same form as that for a point group consisting of the operations E, $C_2(z)$, $C_2(y)$, $C_2(x)$, i, $\sigma(xy)$, $\sigma(xz)$ and $\sigma(yz)$, which is the group D_{2h}. The factor group for the regular polyethylene chain is thus isomorphous with the point group D_{2h}. The *factor group modes*, which include the infrared- and Raman-active modes, will thus have the same types of symmetry species as those of a small molecule belonging to this point group.

The first stage in finding the numbers of factor group modes belonging to the various symmetry species is to write down the total character of each operator of the factor group in the $3n \times 3n$ representation based on a general displacement of the n ($=6$) atoms of any translational repeat unit. The reader is encouraged to check, using the rules already given in subsection 2.4.3 for the character contributed by undisplaced atoms and remembering carefully the definition of the factor group operations, that the total characters are as follows:

oper.	E	$\|C_2(z)\|$	$\|C_2(y)\|$	$\|\bar{C}_2(x)\|$	$\|i\|$	$\|\sigma_g(xy)\|$	$\sigma(xz)$	$\|\sigma(yz)\|$
χ	18	-2	0	0	0	0	2	6

The character table for D_{2h}, expressed in terms of the operations of the factor group, is given in table 3.1, and the use of equation (2.31) leads immediately to the result that the 18×18 representation is equivalent to

$$3A_g + B_{1g} + 2B_{2g} + 3B_{3g} + A_u + 3B_{1u} + 3B_{2u} + 2B_{3u}$$

with $3n = 18$ normal modes. We must now ask which of these modes correspond to simple translations or rotations of the whole polymer molecule.

Each of the three types of translational displacement of a repeat unit represents a translation of the infinite chain as a whole. When, however, we remember that corresponding atoms in all translational repeat units have identical displacements for the factor group modes, we see that only one of the three types of rotational displacement of a unit represents a rotation of the chain as a whole, viz. that around Ox; the other two represent distortions of the chain and are thus associated with genuine

Table 3.1. *Character table for the point group* D_{2h}

D_{2h}	E	$|C_2(z)|$	$|C_2(y)|$	$|\bar{C}_2(x)|$	$|i|$	$|\sigma_g(xy)|$	$\sigma(zx)$	$|\sigma(yz)|$	
A_g	$+1$	$+1$	$+1$	$+1$	$+1$	$+1$	$+1$	$+1$	
B_{1g}	$+1$	$+1$	-1	-1	$+1$	$+1$	-1	-1	
B_{2g}	$+1$	-1	$+1$	-1	$+1$	-1	$+1$	-1	
B_{3g}	$+1$	-1	-1	$+1$	$+1$	-1	-1	$+1$	R_x
A_u	$+1$	$+1$	$+1$	$+1$	-1	-1	-1	-1	
B_{1u}	$+1$	$+1$	-1	-1	-1	-1	$+1$	$+1$	z
B_{2u}	$+1$	-1	$+1$	-1	-1	$+1$	-1	$+1$	y
B_{3u}	$+1$	-1	-1	$+1$	-1	$+1$	$+1$	-1	x

modes of vibration. A regular polymer chain thus has $3n - 4$ factor group modes which correspond to genuine modes of vibration; for polyethylene this number is 14.

It is easily seen that the rotation around Ox and the three translations have the symmetry species shown in the last column of the character table. The genuine vibrations are thus:

$$3A_g + B_{1g} + 2B_{2g} + 2B_{3g} + A_u + 2B_{1u} + 2B_{2u} + B_{3u}$$

and table 3.2 shows symmetry modes corresponding to them. In this table the designations of the symmetry species are given for all possible choices of the orientation of the axes $Oxyz$; those corresponding to the present choice are given in the first row of the table. The descriptions of the symmetry modes usually used and the short forms of these are also shown. It should be noted that the symbols v, δ and γ are widely used to indicate stretching, angle bending and out-of-plane deformation modes, respectively, in molecules of all types. The symbols τ, ω and ρ are sometimes used to indicate twisting, wagging and rocking, respectively, rather than the symbols γ_t, γ_w and γ_r. Modes such as the C—C stretching mode and the C—C—C bending mode are often called *skeletal modes*, since they are essentially vibrations of the main framework of the molecule, rather than of atoms or groups attached to it.

For each of the symmetry species B_{1g}, A_u and B_{3u} of polyethylene there is only one genuine vibrational mode and thus the directions of motion of the atoms in those normal modes are in each case precisely as shown. The amplitudes of motion of the carbon atoms in the mode of B_{3u} species have, however, been exaggerated compared with those of the hydrogen atoms. The precise normal modes for the other symmetry species are mixtures of the symmetry modes shown, although each of them must be very similar to one of the symmetry modes shown. Note that each of the symmetry modes of an isolated —CH_2— group gives rise to two

Table 3.2. Symmetry modes of the regular polyethylene chain

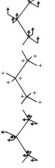

	Ram	Ram	Ram	Ram	inactive	IR	IR	IR
xyz	A_g	B_{1g}	B_{2g}	B_{3g}	A_u	B_{1u}	B_{2u}	B_{3u}
xzy	A_g	B_{2g}	B_{1g}	B_{3g}	A_u	B_{2u}	B_{1u}	B_{3u}
yxz	A_g	B_{1g}	B_{3g}	B_{2g}	A_u	B_{1u}	B_{3u}	B_{2u}
yzx	A_g	B_{3g}	B_{1g}	B_{2g}	A_u	B_{3u}	B_{1u}	B_{2u}
zyx	A_g	B_{3g}	B_{2g}	B_{1g}	A_u	B_{3u}	B_{2u}	B_{1u}
zxy	A_g	B_{2g}	B_{3g}	B_{1g}	A_u	B_{2u}	B_{3u}	B_{1u}
Spectral activity	Ram	Ram	Ram	Ram	inactive	IR	IR	IR

Translation T

Rotation R

C—C stretch ⎱
C—C bend ⎰ v_+

CH$_2$ symmetric stretching	ν_s(CH$_2$)
CH$_2$ antisymmetric stretching	ν_a(CH$_2$)
CH$_2$ bending	δ(CH$_2$)
CH$_2$ wagging	γ_w(CH$_2$)
CH$_2$ rocking	γ_r(CH$_2$)
CH$_2$ twisting	γ_t(CH$_2$)

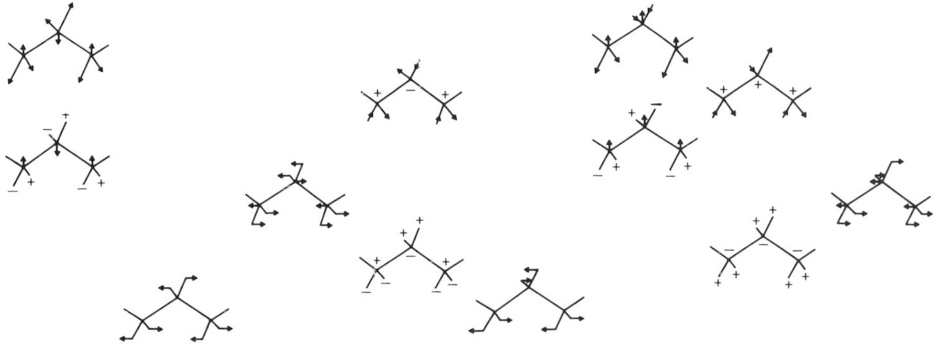

The top part of the table shows the possible orientations of the reference axes with respect to the plane of the carbon backbone and the correspondences between the symmetry species designations for the different choices. The lower part shows, below the appropriate species symbols, symmetry modes corresponding to the 18 allowed normal modes, together with the usual descriptive names and symbols applied to them.

Three CH$_2$ groups are shown in each diagram, with the plane of the carbon atoms rotated slightly around Oz (assuming that the orientation of the axes is as shown in the top row of the table), so that the nearer C—H bond of each CH$_2$ group always appears to the left of the further one. Arrows indicate displacement parallel to Ox, to Oz or to a C—H bond and $+$ and $-$ indicate displacements in the Oyz plane. The y components of displacements corresponding to $+$ and $-$ are of opposite sign and for carbon atoms these displacements are constrained by symmetry to be parallel to Oy. For H atoms the displacement may be taken to be perpendicular to the corresponding C—H bond, bearing in mind that only symmetry modes and not true normal modes are being specified (see also text). A set of symmetry modes of this kind was first illustrated by S. Krimm et al., Journal of Chemical Physics **25** 549 (1956).

Table 3.3. *Some important line groups*

Order	Factor group elements	Isomorphous point group
1	E	C_1
2	E, i	C_i
2	E, σ_h	C_s
2	E, σ_v	C_s
2	$E, \bar{\sigma}_v$	C_s
2	E, C_2'	C_2
n	$E, (n-1)C_n$	$C_n, n \geqslant 2$
n	$E, (n-1)\bar{C}_n^{(k)}$	$C_n, n \geqslant 2$
4	E, σ_v', C_2', i	C_{2h}
4	$E, \bar{\sigma}_v', C_2', i$	C_{2h}
4	E, σ_h, C_2, i	C_{2h}
4	$E, \sigma_h, \bar{C}_2, i$	C_{2h}
4	$E, \sigma_h, \sigma_v', C_2''$	C_{2v}
4	$E, \sigma_h, \bar{\sigma}_v', C_2''$	C_{2v}
4	$E, \bar{\sigma}_v', \sigma_v'', \bar{C}_2$	C_{2v}
4	$E, \bar{\sigma}_v', \bar{\sigma}_v'', C_2$	C_{2v}
$2n$	$E, (n-1)C_n, n\sigma_v$	$C_{nv}, n \geqslant 2$
4	E, C_2, C_2', C_2''	D_2
4	$E, \bar{C}_2, C_2', C_2''$	D_2
8	$E, \sigma_h, \sigma_v', \sigma_v'', C_2, C_2', C_2'', i$	D_{2h}
8	$E, \sigma_h, \bar{\sigma}_v', \sigma_v'', \bar{C}_2, C_2', C_2'', i$	D_{2h}
8	$E, \sigma_h, \bar{\sigma}_v', \bar{\sigma}_v'', C_2, C_2', C_2'', i$	D_{2h}
$2n$	$E, (n-1)\bar{C}_n^{(k)}, nC_2'$	D_n, n odd and >2
$2n$	$E, (n-1)\bar{C}_n^{(k)}, \frac{1}{2}nC_2', \frac{1}{2}nC_2''$	D_n, n even and >2

In this table C_n axes are parallel to the chain axis. C_2' and C_2'' axes are perpendicular to the chain axis and if both are present each C_2' axis bisects the angle between two C_2'' axes and vice versa. σ_h is a mirror plane perpendicular to the chain axis, whereas σ_v, σ_v', σ_v'' and σ_d are mirror planes containing the chain axis. C_2' axes are normal to σ_v' planes and/or parallel to σ_v'' planes, whereas C_2'' axes are normal to σ_v'' planes and/or parallel to σ_v' planes. σ_d planes bisect the angle between C_2' axes. $\bar{\sigma}$ stands for a glide plane and $\bar{C}_n^{(k)}$ means the factor group element corresponding to the screw operation for an n_k helix. $(n-1)P_n$, where P is C, or $\bar{C}^{(k)}$, stands for the $n-1$ different operations that are powers of P_n.

symmetry modes of the factor group, one of type u and the other of type g, and these differ only according to the relative phases of the motions of the two —CH_2— groups.

We have so far considered the line group for a particular simple regular chain, but the line group for any regular chain has a factor group which is isomorphous with one of the point groups and an analysis of the number of modes belonging to the various symmetry species can be performed in exactly the same way. Table 3.3 gives a list of the most important line groups and the point groups with which their factor groups are isomorphous. The vinyl polymers constitute a particularly important class of polymers and their symmetries will now be considered.

3.1.3 *Polyethylene and the vinyl polymers: descent in symmetry*

The structure of the vinyl polymers was discussed in section 1.2. Fig. 3.2 shows the planar zig-zag forms of the isotactic, syndiotactic and atactic configurations and indicates the orientation of the reference axes

Fig. 3.2. Planar zig-zag forms of vinyl polymers, $+CH_2$—$CHX+_n$: (*a*) carbon backbone showing labelling of carbon atoms and orientation of reference axes; (*b*) regular isotactic configuration and orientation of reference axes; (*c*) regular syndiotactic configuration; (*d*) random atactic configuration.

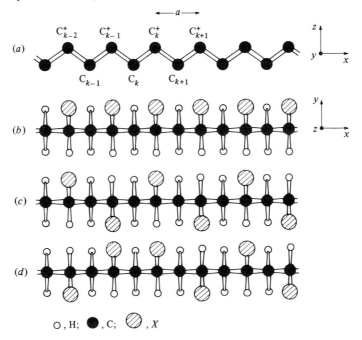

O, H; ●, C; ⊘, X

Table 3.4. *Symmetry elements of polyethylene and stereoregular vinyl polymers*

	$T(x, pa)$	$C_2(z,k)$	$C_2^+(z,k)$	$C_2(z,k+1)$	$C_2^+(z,k+1)$	$C_2(y,k)$	$C_2^+(y,k)$	$C_2(y,k+1)$	$C_2^+(y,k+1)$
Polyethylene	integer	$C_2(z,k)$	$C_2(z,k)$	$C_2(z,k)$	$C_2(z,k)$	$C_2(y,k)$	$C_2(y,k)$	$C_2(y,k)$	$C_2(y,k)$
D_{2h}	E			$\lvert C_2(z)\rvert$				$\lvert C_2(y)\rvert$	
χ	18			-2				0	
Syndiotactic	even	$C_2(z,k)$		$C_2(z,k)$					
C_{2v}	E			$\lvert C_2(z)\rvert$					
χ	36			-2					
Isotactic	integer								
C_s	E								
χ	18								

See text for explanation.

and the labelling of the carbon atoms which will be used in this subsection. The symmetry of these planar structures will now be compared with that of polyethylene, but it must be remembered that in fact the vinyl polymers often take up more complicated conformations, particularly in the isotactic and atactic forms. The orientation of the axes with respect to the carbon backbone is the same as that used in discussing polyethylene.

Each of the structures shown in fig. 3.2 differs from the structure of polyethylene, shown in fig. 3.1, only by the substitution of X atoms or groups for some of the hydrogen atoms of the polyethylene structure, but this removes some of the symmetry elements from the line group. The symmetry groups for the planar zig-zag forms of the vinyl polymers are thus sub-groups of the line group of the planar zig-zag polyethylene molecule. The sub-group for the atactic structure contains only the identity element E and is thus the trivial group C_1 and not a line group at all, since it contains no translational symmetry element. All $3N - 6$ normal modes of vibration of the atactic form thus have the same symmetry species and this form will not be considered further here. For each of the two remaining structures some of the symmetry elements of the polyethylene molecule are retained and the symmetry elements of the three structures are tabulated in table 3.4.

We imagine the isotactic and syndiotactic structures to be derived from the polyethylene structures illustrated in fig. 3.1 by replacing one of the hydrogen atoms on each of the carbon atoms C_k^+ for all values of k by an X atom or group, which we shall at present assume is a single atom. Note that for polyethylene and the isotactic structure the carbon atoms C_k, C_k^+ and the hydrogen or hydrogen and X atoms attached to them form one translational repeat unit of the chain, whereas for the

Table 3.4 (*cont.*)

$\bar{C}_2(x, pa/2)$	$i(k)$	$i^+(k)$	$i(k+1)$	$i^+(k+1)$	$\sigma_g(xy, pa/2)$	$\sigma_g(xz, pa)$	$\sigma(yz, k)$	$\sigma^+(yz, k)$	$\sigma(yz, k+1)$	$\sigma^+(yz, k+1)$
odd $\|\bar{C}_2(x)\|$ 0	$i(k)$ $\|i\|$ 0	$i(k)$	$i(k)$	$i(k)$	odd $\|\sigma_g(xy)\|$ 0	integer $\sigma(xz)$ 2	$\sigma(yz,k)$	$\sigma(yz,k)$ $\|\sigma(yz)\|$ 6	$\sigma(yz,k)$	$\sigma(yz,k)$
						odd $\sigma_g(xz)$ 0		$\sigma^+(yz,k)$ $\|\sigma(yz)\|$ 6		$\sigma^+(yz,k)$
							$\sigma(yz,k)$	$\sigma^+(yz,k)$ $\|\sigma(yz)\|$ 6	$\sigma(yz,k)$	$\sigma^+(yz,k)$

syndiotactic structure the carbon atoms C_k, C_k^+, C_{k+1}, C_{k+1}^+ and the hydrogen or hydrogen and X atoms attached to them form one translational repeat unit. In order to be able to enumerate correctly all the symmetry elements of one translational repeat unit of the syndiotactic chain it is thus necessary to consider all those symmetry elements of the line group of the polyethylene molecule which intersect the Ox axis in two translational repeat units of that molecule, and these are listed at the top of table 3.4. Note that the notation $\sigma_g(xz, pa)$ expresses the fact that in polyethylene the simple mirror plane $\sigma(xz)$ is a special member of the general set of glide operations consisting of reflection in Oxz followed by translation parallel to Ox through any integral multiple pa of a, the translational repeat distance for polyethylene. In the syndiotactic vinyl polymer only odd values of p are allowed; the Oxz plane is a true glide plane, since the translational repeat distance is now equivalent to $2a$. The similar notation $T(x, pa)$ has been used for the translational elements to make explicit here also the magnitude of the translation.

For each structure there are three rows of entries in this table. The symbol for a specific symmetry element, say P, in the first row indicates that the symmetry element at the top of the column is present and that it belongs to the same class as the symmetry element P. The entries 'integer', 'even' and 'odd' indicate that the symmetry element indicated at the top of the column is present in the structure provided that p is an integer, an even integer or an odd integer, respectively. In the second row the symmetry elements of the factor group of the line group are shown. The continuous vertical lines in the table separate those sets of symmetry elements of the line groups which correspond to each of the different elements of the factor group. The symbol for the point group isomorphous with the factor group is shown at the extreme left. The third row gives the total character for one true translational repeat unit of each structure (including polyethylene) for each factor group element.

By using equation (2.31) with the total characters shown in table 3.4 and the characters given in the standard tables for groups C_s and C_{2v} it may be shown that the symmetry species for the genuine vibrational normal modes of the factor group of the planar zig-zag isotactic vinyl chain are

$$9A' + 5A''$$

and that for the planar syndiotactic chain they are

$$9A_1 + 7A_2 + 7B_1 + 9B_2$$

Note that if the Ox and Oy axes are interchanged, as they frequently are in the literature, the designations B_1 and B_2 for the group C_{2v} are also interchanged. Note also that in the standard character table for C_s the normal to the mirror plane is called Oz, and the rotation parallel to the chain thus corresponds to R_z in the table.

The group C_s is a subgroup of the group C_{2v}, which is in turn a subgroup of the group D_{2h}. The full line group of the isotactic chain is not, however, a subgroup of the line group of the syndiotactic chain, although both of them are subgroups of the line group of the polyethylene chain. Because of the *descent in symmetry* $D_{2h} \rightarrow C_{2v} \rightarrow C_s$, the symmetry species of these three groups may be correlated with each other.

For any general descent in symmetry it is easy to deduce the correspondences between the non-degenerate species of the groups of lower and higher symmetry, since the characters of corresponding species must be the same for all the operations that are common to the two groups. This is because the character for a non-degenerate mode shows directly the effect that the corresponding class of operator has on the eigenvector. Each set of degenerate modes of the group of higher symmetry corresponds to an n-dimensional representation of the group for $n > 1$. The appropriate subset of matrices, i.e. those corresponding to the common operations, forms a representation of the subgroup, and their characters, given in the character table for the group of higher symmetry, form the set of total characters for the corresponding n-dimensional representation of the subgroup. Equation (2.31) may thus be used, in conjunction with the character table for the subgroup, to find the corresponding symmetry species for the subgroup. For the example under consideration, where no degenerate species occur, we easily find the correspondences shown in fig. 3.3. Standard *correlation tables for descent in symmetry* are available which show the correlations for all sets of related groups. It is often useful to examine such correlations when comparing the spectra of related molecules, as we shall see later.

So far we have considered the symmetries of the planar zig-zag forms of those stereoregular vinyl polymers for which X is a single atom. If X is a chemical group, the symmetries of either of the two forms can be either the same as when X is a single atom, or lower, depending on the symmetry and orientation of X. If X has reflection symmetry in the Ozy plane, the isotactic molecule retains the symmetry C_s, but otherwise it has no symmetry elements other than $T(x)$, corresponding to factor group C_1. The syndiotactic molecule can have three possible symmetries lower than that already discussed, corresponding to factor groups isomorphous with C_2, C_s or C_1. These are all subgroups of the group C_{2v} and the correlations between their symmetry species are shown in fig. 3.3.

It was pointed out in section 1.2 that stereoregular isotactic vinyl chains often take up helical conformations, and so do many other stereoregular polymer chains. The line groups for helices, and the corresponding factor groups, are thus of particular importance for polymers and will now be considered.

3.1.4 *Helical molecules*

A *regular helical chain* may be defined as a chain which possesses a screw axis, $C_n^{(k)}(z)$, where $n > 1$ and the Oz axis is chosen parallel to the chain axis. The operation corresponding to this is a

Fig. 3.3. Descent in symmetry for point groups D_{2h}, C_{2v}, C_s, C_2 and C_1. In the descent from D_{2h} to C_{2v} the Oz axis is assumed to remain as the C_2 axis in C_{2v} and in the descent from C_{2v} to C_s the Oyz plane is assumed to remain as the mirror plane in C_s. For the group C_{2v}, the B_1 and B_2 species designations are interchanged by interchanging Ox and Oy; table 3.2 shows the effects of different choices of orientation of the axes for D_{2h}.

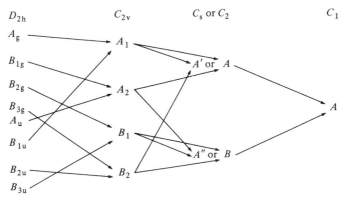

Fig. 3.4. The three-fold helix for an isotactic vinyl polymer. The group X of the structural unit $-CHX-CH_2-$ is shown shaded. The carbon backbone forms a right-handed helix and the right-handed 3_1 and the left-handed 3_2 helices are indicated by the 'ribbons'.

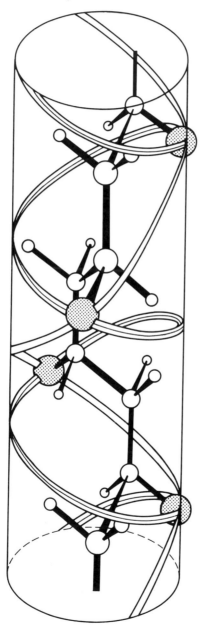

rotation through $2\pi k/n$ about Oz followed by translation through a distance a/n parallel to Oz, where a is the translational period of the chain. The helix may thus be described as consisting of n *physical repeat units* in k turns, which constitute one translational period, and it is called an n_k *helix*. If $k = 1$ it is usually omitted from the symbol for the operation or element. The planar zig-zag polyethylene chain possesses a $C_2^{(1)}$ axis, for which the simpler notation $\bar{C}_2$ was used in the previous subsection, and may thus be considered to be a 2_1 helix, whereas the conformation often assumed by isotactic vinyl polymers is the 3_1 helix illustrated in fig. 3.4. The physical repeat unit is often the same as the chemical repeat unit but need not be.

In specifying a helix, and in deriving its symmetry, several points should be noted. First, n and k should have no common factor, otherwise more than one translational repeat unit is being referred to. Secondly, the value of k is arbitrary to the extent that any integral multiple pn of n may be added to it or subtracted from it, since this will simply insert p extra whole turns per physical repeat unit to give an identical molecule. Thus k should always be less than n to exclude such redundant turns. For greatest simplicity of description $k < n/2$ (or $k = n/2$ for n even), since a helix consisting of n physical repeat units in k clockwise turns can equally well be described as consisting of n physical repeat units in $n - k$ anticlockwise turns. This is illustrated for $n = 3$ in fig. 3.4. The helix with $k < n/2$ is not, however, always of the same handedness, or *chirality*, as the helix described by the bonds in the backbone of the molecule, as it is in fig. 3.4, and the description with the higher value of k may then be preferred. The two helices can, however, differ in chirality only if there is more than one backbone bond per physical repeat unit. For some purposes it may be necessary to specify the chirality of a helix, but not for specifying its line group or the corresponding factor group.

If an n_k helix has no symmetry elements other than products and powers of $T(z)$ and $C_n^{(k)}(z)$, the factor group of the line group is isomorphous with the point group C_n. If the physical repeat unit is sufficiently symmetric there may in addition be two classes of dihedral axes, C_2' and C_2'', which intersect the Oz axis alternately at equal distances $a/(2n)$ along it. The adjacent members of either class make the angle $2\pi k/n$ with each other. The factor group of the line group is then isomorphous with the point group D_n; for n odd the two classes of dihedral axes in the line group gives rise to only one class in the factor group. No higher symmetry is possible, except for $n = 2$, since the operation of reflection reverses the chirality of a helix. Helices with $n = 2$ can have factor groups isomorphous with C_{2v}, C_{2h} or D_{2h}, in addition to C_2 or D_2.

The symmetry species for C_n are $A, E_1, E_2, \ldots, E_{\frac{1}{2}(n-1)}$ for n odd and $A, B, E_1, E_2, \ldots, E_{\frac{1}{2}(n-2)}$ for n even. For D_n they are $A_1, A_2, E_1, E_2, \ldots, E_{\frac{1}{2}(n-1)}$ for n odd and $A_1, A_2, B_1, B_2, E_1, E_2, \ldots, E_{\frac{1}{2}(n-2)}$ for n even (except for D_2, for which they are A, B_1, B_2, B_3). The number of modes belonging to each species may be found using the method already

Fig. 3.5. Symmetry modes corresponding to C—X stretching for a hypothetical 6_1 helix with physical repeat unit —(C—X)— as viewed along the axis of the helix. The degenerate pairs of modes illustrated are orthogonal pairs of standing-wave modes obtained by adding and subtracting the symmetry modes corresponding to waves travelling round the helix axis in opposite directions, so that the amplitudes of motion are given by $a_0 \sin(2\pi mr/6)$ or $a_0 \cos(2\pi mr/6)$, with $m = 1$ for the E_1 mode and $m = 2$ for the E_2 mode. The values of r are shown by the small hexagon for one complete turn of the helix.

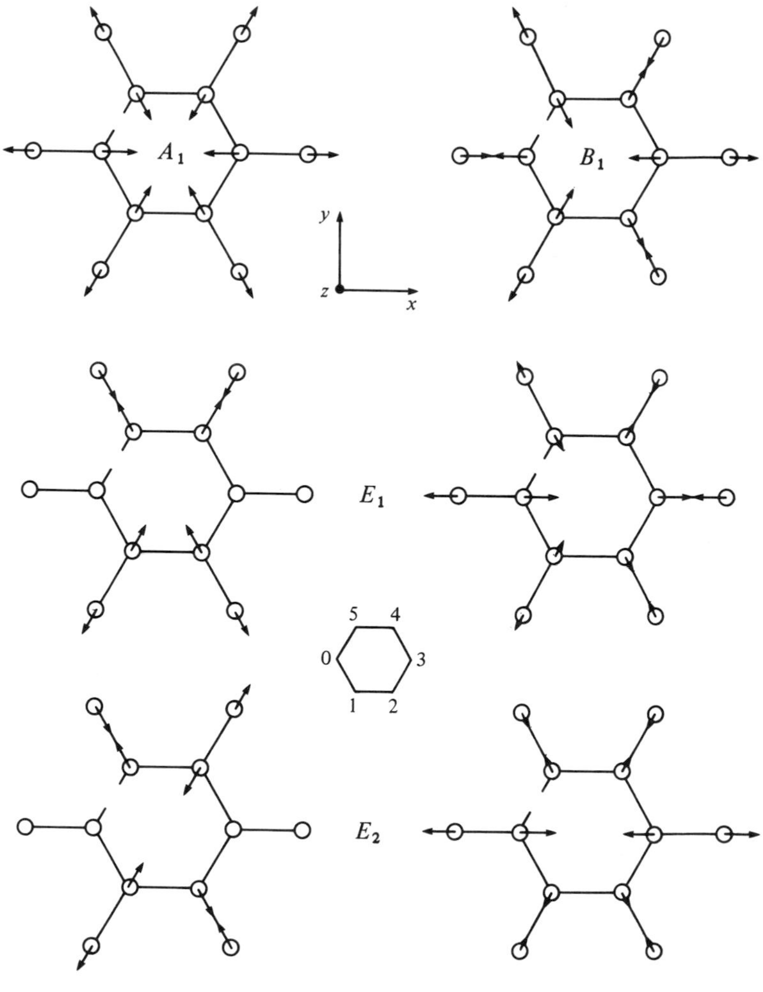

described. In the modes of A, A_1 or A_2 symmetry all physical repeat units vibrate in phase, when related by the screw operation, and in the modes of B, B_1 or B_2 symmetry alternate units vibrate π out of phase (except for D_2, for which B_1 species vibrate in phase and B_2 and B_3 species vibrate π out of phase). The two modes of a degenerate pair of species E_m may be chosen so that there is a phase angle of $\pm 2\pi mk/n$ between the vibrations of adjacent physical units along the helix and a phase angle of $\pm 2\pi m/n$ between consecutive units taken in order round the Oz axis. A more detailed discussion of the degenerate species of point groups C_n and D_n is given in subsection 2.3.2. Symmetry modes corresponding to C—X stretching are shown in fig. 3.5 for a hypothetical 6_1 helix for which the factor group of the line group is isomorphous with D_6.

3.2 The vibrations of polymer crystals

We have so far considered only the vibrations of idealized, isolated polymer chains and we now consider the vibrations of a polymer crystal in which the chains each have essentially the same conformation as the original isolated chain. When the chains are brought together to form a crystal several changes in the nature of the factor group modes take place, since these, like all the normal modes, must now be modes of the crystal as a whole.

A *primitive unit cell* of any crystal is any one of an infinite number of possible choices of parallelepiped of smallest possible volume from which the crystal may be imagined to be built. The crystal is built by the repetition of the contents of this cell in all the volumes which the parallelepiped occupies if it is displaced, without rotation, through all vectors $l\mathbf{a} + m\mathbf{b} + n\mathbf{c}$, where l, m and n take all possible integral values and $\mathbf{a}$, $\mathbf{b}$ and $\mathbf{c}$ are vectors which correspond to the three edges of the parallelepiped. The symmetry elements of a crystal are of similar types to those already considered for single polymer chains, but they no longer meet on a line and they form a *space group*. The set of modes to which the infrared- and Raman-active modes belong are now the modes in which the contents of each primitive unit cell move in phase with those of every other primitive unit cell and the symmetry species of these modes may be obtained by considering the *factor group of the space group*, in a manner analogous to that already used in determining the symmetry species for the corresponding modes of the single chain by considering the factor group of the line group.

From now on the word 'primitive' will be dropped in references to the unit cell but it must be understood to be implied. We shall not discuss the larger unit cells sometimes defined for crystals of certain symmetry classes.

3.2.1   *The nature of the factor group modes*

In a polymer crystal all the chain axes are parallel and one of the edges of the unit cell may be chosen parallel to them. If this edge is defined to be the Oc axis there will be only one chain crossing the Oab plane per unit cell in the simplest type of crystal; more generally there will be m chains per unit cell.

For crystals with only one chain per unit cell there are two possibilities:

(*a*)   The factor group of the space group is the same as that of the line group of the isolated chain.

(*b*)   The factor group of the space group is of lower order than that of the line group of the isolated chain and is a subgroup of it.

If (*a*) holds, the symmetry species of the factor group modes are the same as those of the single chain but there is one more mode, a *lattice mode*, which corresponds to a rotatory vibration, or *libration*, of the whole chain about the chain axis. This mode replaces the free rotation of the isolated chain and is sometimes called an *external mode* to distinguish it from the other *internal modes* of vibration of the chain. It will generally be of rather low frequency, since the intermolecular forces are generally much lower than the intramolecular forces and the moment of inertia of the chain per unit length is usually rather high.

For any symmetry species which contains only one mode the directions of motion of the atoms in the normal mode vibration will be identical to those of the corresponding mode of the single chain, although its frequency will be slightly changed because of the intermolecular forces. The forms of the modes for symmetry species with more than one mode will be changed as well as the frequencies, although the changes will often be small.

If (*b*) holds, not only will there be an additional mode, the lattice mode, but modes which were degenerate for the single chain may no longer be degenerate for the crystal, though in general the frequency splitting will not be very great. Such splitting is called (*static*) *crystal field splitting*, since it is possible to describe the factor group modes as those of a single chain which has been distorted by the forces due to the surrounding chains – the crystal field – so that it has the lower symmetry of the crystal. In addition to the possible crystal field splitting, certain modes inactive in the infrared or Raman spectrum of the single chain may become active for the crystal because of the lowering of the symmetry.

For crystals with more than one chain per unit cell there are also two possibilities:

(a)   The factor group of the space group is of greater order than that of the line group for the isolated chain, which is a subgroup of it.

(b)   The factor group of the space group contains some but not all of the elements of the factor group of the line group for the isolated chain and contains some new elements.

If (a) holds, no crystal field splitting or activation of modes inactive for the single chain occurs, whereas if (b) holds, both effects may occur. For both (a) and (b) a second type of splitting of modes takes place, called *correlation (field) splitting*. Any mode of the single chain (including each of the components of a mode split by crystal field splitting for (b)) is split into m modes which differ, to a first approximation, only in the relative phases of the motions of the m chains in each unit cell; the motions of the m chains are correlated. The form of the motion for each chain is again, to a first approximation, the same as it would be in the isolated chain. In addition, for both (a) and (b), $4m - 3$ new lattice modes appear. Of these, m correspond essentially to rotatory vibrations of the m chains around their axes, each differing from the others primarily in the relative phases of the motions of the m chains in any unit cell, and the remaining $3m - 3$ correspond essentially to vibrations in which whole chains translate bodily with respect to each other. All these lattice modes usually have rather low frequencies.

It is possible to deduce which two symmetry species correspond to the original single symmetry species for modes split by crystal field or correlation splitting by using the ideas discussed in subsection 3.1.3, where descent in symmetry was considered. Note that it may happen, even when there is more than one chain per unit cell, that the factor groups for the single chain and for the crystal are isomorphous with the same point group. The actual elements of the factor groups themselves must, however, be different in spite of this, and the crystal belongs to category (b). (See the following subsection for an example.)

It will be realized, from the description given of the effects of crystallinity on the factor group modes, that changes in the vibrational spectra of a polymer on crystallization can give valuable information about the crystal structure, and some examples of this will be discussed in chapter 6. Some words of caution are, however, appropriate. The treatment just given is an idealized one in two ways. It assumes first that the intermolecular forces are small compared with the intramolecular forces, so that the perturbation of the single-chain vibrations is small. If this is not so, the vibrations of the crystal will not be related in a simple way to those of the single chain. The form of the chain may be severely distorted and the modes will not be clearly divisible into internal and external or lattice modes – there will be no modes in which the individual chains move essentially as rigid units. The treatment also assumes that the vibrations of the crystalline regions can be studied in the absence of non-crystalline material. This is not usually possible for a polymer, although the crystalline content may sometimes be very high, and the presence of the non-crystalline material may cause the observed spectrum to consist of rather broad peaks so that small splittings are obscured.

We now consider briefly, as examples, the factor group modes of crystalline syndiotactic poly(vinyl chloride) (PVC) and of crystalline polyethylene. In particular, we compare the factor group modes of the crystal with those of the single chain for each polymer. The effects of crystallinity on the spectra of these and other polymers are discussed further in chapter 6.

3.2.2   *Factor group modes for crystalline PVC and polyethylene*

The crystal structure of syndiotactic PVC is shown in fig. 3.6, which indicates the conventional crystallographic axes $Oabc$ and their relationship to the axes $Oxyz$ of fig. 3.2. Each chain has the planar zig-zag conformation and the two chains are arranged in such a way that the orthorhombic crystal has new symmetry elements which interchange the two chains, in addition to the symmetry elements of the isolated chain. These new elements are: centres of inversion i; $\bar{C}_2$ screw axes parallel to Oc and passing through the centres of inversion; $\bar{C}_2$ screw axes parallel to Oa, lying in the Oab mirror planes and intersecting the Oc screw axes; glide planes parallel to Oac and passing through the centres of inversion with translational components parallel to Oa. The factor group of the space group is thus isomorphous with the point group D_{2h}. Since the factor group of the single chain is isomorphous with C_{2v}, a subgroup of D_{2h}, we expect no crystal field splitting or activation of modes inactive for the single chain (the A_2 species modes of C_{2v} are inactive in the infrared), but we do expect correlation splitting because of the two chains per unit cell.

Fig. 3.3 shows that the four symmetry species of C_{2v} split into the eight symmetry species of D_{2h} as follows:

$$A_1 \rightarrow A_g + B_{1u}; \ A_2 \rightarrow A_u + B_{1g}; \ B_1 \rightarrow B_{2g} + B_{3u}; \ B_2 \rightarrow B_{3g} + B_{2u}$$

Note that these relationships apply only if the choice of axes $Oxyz$ is the same for the crystal and the single chain and the Oz axis coincides with the C_2 axis for the single chain. Table 3.2 shows the effect of other choices of axes for D_{2h}; for C_{2v} the B_1 and B_2 species designations are interchanged by interchanging x and y. Each factor group mode of the isolated chain gives rise to two factor group modes of the crystal. In the g modes the two chains in the unit cell vibrate in phase with respect to inversion, whereas in the u modes they vibrate π out of phase. The selection rules to be discussed in the next chapter show that for the point group C_{2v} all species except A_2 are both Raman and infrared active; the A_2 species is only Raman active. For point group D_{2h} they show that all u modes except A_u are active in infrared absorption but not in Raman scattering and all g modes are active in Raman scattering but not in infrared absorption. The A_u modes are active in neither effect.

These results show that the infrared and Raman spectra of the crystal would, in the absence of correlation splitting, have many frequencies in

Fig. 3.6. The crystal structure of poly(vinyl chloride). *Oabc* is the set of conventionally chosen crystallographic axes and *Oxyz* corresponds to the choice of axes in fig. 3.2. For clarity, the hydrogen atoms are not shown. The symmetry elements are shown for one complete unit cell.

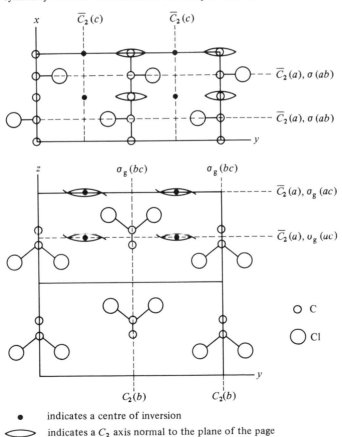

•   indicates a centre of inversion

⬭   indicates a C_2 axis normal to the plane of the page

⬭   indicates a $\bar{C}_2$ axis normal to the plane of the page

common. Because of correlation splitting these frequencies will be slightly different in the two spectra, but neither spectrum on its own will show splitting. The effect is shown for the C—Cl stretching vibrations in fig. 3.7. The splitting is about 6 cm^{-1} for the vibration of B_2 symmetry in the single chain, whereas the vibration of A_1 symmetry has a much smaller splitting. Fig. 3.8 shows schematically the forms of the vibrations of the two molecules in the unit cell for all four C—Cl

Fig. 3.7. The vibrational spectra of poly(vinyl chloride) in the C—Cl stretching region, showing correlation splitting: (*a*) infrared spectrum; (*b*) Raman spectrum.

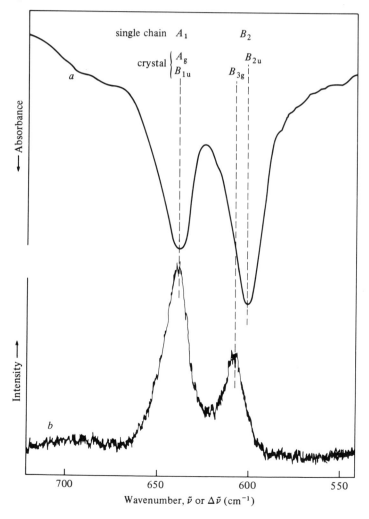

stretching modes. It is important to note that it is the coupling between the vibrations of the two C—Cl pairs in the translational repeat unit of the single chain which causes the gross splitting of the C—Cl stretching frequency into A_1 and B_2 components. These components correspond to in-phase and out-of-phase vibrations, respectively, of the two C—Cl groups in one translational repeat unit. This is a further illustration of the fact that the description of a mode as a simple bond stretching mode is an oversimplification; all modes are modes of the whole molecule or indeed of the whole crystal. Correlation splitting in PVC is discussed in more detail in subsection 6.7.4.

The crystal structure of polyethylene is, like that of PVC, orthorhombic with two chains passing through each unit cell and is shown in fig. 3.9. The factor group of the space group is again D_{2h}. In specifying symmetry species, axes Ox, Oy and Oz have been chosen parallel to Oc, Ob and Oa, respectively. Unlike the two differently oriented molecules in the unit cell of PVC, those in the unit cell of polyethylene are not related by the centres of inversion in the structure, which are located in the individual chains of the polyethylene crystal and half way between adjacent chains in the same orientation. The factor group of the space group is the same as that of the single chain but only some of the symmetry operations of the factor group of the single chain remain symmetry operations for the factor group of the crystal. These are E, $|\sigma(ab)|$, $|C_2(c)|$ and $|i|$, which form a group isomorphous with the point group C_{2h}. This is the factor group of the *site group* for the chains.

Fig. 3.8. The C—Cl stretching modes of crystalline syndiotactic poly(vinyl chloride) as viewed along the chain axis. For clarity, the hydrogen atoms are not shown and the modes are shown as symmetry modes rather than true normal modes. In the true normal modes there are also displacements of the carbon and hydrogen atoms. The centre of inversion which relates the pair of chains within a unit cell is shown; the symmetry of the g modes and the antisymmetry of the u modes with respect to inversion is easily seen.

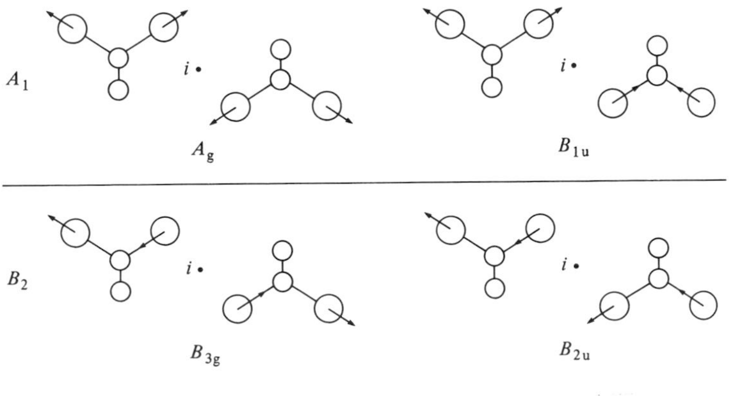

The new operations of the factor group of the space group correspond to two-fold screw axes parallel to the crystallographic Oa and Ob axes and glide planes parallel to the crystallographic Oac and Obc planes. The first of these glide planes is an *axial glide* and has a translational component parallel to Oa of half the translational repeat distance along this axis; the second is a *diagonal glide* with its translational component equal to half the diagonal of the Obc plane of the unit cell. Each of these new symmetry elements of the factor group interchanges the two chains of the unit cell.

Fig. 3.9. The crystal structure of polyethylene. $Oabc$ is the set of conventionally chosen crystallographic axes. The symmetry elements are shown for one complete unit cell.

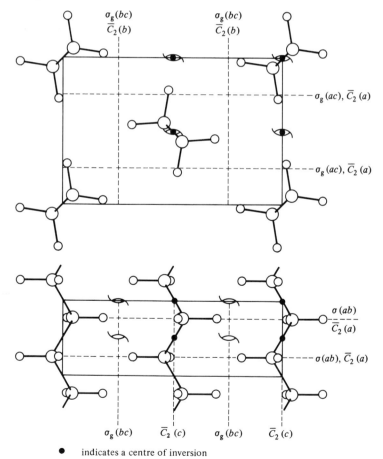

σ indicates a centre of inversion

indicates a $\bar{C}_2$ axis normal to the plane of the figure.

Since the crystal comes into category (*b*) for crystals with more than one chain per unit cell (see subsection 3.2.1), we might at first expect to see crystal field splitting, in addition to correlation splitting and the activation of modes not active for the single chain. The group D_{2h} does not, however, have any degenerate species, so that no crystal field splitting can occur. The site group for the single chain, C_{2h}, is of lower symmetry than that of the isolated chain, D_{2h}. Since centres of inversion remain located in the chains, all u or g modes of the single chain give rise to u or g modes, respectively, of the crystal, and the descent in symmetry is such that A or B_3 species of the factor group of the isolated chain become A species of the factor group of the site group, while B_1 or B_2 species become B species. The correspondence between the symmetry species for the factor groups of the line group of the isolated chain, the site group of the chain in the crystal and the crystal space group are shown in fig. 3.10, which also indicates the spectral activities and numbers of modes of each species. The A_u species of C_{2h} is infrared

Fig. 3.10. Numbers of modes and relationships between symmetry species for single chains and crystals of polyethylene.

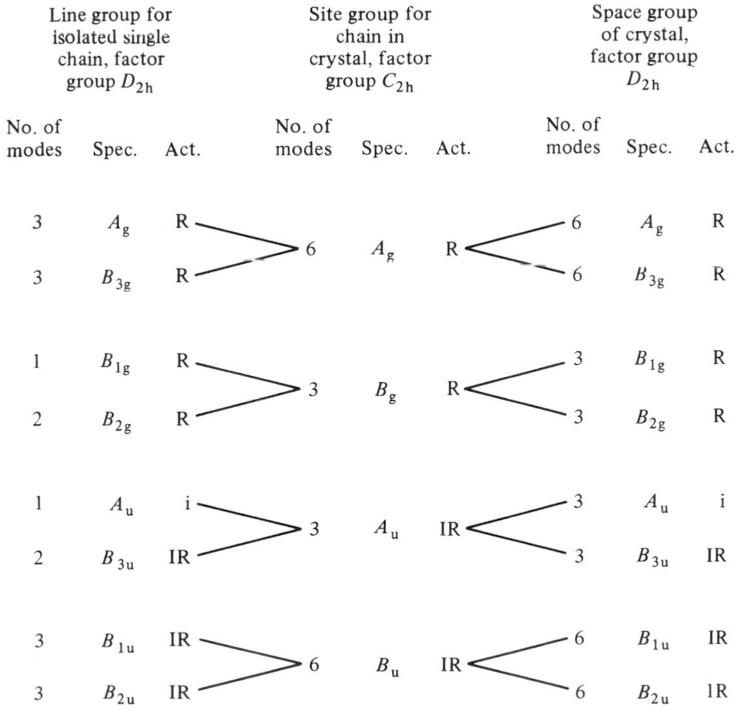

Line group for isolated single chain, factor group D_{2h}			Site group for chain in crystal, factor group C_{2h}			Space group of crystal, factor group D_{2h}		
No. of modes	Spec.	Act.	No. of modes	Spec.	Act.	No. of modes	Spec.	Act.
3	A_g	R				6	A_g	R
3	B_{3g}	R	6	A_g	R	6	B_{3g}	R
1	B_{1g}	R				3	B_{1g}	R
2	B_{2g}	R	3	B_g	R	3	B_{2g}	R
1	A_u	i				3	A_u	i
2	B_{3u}	IR	3	A_u	IR	3	B_{3u}	IR
3	B_{1u}	IR				6	B_{1u}	IR
3	B_{2u}	IR	6	B_u	IR	6	B_{2u}	IR

active, so that the inactive A_u species $-CH_2-$ twisting mode of the isolated single chain (see table 3.2) becomes infrared active for the chain in the crystal. In addition to this crystal field activation of a mode, there is now correlation splitting of all the modes into pairs of modes, the two members of a pair differing according to whether the two chains vibrate in phase or out of phase when related by a particular symmetry operation.

Correlation splitting should not be observable for the A_u species $-CH_2-$ twisting mode, since it gives rise to an infrared active B_{3u} species and an inactive A_u species in the crystal; for all other modes correlation splitting is directly observable in principle. It has been observed in practice for a number of infrared- and Raman-active modes in polyethylene, as discussed in detail in subsection 6.7.2.

In the present section, and elsewhere in this chapter, we have simply stated that certain symmetry species are active or inactive for the IR or Raman spectrum and in the next chapter we shall explain how the activities are derived and how two Raman-active modes of different species or two IR-active modes of different species may be distinguished by the use of polarized radiation, at least in principle. Whether this may be done in practice depends on a variety of factors, some of which will be considered in chapters 5 and 6.

3.3   Further reading

The following books contain further details on the topics of this chapter and some of the topics of chapters 4 and 5: *Infrared Spectroscopy of High Polymers* by R. Zbinden, Academic Press, NY, 1964; *The Theory of Vibrational Spectroscopy and its Application to Polymeric Materials* by P. C. Painter, M. M. Coleman and J. L. Koenig, Wiley 1982.

In addition, the following reviews, among others, may be found helpful: 'Infrared Spectra of High Polymers' by S. Krimm, *Advances in Polymer Science*, vol. 2, p. 51 (1960); 'Vibrational Analysis of Highly Ordered Polymers' by H. Tadokoro and M. Kobayashi in *Polymer Spectroscopy* edited by D. O. Hummel, VCH, 1974. The detailed studies of individual polymers described in the first of these reviews are now somewhat out of date but the introductory section is in large measure still applicable and the review is an important landmark in the subject.

4   Infrared and Raman spectra

4.1   Semiclassical treatment of origins of spectra

4.1.1   *Polarizability tensors and dipole moments*

In this section we explain the origins of infrared absorption and Raman scattering, using as far as possible the ideas of classical physics. Quantum mechanics is introduced only where its use makes the treatment easier or where classical physics is inadequate for explaining a phenomenon or deriving a result. In addition to an explanation of the origins of the effects, an explanation is also given of the *selection rules* which determine which modes of vibration are actually active for infrared or Raman spectroscopy. Since infrared absorption and Raman scattering involve the interaction of molecules with electromagnetic waves, in particular with the oscillating electric field of the waves, it is necessary to consider the electrical properties of the molecules. We begin by defining the *molecular polarizability tensor*.

Imagine first a non-vibrating molecule which belongs to one of the point groups D_2, D_{2h}, D_{2d} or C_{2v}. For such a molecule the intersections of the symmetry elements define a unique set of three mutually perpendicular axes, which will be called $Oxyz$. The molecule consists of a large number of charged particles, the nuclei and the electrons, but is electrically neutral overall. If a steady electric field E is applied to the molecule in a direction parallel to one of the reference axes it will cause a redistribution of the charges in the molecule so that the molecule attains a net electric dipole moment, or undergoes a change in an already existing dipole moment, of amount μ. For low values of E the induced moment μ is proportional to E but its magnitude usually depends on whether E is parallel to Ox, Oy or Oz. The ratio of the induced dipole moment to the applied field is called the *polarizability* and the three values for E parallel to each of the symmetry axes are called the *principal polarizabilities* of the molecule, p_x, p_y, p_z.

If the electric field is applied in a direction which is not parallel to one of the symmetry axes the induced dipole moment will usually not be parallel to E; its components may, however, be deduced by resolving E parallel to the symmetry axes and using the equations

$$\mu_x = p_x E_x, \qquad \mu_y = p_y E_y, \qquad \mu_z = p_z E_z \qquad (4.1)$$

The three principal polarizabilities are said to be the *principal components* of the *polarizability tensor* [p].

Now choose a general set of reference axes $OXYZ$ with some arbitrary but known orientation with respect to $Oxyz$. In general, the application of an electric field parallel to OX will produce components of μ parallel to all three axes OX, OY, OZ and, conversely, a dipole component parallel to OX may be produced by an electric field component parallel to any of the three axes. This means that we can write

$$\mu_X = p_{XX}E_X + p_{XY}E_Y + p_{XZ}E_Z \qquad (4.2a)$$

$$\mu_Y = p_{YX}E_X + p_{YY}E_Y + p_{YZ}E_Z \qquad (4.2b)$$

$$\mu_Z = p_{ZX}E_X + p_{ZY}E_Y + p_{ZZ}E_Z \qquad (4.2c)$$

or, in more compact notation,

$$\boldsymbol{\mu} = \mathbf{p}\mathbf{E} \qquad (4.2d)$$

where $\boldsymbol{\mu}$ and $\mathbf{E}$ are column matrices and $\mathbf{p}$ represents the square matrix of components p_{XX}, p_{XY} etc. These components can be expressed in terms of the principal components p_x, p_y, p_z of [p] and the direction cosines of the angles between the two sets of axes $OXYZ$ and $Oxyz$.

It may be shown that, whether a molecule has any symmetry elements or not, it is always possible to find a set of axes $Ox_0y_0z_0$, called the *principal axes* of the polarizability tensor, for which equations (4.2) reduce to the form of equations (4.1). If the molecule has a principal rotation axis of any order, one of the principal axes of the tensor will coincide with that axis and if the axis is three-fold or greater the principal components of the tensor perpendicular to this axis will be equal. The fact that principal axes exist means that only six of the quantities p_{ij} in equations (4.2) are independent, where i or j may be X, Y or Z, and it may be shown that $p_{ij} = p_{ji}$ for all i, j; the tensor [p] is *symmetric*. For general reference axes two subscripts are required to specify the components of [p] and it is said to be a *second rank* tensor. The reference axes $Oxyz$ for a molecule are usually chosen to be parallel or perpendicular to symmetry elements where possible, with Oz parallel to any principal rotation axis present. The conventional choice of axes with respect to the symmetry elements is usually given in character tables for the point groups, but this leaves ambiguities with respect to the molecule for highly symmetric molecules, which may have, for instance, more than one type of σ_v plane.(See also subsection 2.3.4.)

So far we have considered a steady applied electric field. In infrared and Raman spectroscopy the applied field alternates at the frequency of the incident radiation and this is very much higher for the visible light used in Raman spectroscopy than for the radiation absorbed in infrared spectroscopy, for which it is equal to the frequency of the normal mode of vibration involved. In Raman spectroscopy the polarization of the molecule therefore takes place by distortion of the electronic wave-function, with essentially no effect on the motion of the nuclei, which can only respond at much lower frequencies. The polarizability will, however, depend on the instantaneous positions of the nuclei because they determine the precise form of the electronic wave-function that is distorted by the field. It is the modification of the polarizability by the vibrational motion of the molecule which gives rise to Raman scattering, as discussed in detail in subsection 4.1.3. When infrared radiation of appropriate frequency interacts with the molecule the nuclei and electrons can respond together. It is therefore usual to consider the resulting oscillating electric dipole of the molecule to be produced by the change in magnitude or direction of pre-existing *bond dipoles* and the terms 'polarizability' and 'induced dipole', though strictly applicable, are not used. The mechanism for infrared activity will now be considered in more detail and the selection rule will be deduced.

4.1.2 *Infrared absorption – mechanism and selection rule*

Imagine the effect of irradiating a molecule with radiation of frequency v. In the simplest picture, each atom carries an effective fractional charge which depends on the nature of the atoms to which it is bonded. The electric field acting on these charges causes a distortion of the molecule and in so doing causes a change in the net electric dipole moment of the molecule. Since E varies sinusoidally with frequency v, the molecule tends to be set into vibration at this frequency. If v coincides with the frequency of a suitable mode this mode can be set into resonant vibration and the radiation will then be absorbed. This is the mechanism of infrared absorption according to the classical picture. It is important to realize, however, that not all modes of vibration can absorb radiation in this way.

A mode can only resonate with the oscillatory distortion of the molecule caused by the electric field of the radiation if it belongs to the same symmetry species as the distortion, and this must necessarily be the species to which both the electric field and the vibrational dipole change belong. These are vectors and thus have the same symmetry species as a translation parallel to the electric field. We thus have the following *selection rule* for infrared absorption: a mode of vibration can only absorb radiation polarized parallel to Ox (Oy or Oz) if the mode belongs to the same symmetry species as a translation parallel to Ox (Oy or Oz). We can thus use the column usually given in character tables to indicate the symmetry species of the translations and rotations (see subsection 2.4.3) to deduce which modes will be infrared active for which polarization directions of the incident radiation. It should be noted that, conversely, if the molecule is imagined to be already vibrating in an infrared-active mode the oscillating dipole generated by the vibration will give rise to radiation at the corresponding frequency and with the appropriate direction of polarization. As a specific example of these ideas, consider the bent triatomic molecule belonging to the point group C_{2v} already considered in sections 2.4 and 2.5. In fig. 4.1a the molecule is assumed to consist of a central atom which has an effective negative charge and outer atoms which have an effective, compensating, positive charge. Fig. 4.1b, shows the displacements produced by the application of the field E in the positive Oy direction. This set of displacements clearly belongs to symmetry species B_2 (see subsection 2.4.3) and consists of the superposition of the antisymmetric stretching distortion illustrated in fig. 2.6c and a clockwise rotation (as viewed in the diagram) about an axis parallel to Ox. The antisymmetric stretching mode is thus infrared active for radiation polarized with E parallel to Oy. A similar argument shows that the symmetric stretching mode and the bending mode, illustrated in figs. 2.6a and 2.6b, respectively, are infrared active for radiation polarized with E parallel to Oz.

A more direct way of seeing whether a mode is infrared active and, if so, for which polarization, is to imagine the molecule distorted according to the eigenvector, or symmetry mode vector, for the mode

Fig. 4.1. Infrared activity of the bent triatomic molecule. In this figure the $+$ and $-$ signs represent fractional charges, not displacements: (a) molecule in the absence of an electric field; (b) displacements of charges caused by field E; (c) changes in bond dipoles produced by vibration in the antisymmetric stretching mode and the resultant dipole change.

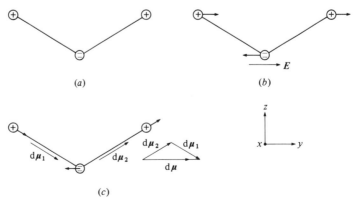

and to consider the changes in the dipole moments due to the individual bonds. These changes are illustrated for the antisymmetric stretching mode in fig. 4.1c and their resultant, the total dipole moment change, μ, is also shown. This is, as expected, parallel to Oy, which confirms that this is the required direction for the E vector for absorption to occur at the frequency of the antisymmetric stretching mode. For a randomly oriented assemblage of molecules, the Oy axes would point randomly in space, so that radiation of any polarization would be absorbed equally, but for oriented systems polarization effects are important and they are considered in section 4.2.

4.1.3 The origin of Raman scattering

We now consider the origin of Raman scattering according to classical physics. Imagine a molecule vibrating in a particular normal mode, m, of frequency v_m and let radiation of frequency v_0 be incident on it. At any particular instant during the vibration of the molecule it will have a polarizability tensor $[p]$, the components of which vary sinusoidally at the frequency v_m because their precise values will depend on the instantaneous distortion of the molecule. Let U_m be the normal coordinate corresponding to the frequency v_m. We can then expand each component p_{ij} of $[p]$ as a Taylor series

$$p_{ij} = p_{ij0} + U_m \frac{\partial p_{ij}}{\partial U_m} + \tfrac{1}{2} U_m^2 \frac{\partial^2 p_{ij}}{\partial U_m^2} + \dots \tag{4.3}$$

where p_{ij0} is the value of p_{ij} in the equilibrium state and the derivatives are understood to be evaluated at $U_m = 0$. We shall neglect all but the first two terms of the series. If the electric field E of the incident radiation is parallel to Ox_j, so that the only non-zero component of E is E_j, the induced dipole μ_i is given by $\mu_i = p_{ij}E_j$, and substitution of the approximate form of p_{ij} from the first two terms of equation (4.3) leads to

$$\mu_i = \left(p_{ij0} + U_m \frac{\partial p_{ij}}{\partial U_m} \right) E_j \tag{4.4}$$

Now

$$E_j = E_{j0} \cos(2\pi v_0 t + \phi_0) \tag{4.5}$$

and

$$U_m = U_{m0} \cos(2\pi v_m t + \phi_m) \tag{4.6}$$

where E_{j0} and U_{m0} are the maximum values of E_j and U_m; ϕ_0 and ϕ_m are unknown phase angles. Thus

$$\mu_i = \left[p_{ij0} + U_{m0} \frac{\partial p_{ij}}{\partial U_m} \cos(2\pi v_m t + \phi_m) \right] E_{j0} \cos(2\pi v_0 t + \phi_0) \quad (4.7)$$

Expanding this expression and rewriting the product of the cosines leads to

$$\mu_i = p_{ij0} E_{j0} \cos(2\pi v_0 t + \phi_0)$$

$$+ \tfrac{1}{2} U_{m0} \frac{\partial p_{ij}}{\partial U_m} E_{j0} \cos[2\pi(v_0 + v_m)t + \phi_0 + \phi_m]$$

$$+ \tfrac{1}{2} U_{m0} \frac{\partial p_{ij}}{\partial U_m} E_{j0} \cos[2\pi(v_0 - v_m)t + \phi_0 - \phi_m] \quad (4.8)$$

The first term in equation (4.8) represents a dipole which oscillates at the frequency v_0 of the incident radiation and which gives rise to scattered radiation at that frequency. This process is called *Rayleigh scattering*. The remaining two terms correspond to dipoles oscillating at frequencies $v_0 \pm v_m$, which give rise to radiation at these frequencies. This is the process called *Raman scattering*. To account for all vibrational modes the right-hand side of equation (4.4) should be summed over all values of m, which leads to a similar summation in equation (4.8). If the electric field of the incident radiation is in an arbitrary direction with respect to the reference axes the right-hand side of equation (4.8) should also be summed over j.

Since the changes in polarizability are usually very much smaller than the polarizability itself, Raman scattering is typically $\sim 10^{-3}$ times less intense than Rayleigh scattering, which is in turn $\sim 10^{-3}$ times less intense than the incident radiation. The radiation scattered with frequency $v_0 - v_m$ is called the *Stokes Raman radiation* and that scattered at frequency $v_0 + v_m$ is called the *anti-Stokes Raman radiation*. The former is usually much more intense than the latter and this may only be understood by considering the quantum-mechanical description of the Raman scattering process.

Each of the modes of oscillation of the molecule may be considered to be a simple harmonic oscillator which oscillates with frequency v_m independently of the oscillators representing the other modes. According to quantum mechanics, a harmonic oscillator of frequency v has equally spaced energy levels of energy E_n given by

$$E_n = (n + \tfrac{1}{2})hv \quad (4.9)$$

where the *quantum number n* is any positive integer or zero. The Raman process consists of the interaction of a photon of the incident light with the molecule in which the molecule loses or gains energy equal to the difference in energy, hv_m, between adjacent energy levels and the photon is scattered with an increase or decrease, respectively, of energy of this amount. For any assembly of harmonic oscillators in thermal equilibrium at temperature T the ratio R_n of the number of oscillators in the state with quantum number n to the number in the ground state with $n = 0$ is given by the Boltzmann factor:

$$R_n = \exp[-(E_n - E_0)/kT] = \exp(-nhv/kT). \qquad (4.10)$$

Thus, unless $hv_m < kT$ there will be many more molecules in the ground state than in any of the excited states and the number of molecules that will be excited to the level $n = 1$ from the ground state by the incident radiation will be significantly greater than the number that are de-excited, since the intrinsic probabilities of these two processes occurring are the same. If $T = 300$ K, then $hv = kT$ for $v = 6 \times 10^{12}$ Hz, a frequency which corresponds to 200 cm^{-1}. Most of the vibrational modes observed in Raman spectroscopy have frequencies higher than this, generally between about 400 and 3500 cm^{-1}, so that $hv_m \gg kT$. A more complete treatment shows that there is also a frequency dependent term $[(v_0 - v_m)/(v_0 + v_m)]^4$ in the ratio of the intensities of the Stokes and anti-Stokes scattering, but the Boltzmann factor is the dominant one so that the Stokes process predominates. It is therefore usual to scan only the Stokes side of the exciting line and Raman spectrometers are usually designed to read wavenumber shifts from the exciting line as positive if the true wavenumber of the detected light is lower than that of the incident light. It should, however, be noted that many lattice mode vibrations in crystals, and some torsional modes in individual molecules, are at lower frequencies and it is sometimes useful to scan the anti-Stokes spectrum when the frequencies of interest are low enough for it to have appreciable intensity.

4.1.4 *Raman scattering and the rule of mutual exclusion*

It follows from the previous subsection that if a particular normal mode with normal coordinate U_m is to be Raman active it is necessary for the partial derivative with respect to U_m of at least one component of the polarizability tensor $[p]$ to be different from zero. If we define α_{ij} by

$$\alpha_{ij} = \frac{\partial p_{ij}}{\partial U_m} \qquad (4.11)$$

then the quantities α_{ij} are the components of a new symmetric second rank tensor $[\alpha]$, the *derivative polarizability tensor*, or *Raman tensor*. It must be remembered that the values of α_{ij} depend on which normal mode they refer to, but for convenience of notation no explicit indication of this is given where such an omission is not confusing. The subscript m will also be dropped from U_m. Whether a mode has non-zero values of α_{ij}, and which of the values are non-zero, depends only on its symmetry species. Before discussing how the selection rules for Raman activity can be derived more generally, we consider specifically how the presence of a centre of inversion affects the spectral activity of a molecule. A molecule which has a centre of inversion is often called a *centrosymmetric molecule.*

It is clear that a molecule with a centre of inversion cannot have a dipole moment in its equilibrium state or at any instant during a normal mode vibration in which it retains the centre of inversion. Such normal modes are thus infrared inactive. Consider, on the other hand, a normal mode in which the centre of inversion is not retained. The effect of the operation i is then to multiply all displacements by -1, which is equivalent to multiplying the normal coordinate U by -1. It may be shown, however, that the operation i leaves all the components of a second rank tensor unchanged. Thus, if any component p_{ij} of $[p]$ is plotted against U for a mode which does not retain the centre of inversion the plot must be similar to that in fig. 4.2a, so that $\alpha_{ij} = \partial p_{ij}/\partial U|_{U=0} = 0$. The mode is thus Raman inactive. These conclusions are summarized in the *rule of mutual exclusion* for centrosymmetric molecules: no mode can be both Raman and infrared active. It should, however, be noted that for some centrosymmetric molecules there are modes which are neither infrared nor Raman active.

Consider as an example the linear X—Y—Z molecule. This molecule belongs to the point group $D_{\infty h}$ and, by a generalization of the method of

Fig. 4.2. Variation of a component p_{ij} of the molecular polarizability tensor with the normal coordinate U: (a) Raman-inactive component; (b) Raman-active component.

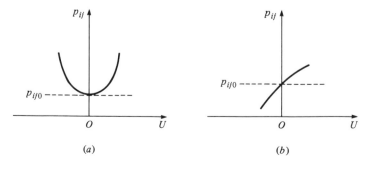

(a)   (b)

subsection 2.4.3, it may be shown that its normal modes of vibration are $A_{1g} + A_{1u} + E_{1u}$, and these are illustrated in fig. 4.3. It is immediately clear from the figure that the A_{1g} mode cannot be infrared active, since it retains the centre of inversion, but the polarizability of the molecule will clearly be different when the bonds are both compressed from when they are both stretched, i.e. the graph of each component of $[p]$ against U will be similar to that in fig. 4.2b, so that α_{ij} is not zero and the mode is Raman active. It is equally clear that the A_{1u} and E_{1u} modes cannot be Raman active, since the polarizability of the molecule is the same for both phases of the oscillation. For the A_{1u} mode there is always one stretched and one compressed bond and for the E_{1u} modes opposite phases are mirror images in either the Oyz or the Oxz planes and the polarizability tensor is unaffected by reflection in one of its principal planes. There is, however, nothing to prevent the molecule from having an oscillating dipole moment in the A_{1u} mode and the type of argument used in subsection 4.1.2, based on bond dipoles, shows that it will be parallel to Oz. It is also readily seen that the mode has the same symmetry as a translation parallel to that axis. Similarly, the two components of the E_{1u} mode will have oscillating dipoles associated with them which are parallel to Ox or Oy. The spectral activity of the molecule is therefore as shown in table 4.1. For this particular centrosymmetric molecule all normal modes are either infrared or Raman active.

Consider as a second example the modes of vibration of the planar square molecule illustrated in fig. 4.4. The molecule belongs to point

Fig. 4.3. Vibrational modes of the linear XYX molecule: (*a*) A_{1g} symmetric stretch; (*b*) A_{1u} antisymmetric stretch; (*c*) and (*d*) degenerate E_{1u} bend. The + and − signs denote displacements up and down, respectively, with respect to the plane of the diagram.

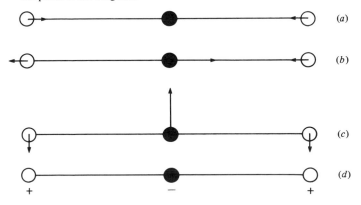

group D_{4h} and analysis by the method of subsection 2.4.3 shows that the vibrational modes are $A_{1g} + B_{1g} + B_{2g} + B_{1u} + E_u$. The spectral activity is shown in table 4.2. For this centrosymmetric molecule one mode, B_{1u}, is active in neither the Raman nor the infrared. The inactivity in the Raman follows from the fact that the centre of symmetry is not retained and it is clear from the form of the vibration that the mode does not belong to the same symmetry species as a translation or, equivalently, that it does not have an oscillating dipole moment associated with it, so that it is not infrared active either.

We turn now to a more formal consideration of the selection rules for Raman activity of molecules of any symmetry and to a discussion of the forms of the Raman tensors for various modes of various point groups.

Table 4.1. *Spectral activity for the linear triatomic molecule XYX*

Mode	i retained?	IR active?	Raman active?
A_{1g}	yes	no	yes
A_{1u}	no	yes	no
E_{1u}	no	yes	no

Fig. 4.4. Vibrational modes of the planar square molecule consisting of four identical atoms. The $+$ and $-$ signs represent displacements perpendicular to the plane of the molecule.

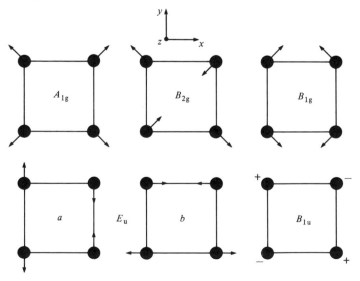

Table 4.2. *Spectral activity for the square molecule* X_4

Mode	i retained?	IR active?	Raman active?
A_{1g}	yes	no	yes
B_{1g}	yes	no	yes
B_{2g}	yes	no	yes
B_{1u}	no	no	no
E_u	no	yes	no

4.1.5 Selection rules for Raman activity: Raman tensors

We have already seen, in subsection 1.4.2, that infrared activity is possible for a particular normal mode only if the symmetry species of the vibration is the same as that of at least one component of a dipole moment or, equivalently, of any vector such as a translation. In an analogous way, Raman activity is possible only if at least one component or combination of components of the Raman tensor $[\alpha]$ belongs to the same symmetry species as the normal mode. Components or combinations which belong to a different symmetry species must be zero. We must thus ask how the components α_{ij} transform under the symmetry operations of the molecule. It is easily shown that these quantities transform in the same way as the corresponding quantities ij, i.e. x^2, y^2, z^2, xy, yz and zx, where (x, y, z) is any general point in the molecule.

Any second order function in x, y and z can be written as a linear combination of the six linearly-independent functions x^2, y^2, z^2, xy, yz and zx. Since any second order function transforms under any operation of any group into another second order function, this set of functions may be used as the basis for a six-dimensional representation of any group. This representation must be reducible, since no molecular point group has symmetry species which have a degree of degeneracy higher than five; in the groups of importance for polymers the degree of degeneracy does not exceed two. Any other set of six linearly-independent second order functions in x, y and z could equally well be used as the basis for a six-dimensional representation. The function $f_0 = x^2 + y^2 + z^2$ is an obvious choice for one of the basis functions, since it is invariant (i.e. its value is multiplied by 1) under every operation of every point group because it is simply the square of the distance r of the point (x, y, z) from the origin. This quantity thus forms a basis for the one-dimensional identity representation belonging to the totally symmetric species of any group.

The most generally suitable choice of five other real basis functions is as follows:

$$f_1 = x^2 - y^2; \quad f_2 = 2z^2 - x^2 - y^2; \quad f_3 = xy; \quad f_4 = yz; \quad f_5 = zx$$

The functions f_1/r, f_2/r, ..., f_5/r are the five real spherical harmonic functions of second order. The functions f_1 to f_5, or the corresponding spherical harmonics, form the most suitable basis set because these functions are not only independent but mutually orthogonal and all the irreducible representations derived for every point group from the five-dimensional representation with f_1 to f_5 as basis can be written with these functions as individual basis vectors, with the exception of the representations corresponding to those doubly degenerate species of non-cubic groups which are reducible in terms of complex coordinates. For this special type of doubly degenerate species it is necessary to use the related set of five complex spherical harmonic functions if individual functions are to be basis vectors.

For the axial groups, that is groups with n-fold axes parallel to Oz and possibly two-fold axes perpendicular to Oz but no other rotational symmetry, z^2 is obviously invariant under all operations of the group, as well as $x^2 + y^2 + z^2$; thus z^2 and $x^2 + y^2$ (and also f_2) become possible bases for the identity representation. For the groups $C_s, C_2, C_{2h}, C_{2v}, D_2$ and D_{2h}, x^2 and y^2 are each independently invariant under all operations of the group and become possible bases for the identity representation. For C_s, C_2 and C_{2h}, xy becomes a fourth possible basis for the identity representation. For the groups C_1, C_i, C_s and all the axial groups, it is in fact possible to choose bases for all irreducible representations from the set of functions $x^2 + y^2, x^2 - y^2, x^2, y^2, z^2, xy, yz$ and zx.

The possible sets of combinations of the functions x^2, y^2, z^2, xy, yz and zx which form bases for the different irreducible representations of the various point groups are usually shown in character tables in a column at the right of the table, as in table 2.4 of subsection 2.3.3, the character table for the point group C_{4v}. If any combination of the functions appears in the row for a particular symmetry species, vibrational modes of that species will be Raman active. The particular combination, or combinations, which appear may be used to derive the explicit form of the Raman tensor or tensors for the vibrational mode, and for some purposes (see, e.g., section 6.10) it is necessary to know this; the method of derivation is now briefly indicated.

We start from the fact that α_{xx} transforms exactly as x^2, α_{xy} as xy and so on. In addition, for a non-degenerate mode, each non-zero component of the tensor must belong to the same symmetry species as the normal mode. Consider a specific example: can α_{xx} be non-zero for the A_1 species of point group C_{4v}? Now $\alpha_{xx} = (\alpha_{xx} + \alpha_{yy})/2 + (\alpha_{xx} - \alpha_{yy})/2$ and the last column of table 2.4 shows that $x^2 + y^2$, and thus $\alpha_{xx} + \alpha_{yy}$, belongs to the A_1 species whereas $x^2 - y^2$, and hence $\alpha_{xx} - \alpha_{yy}$, does not. Thus α_{xx} can only belong to this species if $\alpha_{xx} = \alpha_{yy}$. In a similar way it is possible, by setting equal to zero the combinations of tensor components corresponding to any of f_0 to f_5 which do not belong to that particular species, to deduce any relationships which must exist between the Raman tensor components of any non-degenerate species and also whether any components must be zero. The following relationships can thus be found between expressions in the column of the character table relating to second order functions and the Raman tensors for the corresponding non-degenerate species:

$$x^2 + y^2 \text{ implies } \alpha_{xx} = \alpha_{yy}$$

$$x^2 + y^2 + z^2 \text{ implies } \alpha_{xx} = \alpha_{yy} = \alpha_{zz}$$

$$x^2 - y^2 \text{ implies } \alpha_{xx} = -\alpha_{yy}$$

The entry x^2, y^2, z^2, xy, yz or zx implies simply that the corresponding component α_{xx} or α_{yy}, etc. may be non-zero. For C_{4v} the forms of the Raman tensors corresponding to non-degenerate modes are thus of the form

$$A_1: \begin{bmatrix} a & . & . \\ . & a & . \\ . & . & b \end{bmatrix} \qquad B_1: \begin{bmatrix} c & . & . \\ . & -c & . \\ . & . & . \end{bmatrix} \qquad B_2: \begin{bmatrix} . & d & . \\ d & . & . \\ . & . & . \end{bmatrix}$$

where the dots signify zero components, and the A_2 mode is Raman inactive.

Similarly, for the pairs of tensors corresponding to doubly degenerate species of axial point groups (other than those for which the individual complex spherical harmonic functions form bases for two separate complex one-dimensional representations) we can deduce the simplest forms of each of the individual tensors in the following way:

(i)   First set equal to zero any of f_0' to f_5' which does not belong to that symmetry species, where f_i' is the combination of tensor components corresponding to f_i.

(ii)   For any tensor of the degenerate multiplet which contains f_i', set to zero all those f_1' to f_5' which do not have the same symmetry as f_i' with respect to any C_2 or σ_v element parallel to Ox or Oy.

(iii)   Use the condition that the total intensity due to any multiplet must be invariant for any operation of the group to relate the components of the tensors of a pair quantitatively to each other.

Consideration of these three point shows, for instance, that the pair of tensors corresponding to the degenerate E_g modes for the point group D_{3d} are of the form

$$\begin{bmatrix} c & \cdot & \cdot \\ \cdot & -c & d \\ \cdot & d & \cdot \end{bmatrix} \qquad \begin{bmatrix} \cdot & -c & -d \\ -c & \cdot & \cdot \\ -d & \cdot & \cdot \end{bmatrix}$$

For those doubly degenerate species so far excluded it may be shown that the entries $(x^2 - y^2, xy)$ or (yz, zx) imply tensor pairs of the forms

$$\begin{bmatrix} a & b & \cdot \\ b & -a & \cdot \\ \cdot & \cdot & \cdot \end{bmatrix} \begin{bmatrix} b & -a & \cdot \\ -a & -b & \cdot \\ \cdot & \cdot & \cdot \end{bmatrix} \quad \text{or} \quad \begin{bmatrix} \cdot & \cdot & c \\ \cdot & \cdot & d \\ c & d & \cdot \end{bmatrix} \begin{bmatrix} \cdot & \cdot & -d \\ \cdot & \cdot & c \\ -d & c & \cdot \end{bmatrix}$$

respectively. If both pairs of functions belong to the same doubly degenerate species then the possible pairs of tensors have only $\alpha_{zz} = 0$.

4.2   Polarization effects
4.2.1   *Polarized infrared spectroscopy – dichroic ratios*
If a polymer is allowed to solidify from the melt or from solution, to give a rubbery, glassy or partially crystalline solid, the axes of the molecular segments or crystallites are usually randomly or nearly randomly oriented, i.e. there is no tendency for the axes to be parallel to any particular direction in the solid rather than another. It is difficult to produce perfectly random orientation, but if there is no preferred orientation the material is isotropic in its physical properties. In particular, if an infrared spectrum is obtained using polarized infrared radiation incident normally on a thin film sample produced in such a way that its molecules are randomly oriented the spectrum will be the same for any orientation of the polarization vector within the plane of the film. If, however, the film is stretched before the spectrum is obtained it will generally be found that the spectrum is no longer independent of the orientation of the polarization vector. This is because the stretching causes a partial alignment of the axes of the molecules, usually towards the direction of stretching, so that the material is no longer isotropic.

The anisotropy of the sample manifests itself in the infrared absorption spectrum because the contribution that any particular molecular segment makes to the total absorption of the sample at the frequency of a particular normal mode depends on the angle between the electric field vector of the incident radiation and the oscillating dipole moment vector of that molecular segment. Only the component of the electric field parallel to the dipole moment vector is effective and thus the contribution is greatest when the electric field and the dipole moment are parallel and zero when they are perpendicular to each other. Producing a preferred alignment of the molecular axes by stretching the sample also produces a certain degree of preferred alignment of the dipole moments. The absorbance of the sample for radiation polarized parallel to the stretching direction, $A_\parallel$, will in general now be different from that for radiation polarized perpendicular to the stretching direction, $A_\perp$, for any absorption peak. Fig. 4.5 shows an example. The quantity $D = A_\parallel / A_\perp$ is called the *dichroic ratio* of the sample for the corresponding mode. If $D > 1$ the mode is said to be a *parallel mode*, or π *mode*; otherwise it is a *perpendicular mode* or σ mode. It can be shown

Fig. 4.5. Part of the infrared spectrum of oriented poly(ethylene terephthalate). —— Electric field vector parallel to the draw direction; – – – – electric field vector perpendicular to the draw direction.

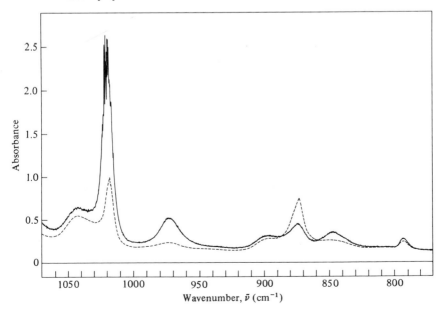

that, for the simplest kind of distribution of molecular orientations produced by stretching, π modes are those for which the angle θ between the dipole moment due to any molecular segment and the axis of that segment is less than $54.7°$ and that σ modes are those for which $\theta > 54.7°$. Information about whether a particular mode is a π or σ mode may help to identify the origin of the mode, as will be explained in subsection 4.3.2. Measurements of *infrared dichroism* can also give useful information about the degree of molecular orientation in polymer samples and their use for this purpose is discussed in chapter 6.

4.2.2   *Polarized Raman spectroscopy – depolarization ratios*

Two beams of radiation are necessarily involved in the recording of a Raman spectrum; the incident beam and the scattered beam. The fact that the state of polarization of both beams must be considered makes a discussion of polarization effects in Raman spectroscopy necessarily more complicated than that just given for infrared spectroscopy. It also has the effect of making Raman spectroscopy a more powerful tool than infrared spectroscopy for studying both oriented and isotropic systems of molecules because more information can be obtained about them. In this section we shall consider only the information obtainable about isotropic systems; oriented systems are discussed in chapter 6.

Equations (4.2d), (4.8) and (4.11) show that the amplitude of the induced dipole moment which gives rise to Raman scattering is given by

$$\boldsymbol{\mu} = \text{const} \cdot \alpha \mathbf{E}_0 \tag{4.12}$$

Let polarized light be incident on a single molecule with the electric field of the light wave parallel to the axis Ox_j of a set of axes $Ox_1x_2x_3$ fixed in space, so that the only non-zero component of the electric field is E_j. If the scattered radiation is observed through an analyser which allows only light with its electric vector parallel to Ox_i to pass, the amplitude transmitted by the analyser is proportional to μ_i. Thus, the observed intensity is proportional to α_{ij}^2. The constant of proportionality depends on E_{j0}, i.e. on the incident light intensity, and on instrumental factors which govern the efficiency with which the Raman scattered light is collected. Provided that the observed intensities have been corrected for any polarization sensitivity of the spectrometer, we can thus write

$$I_{ij} = I_0 \alpha_{ij}^2 \tag{4.13}$$

where I_{ij} is the intensity observed for incident and scattered light polarization vectors $\boldsymbol{E}$ parallel to Ox_j and Ox_i, respectively, and I_0 is a constant independent of i and j.

We now evaluate the ratio $\rho_l = I_\perp/I_\| = I_{ij}/I_{ii}$, with $i \neq j$, for an assembly of randomly oriented molecules, where $I_\|$ is the intensity measured for parallel polarization vectors and $I_\perp$ that for perpendicular polarization vectors of the incident and scattered light, respectively. The quantity ρ_l is called the *linear depolarization ratio*, linear because the incident and the scattered radiation are linearly polarized. Thus

$$\rho_l = \frac{I_\perp}{I_\|} = \frac{\sum \alpha_{ij}^2}{\sum \alpha_{ii}^2} = \frac{\langle \alpha_{ij}^2 \rangle}{\langle \alpha_{ii}^2 \rangle} \tag{4.14}$$

where the summations extend over all the molecules which contribute to the observed scattered light and the brackets $\langle \ \rangle$ denote values averaged over the random distribution of these molecules.

In order to perform the averages we introduce the principal components α_1, α_2 and α_3 of $[\alpha]$ and use the expression

$$\alpha_{ij} = \sum_k a_{ik} a_{jk} \alpha_k \tag{4.15}$$

where a_{ij} are the elements of the direction cosine matrix which relates the principal axes of $[\alpha]$ to the axes $Ox_1 x_2 x_3$ fixed in space. Thus

$$\rho_l = \frac{\langle \sum_{kl} a_{ik} a_{jk} a_{il} a_{jl} \alpha_k \alpha_l \rangle}{\langle \sum_{kl} a_{ik}^2 a_{il}^2 \alpha_k \alpha_l \rangle}, \quad i \neq j$$

$$= \frac{\sum_{kl} \alpha_k \alpha_l \langle a_{ik} a_{jk} a_{il} a_{jl} \rangle}{\sum_{kl} \alpha_k \alpha_l \langle a_{ik}^2 a_{il}^2 \rangle}, \quad i \neq j$$

$$= \frac{\sum_k \alpha_k^2 \langle a_{ik}^2 a_{jk}^2 \rangle + 2 \sum_{k<l} \alpha_k \alpha_l \langle a_{ik} a_{jk} a_{il} a_{jl} \rangle}{\sum_k \alpha_k^2 \langle a_{ik}^4 \rangle + 2 \sum_{k<l} \alpha_k \alpha_l \langle a_{ik}^2 a_{il}^2 \rangle}, \quad i \neq j \tag{4.16}$$

It can be shown that the values of the averages are
$$\langle a_{ik}^2 a_{jk}^2 \rangle = \langle a_{ik}^2 a_{il}^2 \rangle = \tfrac{1}{15} \quad (i \neq j, k \neq l)$$
$$\langle a_{ik} a_{jk} a_{il} a_{jl} \rangle = -\tfrac{1}{30} \quad (i \neq j, k \neq l); \quad \langle a_{ik}^4 \rangle = \tfrac{1}{5}$$

Hence

$$\rho_l = \frac{\alpha_1^2 + \alpha_2^2 + \alpha_3^2 - \alpha_1\alpha_2 - \alpha_2\alpha_3 - \alpha_3\alpha_1}{3(\alpha_1^2 + \alpha_2^2 + \alpha_3^2) + 2(\alpha_1\alpha_2 + \alpha_2\alpha_3 + \alpha_3\alpha_2)} = \frac{3\beta^2}{45\bar{\alpha}^2 + 4\beta^2}$$

where
$$\tag{4.17}$$

$$\bar{\alpha} = \tfrac{1}{3}(\alpha_1 + \alpha_2 + \alpha_3) \tag{4.18a}$$

$$2\beta^2 = (\alpha_1 - \alpha_2)^2 + (\alpha_2 - \alpha_3)^2 + (\alpha_3 - \alpha_1)^2 \tag{4.18b}$$

The quantity $\bar{\alpha}$ is the mean value of the three principal components of $[\alpha]$ and is often called the *spherical part* of the tensor. The quantity β

measures the *anisotropy* of the tensor, since it is only zero if all three principal components are equal, in which case $[\alpha]$ has spherical symmetry.

A second depolarization ratio ρ_n is sometimes defined for illumination with unpolarized or natural light when the scattered light is observed in a direction at right angles to that along which the incident light propagates:

$$\rho_n = \frac{I'_\perp}{I'_\parallel} \tag{4.19}$$

where $I'_\parallel$ and $I'_\perp$ are the intensities for the observed light polarized parallel or perpendicular, respectively, to the propagation direction of the incident light. It may be shown, by a method similar to the above, that

$$\rho_n = \frac{6\beta^2}{45\bar{\alpha}^2 + 7\beta^2} = \frac{2\rho_l}{1 + \rho_l} \tag{4.20}$$

Since the laser beams used nowadays for exciting Raman spectra are usually polarized, ρ_n is not often quoted and will not be considered further. The importance of the depolarization ratio ρ_l is that its value gives information about the symmetry species to which a mode belongs and thus the determination of depolarization ratios provides information about the symmetry of the molecule.

It is a property of any second rank tensor $[T]$ that the quantity $T_{11} + T_{22} + T_{33}$ is independent of the choice of reference axes, i.e. it is a *tensor invariant*. Let δ_{ij} be the change in the component p_{ij} of the molecular polarizability tensor produced by any distortion of the molecule from its equilibrium state, so that any component α_{ij} of the Raman tensor $[\alpha]$ is equal to the value of δ_{ij} for unit positive vibrational distortion of the molecule from its equilibrium state according to the form of the eigenvector for the mode in question. Let this be a non-degenerate mode not belonging to the totally symmetric species, so that there must be at least one symmetry operation which has character -1. Consider the molecule with the unit positive distortion and imagine a rotation of axes, with respect to the fixed molecule, which produces the same relative reorientation of the molecule with respect to the axes as would a particular operation with character -1, where the operation moves the molecule and the axes are regarded as fixed. Since it is a tensor invariant, the quantity $\delta_{11} + \delta_{22} + \delta_{33}$ is unchanged by the rotation of axes. The operation on the molecule for fixed axes would, however, reverse the displacements of all the atoms and hence would reverse the signs of all the δ_{ij} (see fig. 4.2b). It follows that $\delta_{11} + \delta_{22} + \delta_{33}$ must be zero for this mode and so therefore must $\frac{1}{3}(\alpha_{11} + \alpha_{22} + \alpha_{33}) =$

$\frac{1}{3}(\alpha_1 + \alpha_2 + \alpha_3) = \bar{\alpha}$. A slightly more difficult argument shows that $\bar{\alpha}$ must also be zero for degenerate modes, so that $\bar{\alpha} = 0$ for all modes which do not belong to totally symmetric species. Thus the depolarization ratio takes the following values:

for totally symmetric species, $0 \leqslant \rho_l \leqslant 3/4$ (4.21a)

for all other species, $\rho_l = 3/4$ (4.21b)

If $\rho_l = 0$, the Raman peak, or the corresponding mode, is said to be *totally polarized*, and β must be zero; the polarizability is isotropic at all instants during the vibration so that light is scattered only with the same polarization as that of the incident light. The only species for which $\rho_l = 0$ are the totally symmetric species of cubic or icosahedral symmetry, so that this value does not occur for the vibrational modes of polymers. If $\rho_l < \frac{3}{4}$ the peak or mode is said to be *polarized* or *partially polarized* (p) and if $\rho_l = \frac{3}{4}$ the line is said to be *depolarized* (d). Modes with $\rho_l = \frac{3}{4}$ are sometimes said to be 'completely depolarized' but this is a usage best avoided since, by definition, completely depolarized scattering would have $\rho_l = 1$.

It should be noted that measurements of depolarization ratios, particularly on polymers, often give results that are higher than would be given by an ideal sample because of polarization scrambling caused by inhomogeneities in the sample, and care must be taken in interpreting them. Care must also be taken to avoid molecular orientation in the sample.

4.3 Vibrational assignments

4.3.1 *Introduction*

The observed vibrational spectrum of any polymer consists of a large number of peaks and any peak is said to have been *assigned* when its origin is understood. This understanding, or *assignment*, may, in principle, be at one of several levels of refinement. In practice, the process of determining the assignment may involve tackling the problem at several levels simultaneously. The levels at which the assignment may be known for a polymer are, in order of increasing detail:

(1) to a chemical species or molecular configuration;
(2) to the crystalline or amorphous region;
(3) to a specific molecular conformation;
(4) to a symmetry species;
(5) to a specific normal mode, i.e. to a specific eigenvector.

It is clear that assignment to a chemical species or molecular configuration is required only if more than one species or configuration is present in the sample, but this is often so, particularly in commercial polymers. If samples are available in which the concentrations of the species or configurations vary sufficiently it should be possible, particularly for a simple binary mixture, to divide the observed peaks in the spectrum into two sets such that the intensities of the peaks in one set all increase with respect to those of the peaks in the other set when the concentration is changed. Each subspectrum corresponding to a particular set can then be treated as a separate spectrum for further assignment. Some peaks may not be sensitive to concentration and must be treated as belonging partially to both subspectra, since such a peak must be the result of the overlap of peaks belonging to each of the two components.

In a similar manner, if the crystallinity of a sample can be varied substantially, for instance by suitable heat treatment, it should be possible to assign some peaks to the crystalline regions and some to the amorphous regions. It should be noted that heat treatment of a sample may produce changes of molecular conformation in the non-crystalline regions and spectral changes will generally accompany these in addition to any due to changes in crystallinity or crystalline perfection.

Examples of these kinds of assignment, which are at the first two levels of refinement, and of their usefulness in providing information about the structure of polymers will be discussed in subsequent chapters. The practical difficulties that arise because of the finite widths and consequent overlapping of the peaks will be discussed there. In the remainder of this section we shall consider various methods which may be used to give information about assignments at the higher levels of refinement, or at the lower levels when simpler methods fail. We first indicate how the ideas of symmetry, selection rules and polarization effects discussed in earlier sections may be used in principle to produce assignments for the spectrum of a polymer consisting of a single chemical species with a single configuration and a regular conformational structure. We then consider three related techniques: the use of group frequencies, model compounds and isotope substitution. A more theoretical approach to detailed vibrational assignment and to the overall understanding of molecular dynamics involves the attempt to calculate vibrational mode frequencies and a brief introduction to vibrational calculations is given in section 4.4.

4.3.2   *Factor group analysis – chain or local symmetry*
    Consider a monosubstituted vinyl polymer consisting largely of molecules of only one tacticity. The first column of table 4.3 shows all the reasonably likely structures for each tacticity, together with the

Table 4.3. *Selection rules for monosubstituted vinyl polymers*

Structure	Symmetry of factor group	R / IR	p / π	p / σ	d / π	d / σ	p / i	d / i	i / π	i / σ
Syndiotactic										
n_k helix, $n>3$	D_n					✓	✓	✓		✓
3_1 helix	D_3					✓	✓			✓
2_1 helix	D_2				✓	✓	✓			
planar zig-zag	C_{2v}				✓	✓	✓		✓	
Isotactic										
n_k helix, $n>3$	C_n		✓		✓				✓	
3_1 helix	C_3		✓		✓					
planar zig-zag	C_s		✓	✓						

[a] A tick indicates that the factor group has a symmetry species with the corresponding spectral activities.
Adapted from fig. 12 of J. L. Koenig, *Applied Spectroscopy Reviews* **4** 223 (1979), by courtesy of Marcel Dekker, Inc., NY.

point group which is isomorphous with the factor group of the line group for each structure. The remaining columns show, for each structure, the types of spectral activity which may be exhibited by an optically active mode. The symbols p, d, σ and π have the significance explained in section 4.2 and i stands for inactive.

Examination of table 4.3 shows that no two structures have the same set of permitted activities if both Raman and infrared spectra can be obtained using polarized light. The infrared spectra must, of course, be obtained on oriented samples but the Raman spectra must be obtained on unoriented samples. It is thus possible, at least in principle, to determine which of the seven structures is taken up by the vinyl polymer simply by noting the types of spectral activity shown, and to assign the observed vibrational modes to their symmetry species. Note, however, that if oriented samples are not available for the infrared studies the structures C_{2v} and C_n cannot be distinguished simply by the types of spectral activity shown, nor can the structures C_3 and C_s. If only infrared spectra or Raman spectra, but not both, are available, none of the structures can be distinguished solely on the basis of the allowed types of spectral activity, even if oriented samples are used for the infrared spectra. This is one illustration of the great advantages that the combination of infrared and Raman spectroscopies has over the sole use of either technique.

In the example just discussed, the number of different symmetry species varies from two to four for the different possible structures. As described in subsection 3.2.2, one of the commonest vinyl polymers, PVC, has the planar syndiotactic structure in its crystallites, with 12 atoms per physical repeat unit, and this means (neglecting correlation splitting) that there are 32 true vibrational modes distributed among four different symmetry species. Assignment to the correct symmetry species is thus a big step towards the full assignment of a particular vibrational mode. Consider, in contrast, another very common polymer, the polyester poly(ethylene terephthalate) or PET, which illustrates the inadequacy of this direct approach.

In the crystalline regions PET has the structure shown in fig. 4.6. The repeat unit has 22 atoms and there are thus 62 normal modes. Even if the hydrogen atoms of the $\diagdown CH_2$ groups are neglected, the molecule is not quite planar and it will thus be observed that the chain has no symmetry elements except for centres of inversion at points corresponding to i_1 and i_2, the centres of the terephthaloyl and glycol residues, respectively. The factor group of the line group is thus C_i and the 62 normal modes therefore belong to only two different symmetry species, A_g and A_u. Even if the approximation is made that the molecule is planar (apart from the H atoms of the $\diagdown CH_2$ groups), the only additional symmetry elements introduced are C_2 axes, normal to the plane of the chain and passing through points such as i_1 and i_2, and a mirror plane, C_h, coincident with the plane of the molecule. The corresponding factor group is C_{2h} and there are still only four different symmetry species: A_g, B_g, A_u and B_u.

If, however, the paradisubstituted benzene ring (i.e. the ring and the two attached carbon atoms) could be considered on its own it would be seen to have two additional two-fold rotation axes and two additional mirror planes, so that it would have the higher symmetry of D_{2h}. Thus, if any of the vibrational modes of the whole molecule are essentially localized to the benzene ring they may be discussed, to a first approximation, as if they were modes of a molecule with the *local symmetry*, D_{2h}, of the ring and assigned to the appropriate local symmetry species in addition to the appropriate symmetry species of the whole molecule. The relationships between the symmetry species of the groups C_i, C_{2h} and D_{2h} are shown in fig. 4.7, which also indicates their relationship to those of the group D_{6h} to which the benzene molecule belongs. The spectral activities of the various species are also shown.

In using such local symmetries one must be aware that the corresponding selection rules will not be obeyed exactly, e.g. modes which are
predicted to be inactive may be weakly active because the symmetry of
the whole molecule is not the same as the local symmetry. The local
symmetry method is usually combined with the use of the idea of group

Fig. 4.6. The chain conformation in poly(ethylene terephthalate) crystallites.
(Reproduced by permission from R. de P. Daubeny *et al.*, *Proceedings of the
Royal Society* **A226** 531 (1954).)

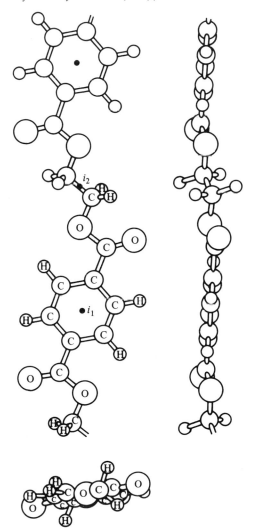

vibrations and the study of the spectra of model compounds. The spectral assignments of PET are discussed in chapters 5 and 6, where the applicability of these ideas to both PET and polystyrene is considered.

If assignments can be made at the level of symmetry species for a particular conformation of a known molecular structure, the final level of assignment consists in principle of associating a specific eigenvector with the mode since, for most molecules, including polymers, there is usually more than one mode belonging to a given symmetry species. It is not generally possible to determine the precise form of the eigenvector corresponding to any vibrational mode and the closest approach to such knowledge is obtained from the results of vibrational calculations, which will be discussed in section 4.4. For some purposes, however, it is sufficient to have approximate descriptions of the eigenvectors, which serve to distinguish between modes of the same symmetry species, and these descriptions are often given in terms of internal symmetry

Fig. 4.7. Descent in symmetry $D_{6h} \rightarrow D_{2h} \rightarrow C_{2h} \rightarrow C_i$. It is assumed that one of the σ_v planes of D_{6h} becomes the $\sigma(y, z)$ plane of D_{2h}. The axis normal to the ring is Oz for D_{6h}, D_{2h} and C_{2h}.

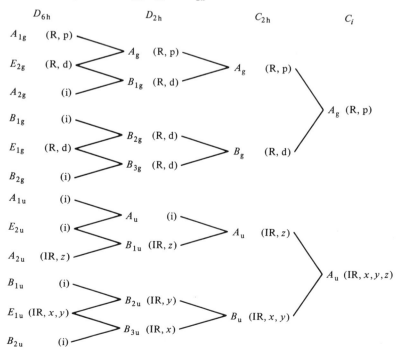

coordinates of the type discussed in section 2.5 and shown for the regular polyethylene chain in table 3.2. Symmetry modes are often chosen on the basis that specific atoms or groups of atoms vibrate essentially independently of the other atoms in the molecule; such vibrations are called *group vibrations* or *group modes* and analysis in terms of such vibrations is an extreme form of the use of local symmetry.

4.3.3 Group vibrations

It is important to remember that every vibration of a molecule or crystal involves, in principle, the motion of every atom in the molecule or crystal, except where the amplitude of vibration of an atom or atoms is zero by symmetry. Nevertheless, in section 3.2 we divided the vibrations of polymer crystals, to a first approximation, into two sets. The first set consisted of lattice modes, i.e. vibrations of whole chains relative to each other, the frequencies of which are not related to those of isolated chains; the second set consisted of internal vibrations of the individual chains, the frequencies of which are often not very different from those of isolated chains. In a similar way, it is sometimes possible to divide the vibrations of a molecule into two sets, those which involve mainly the coupled internal vibrations of particular groups of atoms and those which involve mainly the vibrations of these groups of atoms with respect to each other. A simple example of this was given in fig. 2.9, section 2.5. The actual vibrations which correspond to figs. 2.9b and 2.9c involve, to a first approximation, only internal vibrations of the $=CH_2$ groups, whereas the actual vibration which corresponds to fig. 2.9a is mainly a vibration of the two $=CH_2$ groups with respect to each other.

It should be noted, however, that the vibration which corresponds to fig. 2.9b is only one of two possible modes in which each of the $=CH_2$ groups shows a symmetric stretch (see figs. 4.8a and 4.8b). The two modes have slightly different frequencies, 3026 and 2989 cm^{-1}, respectively, because the vibrations of the $=CH_2$ groups are coupled through the $C=C$ bond. The difference of only 1 % in their frequencies shows that the $=CH_2$ symmetric stretch gives rise to a good *group frequency*. The two vibrations which correspond essentially to antisymmetric $=CH_2$ stretching, shown in figs. 4.8c and 4.8d, have frequencies 3106 and 3103 cm^{-1}, respectively, with a difference of about 0.1 %, so that the antisymmetric $=CH_2$ stretching frequency is an excellent group frequency. The four vibrations which involve to a first approximation only the stretching of C—H bonds of $=CH_2$ groups have a total spread of frequencies from 2989 to 3106 cm^{-1}, or about 4 %, so that the =C—H stretching frequency is also a moderately good group frequency, but not such a good one as either the symmetric or antisymmetric stretches of the =CH$_2$ groups as a whole.

These ideas may be extended by considering the vibrations of a particular group of atoms common to several different molecules. If in each of two different molecules which include the same group of atoms, e.g. a =CH$_2$ group, there is a vibrational mode which involves to a good approximation only one of the internal modes of vibration of that group of atoms, the frequencies of the two modes will be very closely the same and the molecules are said to show this particular group frequency of the =CH$_2$ group. Fig. 4.9 shows some of the more important general types of group frequencies observable in polymers and the approximate ranges of frequency over which they are observed.

For a molecule to have a normal mode which is a group mode, to a first approximation, one of two conditions must be satisfied. Either there must be no other group mode for the molecule which is both at a similar frequency and of the same symmetry species (with respect to the symmetry of the whole molecule) as the first group mode considered or,

Fig. 4.8. The C—H stretching vibrations of ethylene, CH$_2$=CH$_2$. The displacements shown correspond to symmetry modes rather than true normal modes; the displacements of the carbon atoms would, however, be small compared with those of the hydrogen atoms in the true normal modes.

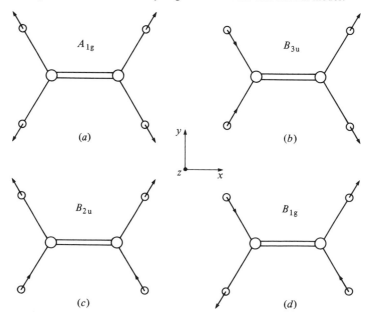

if there is, the coupling between the two groups must be weak. Weak coupling generally implies separation by several other atoms or groups of atoms. It should be clear from this discussion that even if some of the modes for a particular group are observed in the vibrational spectrum of a molecule or polymer, this does not mean that all the other vibrations of the group will necessarily be observable as clear group vibrations.

The occurrence of group frequencies has been and still is of great importance in vibrational spectroscopy for several reasons. First, they are useful in chemical analysis; if the vibrational spectrum of an unknown substance shows one or more modes with frequencies typical of a particular group of atoms, the presence of this group in the molecule is strongly suggested. Secondly, they are useful in making vibrational assignments at the level of specifying an approximate eigenvector for a mode. In polymers, in particular, the first use may be extended to determining the presence or absence of a particular configurational or conformational isomer in the polymer or to a more quantitative assessment of the relative amount, as discussed in chapters 5 and 6, and the second may be extended to obtaining information about the distribution of orientations of particular groups of atoms in an oriented polymer, as explained in section 6.10.

Fig. 4.9. Group frequencies. The upper part of the diagram shows the regions of the vibrational spectrum in which some general categories of vibration may be observed. The lower part indicates some of the subregions in which vibrations of more specific examples of these general types may be observed. A more detailed subdivision is given in the next chapter, fig. 5.12.

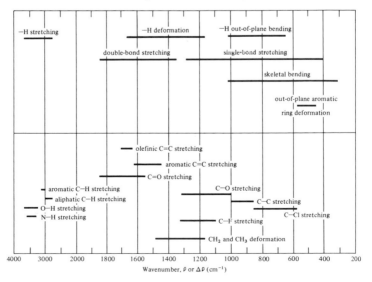

4.3.4   *Model compounds*

As indicated above, not all the vibrational modes of a molecule or polymer chain can be described as simple group modes and, even when they can, there may be similar groups in different environments which have significantly different frequencies, such as the C—X groups in an atactic or conformationally irregular isotactic or syndiotactic vinyl polymer. A useful approach to assignment may then be the use of *model compounds*. These compounds have small molecules with certain features in common with the polymer chain. The assumption is made that many of their vibrational modes will have frequencies which are very close to those of the corresponding part of the polymer chain. The polymer modes can accordingly be assigned to vibrations of the appropriate part of the molecule.

Good examples of this approach, which will be discussed further in chapters 5 and 6, are the use of compounds such as dimethyl tereph-

thalate, $CH_3.O.CO$—⟨O⟩—$CO.O.CH_3$, and other compounds

containing the terephthaloyl group, as model compounds for the terephthaloyl regions of PET and the use of various secondary chlorides, such as $CH_3.CHCl.CH_3$ and $CH_3.CHCl.(CH_2)_2.CHCl.CH_3$, as model compounds for various specific configurational and conformational isomers which may be present in an atactic PVC chain. This approach to assignment has been and remains a useful one, but it must be made with caution, because any structural difference between two molecules affects the frequency and eigenvector of every mode of vibration to some extent. It is not always easy to predict what the effect will be without detailed calculation, which presupposes a knowledge of the relevant force constants and precise molecular geometry. The model compound method is therefore best used not in isolation, but in combination with other methods.

4.3.5   *Isotope substitution*

The frequency of a vibrational mode depends on the masses of the atoms as well as on the geometry of the molecule and the force constants. If the mass of one type of atom can be changed by substituting a different isotope from the normally predominant one, the other factors which determine the frequencies of the modes remain constant, and certain simple predictions can be made about the changes in frequency which will be observed. Clearly, if the type of atom for which the substitution is made does not move in a particular mode, the frequency of that mode will be unchanged, whereas if the mode involves predominantly the movement of this type of atom the change of frequency may be quite large. Isotope substitution may thus be a useful aid in making vibrational assignments.

Consider, for instance, the totally symmetric 'breathing mode' of the methane molecule, in which all four hydrogen atoms move radially in and out in phase with each other. In the simplest model the frequency of this mode is given by

$$v = \frac{1}{2\pi} \sqrt{\frac{k}{m_H}} \tag{4.22}$$

where m_H is the mass of the hydrogen atom and k is the force constant for C—H stretching, since in this mode the carbon atom does not move and the vibration has the same frequency as that of a diatomic molecule consisting of a hydrogen atom bonded to an atom of infinite mass by a bond with force constant k. Since the frequency of this mode is independent of the mass of the carbon atom it would not change if the mass of the carbon atom were changed from the usual 12 to 13 amu (atomic mass units) by isotope substitution, whereas if the masses of all the hydrogen atoms were changed from 1 to 2 amu by the substitution of deuterium the frequency would, in the harmonic approximation, be reduced by a factor $\sqrt{2} = 1.414$.

This example is extreme, but illustrates that the substitution of deuterium for hydrogen is a very useful aid in vibrational assignment because the shifts produced may be very large if the motion involved in a particular mode is predominantly that of hydrogen atoms. Even if the carbon atom to which a hydrogen atom is attached did move, and behaved as if it were completely free except for its bond to the hydrogen atom, the effective masses entering the equations for C—H or C—D modes would be the reduced masses $m_C m_H/(m_C + m_H)$ and $m_C m_D/(m_C + m_D)$, so that the ratio of the frequencies would be the square root of the ratio of these quantities, i.e. $\sqrt{(2 \times 13/14)} = 1.363$. For a vibration in which the C of the C—H bond is strongly involved but which is not predominantly a C—H vibration, the H and C atoms move together to a first approximation and thus the frequency will only be changed by a factor of approximately $\sqrt{[(m_C + m_D)/(m_C + m_H)]} = \sqrt{(14/13)} = 1.038$.

The effect of isotope substitution on the frequency of any particular normal mode could only be predicted accurately if the precise molecular geometry and force constants were known, and in practice only approximate shifts can be calculated. There is however, an important rule which gives an exact relationship between all the vibrations of a given symmetry species of two molecules which differ only by isotope substitution. This rule, which was derived independently by Teller and Redlich and is generally known as the *Teller–Redlich product rule*, states

$$\left(\frac{v_1'}{v_1}\right)\left(\frac{v_2'}{v_2}\right)\cdots\left(\frac{v_n'}{v_n}\right) = \sqrt{\left(\frac{m_1}{m_1'}\right)^a\left(\frac{m_2}{m_2'}\right)^b\cdots\left(\frac{M'}{M}\right)^t f(I,I')} \qquad (4.23)$$

where all primed symbols refer to the molecule with isotope substitution and the unprimed symbols refer to the same quantities for the 'original' molecule. The quantities $v_1, v_2, \ldots, v_n$ are the frequencies of the n normal vibrations of the given symmetry species, the quantities $m_1, m_2, \ldots$ are the masses of an individual member of each of the symmetry-related sets of atoms which make up the molecule. M is the total mass of the molecule and t is the number of translations of the given symmetry species. The function $f(I, I')$ is the product of the quantities I_i'/I_i for those values of $i = x$, y or z for which rotation about Oi belongs to the given symmetry species, where I_i is the moment of inertia of the molecule about Oi. The powers $a, b, \ldots$ are the numbers of vibrational modes (including non-genuine ones) that each symmetry-related set of atoms contributes to the symmetry species, i.e. the numbers which would be calculated from equation (2.31) if only the atoms of each set were used in the calculation.

The frequencies which occur in equation (4.23) are strictly speaking those which would be observed if the anharmonic contributions to the forces between the atoms were zero, but the equation will hold to a good approximation for the observed fundamental frequencies. It may be show that if the prime indicates the heavier isotope, the left-hand side of equation (4.23) should be slightly greater than the right-hand side if the observed, slightly anharmonic, fundamental frequencies are used.

The simplest application of the Teller–Redlich rule occurs if there is only one genuine vibration of a particular symmetry species, when equation (4.23) gives directly the ratio of the frequencies, and it is then a simple matter to check a proposed assignment to a symmetry species. More generally, assignments of all the modes belonging to a given species must be attempted before the rule can be used to check whether the assignments constitute a possibly correct set or not. As an example of the simpler type of application, consider the CH_2 twisting mode of B_{1g} symmetry and the CH_2 wagging mode of B_{3u} symmetry for polyethylene, which each belong to symmetry species with only one genuine normal vibration (see table 3.2). For polyethylene all hydrogen atoms belong to the same symmetry-related set and all carbon atoms belong to the same set. Neither of the symmetry species considered has a rotation as a non-genuine vibration, so that $f(I, I') = 1$. There is one translation belonging to the B_{3u} species but none belonging to the B_{1g} species. Hence, with fully deuterated polyethylene as the isotope substituted molecule and ordinary polyethylene as the 'original' molecule, we have

for B_{1g}:
$$\frac{v'}{v} = \sqrt{\left(\frac{1}{2}\right)^1 \left(\frac{12}{12}\right)^0 \left(\frac{32}{28}\right)^0} \times 1 = 0.707 = 1/1.414$$

for B_{3u}:
$$\frac{v'}{v} = \sqrt{\left(\frac{1}{2}\right)^1 \left(\frac{12}{12}\right)^1 \left(\frac{32}{28}\right)^1} \times 1 = 0.756 = 1/1.323$$

since for each species the H atoms contribute one vibrational mode and the carbon atoms contribute one vibrational mode to the B_{3u} species and none to the B_{1g} species. The experimental results are in extremely good agreement with the predicted ratios. The result for the B_{1g} species is, of course, easily obtained by the physical argument that the carbon atoms do not move in the corresponding CH_2 twisting vibration and the result for the B_{3u} species by the argument that two hydrogen atoms move in the opposite direction to each carbon atom, so that the ratio of the frequencies will be the square root of the ratio of the reduced masses, $(m_H/m_D)(m_C + 2m_D)/(m_C + 2m_H)$. Such simple arguments are, however, not always available. For polymers there are generally many modes belonging to a given symmetry species, as pointed out in subsection 4.3.2, so that the Teller–Redlich rule is difficult to apply. It has, however, been used to check assignments in a number of polymers, including poly(ethylene terephthalate). A simpler approximate expression has been given by Krimm, which applies to a single vibrational frequency. (See the review by Krimm cited in section 3.3.)

4.4 Force fields and vibrational calculations

So far we have explained how to calculate the numbers of normal modes of each symmetry species for any molecule, but the frequencies of the modes have been taken as quantities to be determined empirically, using the methods of assignment outlined in the previous section. Often such methods will allow many modes to be correctly assigned at the level of specific symmetry modes of the molecule or particular groups within it. There are, however, often other modes which resist such assignments because they are mixtures of simple symmetry or group modes of the same symmetry species in which no single component mode is dominant. Further information on the nature of these modes can, in principle, be obtained by calculating the normal mode frequencies and eigenvectors from the geometry of the molecule and the interatomic force constants. Such calculations can also show to what extent assignments made to symmetry or group modes are really approximations.

We give in this section an elementary introduction to the methods used in vibrational calculations. Sufficient detail is included to cover the general principles of the methods and the terminology, so that with this section as background the reader should be able to understand the references to vibrational calculations in chapters 5 and 6 and in the general literature. The subject is, however, highly technical and the reader is referred to the books and papers indicated in section 4.7 for further details. Many practising polymer spectroscopists will never perform a full vibrational calculation, but it is important to understand the assumptions and approximations made in such calculations and to develop a critical approach to the conclusions drawn from them. The discussion is given first in terms of small molecules; the modifications required to apply it to polymers are explained in subsection 4.4.7.

4.4.1 *The vibrational problem in Cartesian coordinates*

We start from equation (2.4), viz.

$$m_a \frac{d^2 x_{ai}}{dt^2} = -\sum_{bj} k_{ai,bj}(x_{bj} - x_{bj}^0)$$

and first simplify the notation. Since x_{ai}^0 is independent of t, x_{ai} can be replaced by $(x_{ai} - x_{ai}^0)$ on the left-hand side of the equation; an immediate simplification then results by replacing the quantities $(x_{ai} - x_{ai}^0)$, which are the displacements from the equilibrium positions, by x_{ai}. This simply moves the origin of the coordinates referring to any given atom to the equilibrium position of that atom. If we further replace the labels ai by i, where the new set of values of i goes from 1 to $3N$, so that the old $a1, a2, a3$, for example, become simply three different values of the new i, the equation reduces to

$$m_i \frac{d^2 x_i}{dt^2} = -\sum_j k_{ij} x_j \tag{4.24}$$

where, for consistency, m_a has been replaced by m_i.

Equation (4.24) holds for all values of i. Assume that a solution of this set of $3N$ simultaneous equations can be found which corresponds to simple harmonic vibrations of all atoms about their equilibrium positions with a common angular frequency $\omega = 2\pi\nu$. For such a solution the following equation must hold for all values of j from 1 to

$3N$:

$$x_j = X_j e^{i\omega t} \tag{4.25}$$

where the quantities X_j represent the amplitudes of vibration. Substituting in equation (4.24) leads to

$$\omega^2 m_i X_i = \sum_j k_{ij} X_j \tag{4.26}$$

This set of simultaneous equations can be represented in matrix form by

$$\omega^2 \mathbf{MX} = \mathbf{KX} \tag{4.27}$$

where

$$\mathbf{M} = \begin{pmatrix} m_1 & & & & \\ & m_2 & & 0 & \\ & & m_3 & & \\ & 0 & & \ddots & \\ & & & & m_{3N} \end{pmatrix} \quad \text{and} \quad \mathbf{X} = \begin{pmatrix} X_1 \\ X_2 \\ X_3 \\ \vdots \\ X_{3N} \end{pmatrix} \tag{4.28}$$

The equations can only be solved for non-zero values of the amplitudes X_i if the value of ω satisfies the *secular equation*

$$|\mathbf{K} - \omega^2 \mathbf{M}| = 0 \tag{4.29}$$

or

$$|\mathbf{M}^{-1}\mathbf{K} - \mathbf{E}\lambda| = 0 \tag{4.30}$$

where $\mathbf{E}$ is the unit $3N \times 3N$ matrix and $\lambda = \omega^2$. For any solution of equation (4.30), say λ_k, corresponding to the normal mode k, it is possible to find a column matrix $^k\mathbf{X}$ which satisfies equation (4.27). A suitably normalized set of displacements $^k\mathbf{X}$ is the eigenvector for the normal mode k.

In section 2.5 other types of coordinate system than the Cartesian one used here were introduced, viz. those based on symmetry coordinates or internal coordinates, and we now consider how the use of other coordinate systems can simplify the calculation of the normal modes. We consider first a generalized set of coordinates.

4.4.2 *Generalized coordinates – the Wilson $\mathbf{G}$ and $\mathbf{F}$ matrices*

Suppose that a general system of coordinates z_i is related to the set x_i by the transformation

$$\mathbf{z} = \mathbf{Rx} \tag{4.31}$$

We shall assume to start with that z represents a complete independent set of $3N$ coordinates, so that $\mathbf{R}$ is a $3N \times 3N$ matrix. Since the set is complete it must be possible to write the variable part of the potential energy, in the harmonic approximation, in the form

$$V = \tfrac{1}{2} \sum_{ij} F_{ij} z_i z_j = \tfrac{1}{2} \tilde{\mathbf{z}} \mathbf{F} \mathbf{z} \tag{4.32}$$

where the matrix $\mathbf{F}$ corresponds to the force constant matrix $\mathbf{K}$ in the treatment in Cartesian coordinates. Now consider the kinetic energy T. This is given by

$$T = \tfrac{1}{2} \sum_i m_i \dot{x}_i^2 = \tfrac{1}{2} \tilde{\dot{\mathbf{x}}} \mathbf{M} \dot{\mathbf{x}} \tag{4.33}$$

in terms of the Cartesian coordinates. It follows from equation (4.31) that

$$\mathbf{R}^{-1}\dot{\mathbf{z}} = \dot{\mathbf{x}} \quad \text{and} \quad \tilde{\dot{\mathbf{z}}}\tilde{\mathbf{R}}^{-1} = \tilde{\dot{\mathbf{x}}} \tag{4.34}$$

Hence

$$T = \tfrac{1}{2}\tilde{\dot{\mathbf{z}}}\mathbf{G}^{-1}\dot{\mathbf{z}} = \tfrac{1}{2} \sum_{ij} G_{ij}^{-1} \dot{z}_i \dot{z}_j \tag{4.35}$$

where G_{ij}^{-1} means the ij component of $\mathbf{G}^{-1}$ and

$$\mathbf{G} = \mathbf{R}\mathbf{M}^{-1}\tilde{\mathbf{R}} \tag{4.36}$$

Substituting V and T into Lagrange's equation of motion

$$\frac{\partial V}{\partial z_i} + \frac{\mathrm{d}}{\mathrm{d}t} \frac{\partial T}{\partial \dot{z}_i} = 0 \tag{4.37}$$

leads, for all i, to

$$\sum_j (F_{ij} z_j + G_{ij}^{-1} \ddot{z}_j) = 0 \tag{4.38}$$

Assuming a solution to the vibrational problem of the form of equation (4.25) with z replacing x and $\mathbf{Z}$ replacing $\mathbf{X}$ leads to

$$\sum_j (F_{ij} - \omega^2 G_{ij}^{-1}) Z_j = 0 \tag{4.39}$$

for all i. Equation (4.39) is a set of simultaneous equations which represent a generalization of the set in equation (4.26). It follows that the generalized secular equation is

$$|\mathbf{G}\mathbf{F} - \mathbf{E}\lambda| = 0 \tag{4.40}$$

This is the most important equation of the **GF** matrix method, first given by Wilson, for solving the vibrational problem. We now consider some special coordinate systems.

4.4.3 *Mass-weighted Cartesian coordinates and normal coordinates*

Let **R** in equation (4.31) be the diagonal matrix with elements $R_{ii} = \sqrt{m_i}$ and let the corresponding specific new set of coordinates be **y**. Then

$$y_i = x_i\sqrt{m_i} \quad \text{and} \quad F_{ij} = k_{ij}/\sqrt{(m_i m_j)} \tag{4.41}$$

and the new coordinates y_i are called *mass-weighted coordinates*. It follows from equation (4.36) that

$$\mathbf{G} = \mathbf{RM}^{-1}\tilde{\mathbf{R}} = \mathbf{E} = \mathbf{G}^{-1} \tag{4.42}$$

Equation (4.39) now becomes

$$\omega^2 Y_i = \sum_j F_{ij} Y_j \tag{4.43}$$

and for the normal mode k

$$\omega_k^2 Y_{ki} - \sum_j F_{ij} Y_{kj} \tag{4.44}$$

(Note that when two subscripts are used on the symbol for a column vector the first will refer to the mode and the second will identify the component, i.e. $Y_{ki} = {}^kY_i$.) Multiplying both sides by Y_{li} and summing over i leads to

$$\omega_k^2 \sum_i Y_{ki} Y_{li} = \sum_{ij} F_{ij} Y_{kj} Y_{li} = \sum_{ij} F_{ji} Y_{ki} Y_{lj} \tag{4.45}$$

Using $F_{ji} = F_{ij}$ and the first part of equation (4.45) with k and l interchanged leads to

$$\omega_l^2 \sum_i Y_{li} Y_{ki} = \sum_{ij} F_{ji} Y_{lj} Y_{ki} \tag{4.46}$$

Equations (4.45) and (4.46) show immediately that if $\omega_k \neq \omega_l$, then

$$\sum_{ij} F_{ij} Y_{kj} Y_{li} = \sum_i Y_{ki} Y_{li} = 0 \text{ for all } l, k \quad (l \neq k) \tag{4.47}$$

The second part of equation (4.47) expresses the orthogonality of the eigenvectors of non-degenerate modes. Orthogonal eigenvectors can always be chosen for the modes of any degenerate set and we shall assume from now on that the ${}^kY$ represent a complete orthogonal set of eigenvectors. We shall further assume that they have been normalized so that

$$\sum_i Y_{ki}^2 = 1 \quad \text{for all } k \tag{4.48}$$

Putting $l = k$ in equation (4.45) now shows that

$$\sum_{ij} F_{ij} Y_{kj} Y_{ki} = \omega_k^2 \sum_i Y_{ki}^2 = \omega_k^2 \tag{4.49}$$

If we define new coordinates U_i by the equation

$$\mathbf{y} = \mathcal{Y}\mathbf{U} \tag{4.50}$$

where $\mathcal{Y}$ is the square matrix with columns $^k\mathbf{Y}$, so that $\mathcal{Y}_{ik} = Y_{ki}$ we have

$$y_i = \sum_k \mathcal{Y}_{ik} U_k = \sum_k Y_{ki} U_k \tag{4.51}$$

This equation indicates that the overall distortion of the molecule given by the coordinates y_i can be regarded as a sum over distortions each of the form of one of the individual normalized eigenvectors, the coefficient of the contribution from the mode k being U_k. The coefficients U_k are therefore called the *normal coordinates* and have already been referred to in earlier sections, e.g. subsection 2.4.2. Here $\mathbf{y}$ replaces $\mathbf{I}$ and $^k\mathbf{Y}$ replaces $\mathbf{D}_k$ in equation (2.27).

Since $\mathbf{G}^{-1} = \mathbf{E}$, substitution of y for z in equation (4.35) and use of equation (4.51) leads to

$$T = \tfrac{1}{2} \sum_{kl} \left(\dot{U}_k \dot{U}_l \sum_i Y_{ki} Y_{li} \right) \tag{4.52}$$

Use of equations (4.47) and (4.48) now shows that

$$T = \tfrac{1}{2} \sum_k \dot{U}_k^2 = \tfrac{1}{2} \tilde{\dot{U}} \dot{U} \tag{4.53}$$

Similarly, equation (4.32) becomes

$$V = \tfrac{1}{2} \sum_{kl} \left[U_k U_l \left(\sum_{ij} F_{ij} Y_{ki} Y_{lj} \right) \right] \tag{4.54}$$

Equations (4.47) and (4.49) show that this reduces to

$$V = \tfrac{1}{2} \sum_k \omega_k^2 U_k^2 = \tfrac{1}{2} \tilde{U} \Lambda U \tag{4.55}$$

where the matrix Λ is defined by

$$\Lambda_{ij} = 0 \quad \text{for } i \neq j, \qquad \Lambda_{ii} = \omega_i^2 = \lambda_i \tag{4.56}$$

Equations (4.53) and (4.55) show that both the kinetic and potential energies are simply sums over separate contributions from each of the normal mode vibrations, with no cross terms.

4.4.4 The solution in principle; potential energy distributions

Let any general complete set of coordinates be related to the normal coordinates by

$$\mathbf{z} = \mathbf{LU} \tag{4.57}$$

The displacements z_i in the eigenvector of normal mode k are now given by L_{ik} ($U_k = 1$, all other $U_i = 0$), so that the columns of $\mathbf{L}$ represent the normal mode eigenvectors expressed in terms of the set of coordinates $\mathbf{z}$ and are thus proportional to the solutions $^k\mathbf{Z}$ of equation (4.39). It follows from equations (4.32) and (4.35) that

$$V = \tfrac{1}{2}\mathbf{\tilde{U}\tilde{L}FLU} \tag{4.58a}$$

$$T = \tfrac{1}{2}\mathbf{\dot{\tilde{U}}\tilde{L}G^{-1}L\dot{U}} \tag{4.58b}$$

Comparison of equations (4.58) with equations (4.55) and (4.53) shows that

$$\mathbf{\tilde{L}FL} = \mathbf{\Lambda} \tag{4.59a}$$

$$\mathbf{\tilde{L}G^{-1}L} = \mathbf{E} \tag{4.59b}$$

and thus

$$\mathbf{L\tilde{L}} = \mathbf{G} \tag{4.60}$$

Equations (4.59a) and (4.60) yield immediately the matrix equivalent of equation (4.39), viz.

$$\mathbf{GF} = \mathbf{L\Lambda L}^{-1} \tag{4.61}$$

The relationship expressed in equation (4.60) replaces the orthogonality and normality of the eigenvectors $\mathbf{Y}$ expressed by equations (4.47) and (4.48).

It has so far been tacitly assumed that the coordinates are real. If complex coordinates are chosen, as is sometimes convenient, the treatment given in subsections 4.4.2, 4.4.3 and the present subsection still applies, provided that the transposed matrices that appear are replaced by their complex conjugates.

The complete solution of the vibrational problem expressed in terms of any set of coordinates $\mathbf{z}$ now reduces in principle to the following steps, although more sophisticated methods of solution are generally used for molecules consisting of more than a few atoms:

(i)   Write down expressions for the **F** and **G** matrices appropriate to the chosen set of coordinates. (See subsections 4.4.5 and 4.4.6.)

(ii)   Solve the equations (4.40) to find the possible values of the quantities λ_k and hence the normal mode (angular) frequencies, $\omega_k = \sqrt{\lambda_k}$.

(iii)   Substitute any one of these values λ_k into the simultaneous equations (4.39) and solve for ${}^k\mathbf{Z}$.

(iv)   Normalize the eigenvectors ${}^k\mathbf{Z}$ so that the matrix **L** which has ${}^k\mathbf{Z}$ as its columns satisfies equation (4.59b) or equation (4.60).

Although the best description of the form of a normal mode k is provided by specifying the components of its eigenvector ${}^k\mathbf{Z}$ with respect to some suitable system of coordinates, because this description is complete, a less complete description is often given in terms of a so-called *potential energy distribution*, derived as follows. Equation (4.55) shows that the potential energy associated with a single mode k is $\frac{1}{2}\omega_k^2 = \frac{1}{2}\lambda_k$ for unit displacement ($U_k = 1$). Equation (4.59a) shows that

$$\lambda_k = \sum_{ij} L_{ik}L_{jk}F_{ij} \tag{4.62}$$

If we assume that the off-diagonal terms ($i \neq j$) of **F** are small compared with the diagonal terms, then

$$\lambda_k = \sum_i L_{ik}^2 F_{ii} \tag{4.63}$$

and to this approximation the potential energy associated with the mode k is the sum of terms each of which is associated with only one of the coordinates z_i. The set of quantities

$$V_{ki} = L_{ik}^2 F_{ii}/\lambda_k \tag{4.64}$$

for all i is called the *potential energy distribution* for the mode k and gives an indication of the importance of each of the coordinates z_i to the mode. Usually the V_{ki} are expressed as percentages. It is clear from the definition that the term 'potential energy distribution' is confusing, since the sum of the values V_{ki} over all coordinates i does not have to be unity (or 100%). In addition, contributions less than (say) 10% are often neglected, so that quoted values may add up to more or less than unity (or 100%). Nevertheless they give a useful simple indication of the nature of a normal mode, particularly if the coordinates **z** are chosen appropriately.

We now turn to a discussion of some further simplifications which can be made to the problem of determining the frequencies and eigenvectors of the normal modes by choosing suitable coordinate systems.

4.4.5 *Symmetry coordinates and internal symmetry coordinates*

It was shown in chapter 2 that every normal mode or set of degenerate normal modes belongs to a symmetry species and the idea of symmetry modes, symmetry mode vectors and symmetry coordinates was introduced. Any eigenvector of a given symmetry species can be expressed as a sum over any complete set of independent symmetry mode vectors of the same symmetry species. The coefficients in this sum are the values of the symmetry coordinates for the eigenvector. To be complete the set of symmetry modes must clearly contain the same number of members as there are normal modes of the given species. By comparison with equation (4.57) it follows that the symmetry coordinates ^aS of any general set of displacements of a particular symmetry species a are given by

$$^aS = {}^a\mathbf{l}\,{}^a\mathbf{U} \tag{4.65}$$

where ^aS_i represents a particular symmetry coordinate of symmetry species a, $^al_{ij}$ represents the component i of the eigenvector j of species a with respect to the symmetry coordinates and ^aU_j represents the contribution of the eigenvector j to the general set of displacements considered. Any totally general set of displacements of the molecule can be written in terms of a set of symmetry coordinates $\mathbf{S}$ composed of the subsets appropriate to each of its symmetry species and we can thus write

$$
\mathbf{S} = \begin{pmatrix} {}^a\mathbf{S} \\ {}^b\mathbf{S} \\ \cdot \\ \cdot \\ \cdot \end{pmatrix} = \begin{pmatrix} {}^a\mathbf{l} & & & \\ & {}^b\mathbf{l} & & 0 \\ & & \cdot & \\ & 0 & & \cdot \\ & & & \cdot \end{pmatrix} \begin{pmatrix} {}^a\mathbf{U} \\ {}^b\mathbf{U} \\ \cdot \\ \cdot \\ \cdot \end{pmatrix} \tag{4.66a}
$$

or

$$\mathbf{S} = \mathbf{L_s U} \tag{4.66b}$$

where the diagonal block matrix $\mathbf{L_s}$ is the equivalent of the more general matrix $\mathbf{L}$ in equation (4.57).

Equation (4.61) becomes

$$\mathbf{G_s F_s} = \mathbf{L_s \Lambda L_s^{-1}} \tag{4.67}$$

Now the inverse of a block matrix is a block matrix consisting of the inverses of the individual blocks in the original matrix and multiplying two block matrices of the same form leads to a further matrix of the same form. Equation (4.67) thus shows that G_sF_s is of block diagonal form. This means that equations (4.39) split up into separate sets, one for each block of G_sF_s, i.e. one set for each symmetry species. It may be shown that a special method of choice of the symmetry coordinates produces a further simplification whereby the block for an n-dimensionally degenerate species splits up into n equal blocks. The method of solution of the vibrational problem listed in steps (i) to (iv) of subsection 4.4.4 can now be applied separately to the modes of each symmetry species in turn, which greatly simplifies the problem.

In the argument just given the vital factor leading to the diagonal block nature of G_sF_s was the separation of both the normal modes and the reference coordinates into sets in such a way that each set of normal modes corresponded to a different set of reference coordinates. The sets were actually chosen so that each set contained all modes of a given symmetry species. Suppose, however, that we had used only two sets of normal modes, one corresponding to all genuine modes of vibration and one corresponding to all translations and rotations. The corresponding sets of coordinates would then have been one consisting only of internal coordinates, for which we shall later use the symbol q, and one consisting only of translational or rotational coordinates. The argument would then have led to a separation of equations (4.39) into two sets, one corresponding to the true vibrations and the other to the translations and rotations.

Use of internal symmetry coordinates will clearly give the greatest simplification of the vibrational problem. Special techniques may be used to generate suitable symmetry coordinates from internal coordinates; it is then easy to generate the G_s and F_s matrices from the corresponding matrices for the internal coordinates. We now consider the nature of the force fields and force constants which must be specified in order to calculate the F matrix.

4.4.6   *Force fields and force constants*

Since there are $3N - 6$ internal coordinates for a molecule consisting of N atoms it is necessary, in principle, to specify $\frac{1}{2}(3N - 6)(3N - 5)$ independent force constants if the molecule has no symmetry elements. The molecule has only $3N - 6$ genuine normal modes of vibration, so that the number of force constants required usually exceeds the number of modes of vibration even when the symmetry of the molecule is taken into account. The previous subsection shows that the vibrational problem can be solved independently for the set of internal symmetry coordinates corresponding to any chosen symmetry species and it follows that corresponding to any symmetry species with n normal vibrational modes there are $\frac{1}{2}n(n + 1)$ independent force constants. This number is always greater than n unless $n = 1$. In order to reduce the complexity of the calculation, and also to render more obvious the physical significance of the force constants, various types of approximate *force field* are used. These are based on four types of internal coordinates which are illustrated in fig. 4.10.

In the *central force field* the only forces assumed to exist are those acting along the lines joining pairs of atoms, so that the **F** matrix is diagonal for a set of internal coordinates consisting of any appropriate set of $3N - 6$ interatomic distances. This force field, the *general central force field*, or GCFF, is usually simplified by assuming that only atoms within a certain distance of each other interact, so that the number of force constants is substantially reduced for structures consisting of large

Fig. 4.10. Internal coordinates: (*a*) change of bond length, Δr; (*b*) change of bond angle, $\Delta\phi$; (*c*) change of internal rotation angle, $\Delta\tau$; (*d*) out-of-plane displacement angle, $\Delta\theta$.

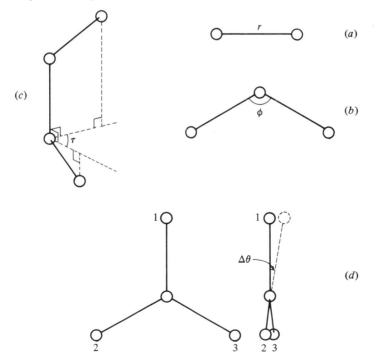

number of atoms. This force field is not very realistic for covalently bonded molecules because it includes no forces which tend specifically to maintain the angles between the covalent bonds at their equilibrium values. The *valence force field*, or VFF, is more realistic for such molecules, including, of course, polymers. In this force field there are forces which resist the changes of any of the internal coordinates of the types shown in fig. 4.10.

In the simplest version of the VFF, the *simple valence force field*, or SVFF, the **F** matrix has only the diagonal terms, F_{ii}, not equal to zero. Such a simple force field is usually incapable of giving good agreement with observed vibrational frequencies; it is necessary to allow for various interactions between the internal coordinates, such as interactions between the coordinates corresponding to the stretching of two bonds to the same atom and to the bending of the angle between them. If two internal coordinates q_i and q_j interact, the element F_{ij} of the **F** matrix will be non-zero and such force constants are called *interaction force constants*. For the GVFF all F_{ij} are non-zero. A further force field related to the VFFs is the *Urey–Bradley–Shimanouchi force field*, or UBSFF, in which the force constants of the SVFF are supplemented by repulsive forces between non-bonded atoms rather than by the explicit interaction force constants of the GVFF. These repulsive forces simulate the van der Waals forces between the non-bonded atoms. The interatomic distances for the non-bonded atoms depend on the bond lengths and angles which are used to specify the internal coordinates in the VFFs and do not constitute additional internal coordinates. It may be shown that when this dependence is taken into account the introduction of the forces between non-bonded atoms leads to non-zero off-diagonal terms in the **F** matrix. The total number of independent force constants required is, however, less than for the GVFF and this is an additional reason for using the UBSFF. Further reductions in the number of independent force constants can be made by neglecting the interaction between any pair of atoms unless they are bonded to a third common atom and by assuming a specific form for the repulsive interaction, such as the form $V = k/r^9$ assumed by Shimanouchi.

Once an approximate force field has been chosen it is necessary to choose the numerical values of the force constants before the frequencies can be calculated. This usually involves making use of force constants already available for similar, usually smaller, molecules which contain the same chemical groups, i.e. it is usually assumed that to a first approximation force constants can be *transferred* from one molecule to another. For very small molecules it is often possible to deduce force constants directly or almost directly from frequencies; for instance the frequencies of the totally symmetric modes of CCl_4 or CH_4, in which the carbon atom does not move, give directly the force constant for stretching of the C—Cl or C—H bond, respectively, in the SVFF model. Since the time when vibrational calculations were first made, a great deal of information has gradually been built up about the values and *transferability* of force constants for small molecules and this has been increasingly applied to larger molecules, including polymers.

The way in which a vibrational calculation is now usually made is, in outline, as follows:

(i)   A set of force constants is chosen, using data in the literature, and the vibrational frequencies are calculated by the procedures outlined in subsections 4.4.1 to 4.4.5.

(ii)   The difference between each calculated frequency and the corresponding observed frequency is found.

(iii) These differences are used with the partial derivative of every frequency with respect to every force constant to find a better set of force constants, i.e. a set which predicts the observed frequencies better.

(iv)   Steps (ii) and (iii) are repeated until an acceptably good fit is found between the observed and calculated frequencies.

It is not necessary to use all frequencies or force constants in the refinement process provided that the number of frequencies used is greater than or equal to the number of force constants that are allowed to vary. It is this aspect of the procedure which makes a vibrational calculation potentially predictive, and hence an aid to assignment, rather than simply a fitting exercise from which a set of force constants is obtained.

This procedure sounds straightforward until one realizes that at each step there is a degree of arbitrariness. Step (i) already implies a choice of force field, an approximation in itself. Step (ii) implies either that the calculated and observed frequencies agree so well with each other that no doubt exists as to which are corresponding frequencies in the two sets or that the observed frequencies have been assigned unambiguously at least to a symmetry species and that this eliminates the possibility of incorrect assumptions about correspondences. In practice it is not usually possible to eliminate ambiguities completely. Steps (iii) and (iv) assume the convergence of an iterative process, which is only guaranteed if the force field chosen is able to describe well enough the real interactions in the molecule and if the initial set of force constants is reasonably close to the most appropriate ones. We conclude this subsection with a quotation, reprinted by permisson, from *Introduction to the Theory of Molecular Vibrations and Vibrational Spectroscopy* by L. A. Woodward, Oxford University Press, 1972, p. 228:

the calculation of force fields from observed frequency data has become almost as much an art as a science. There is indeed little that is reliable to go upon in choosing the type of approximate field to be used, for approximate fields that are physically incompatible with one another may turn out to give equally good 'best fits' with the observed frequency data. Nor is there any firm principle upon which to base the guess of the initial set of force constants which are to be subjected to improvement. Lastly, in difficult cases it may be necessary to impose constraints in order to obtain the desired convergence; and there is a measure of arbitrariness as to how this shall be done. It should not be forgotten that in the end the force constants that are obtained are at best only the most satisfactory set that can be constructed with the particular type of approximate force field that was arbitrarily adopted as the basis of the calculation. If the constants happen to give an excellent fit between calculated and observed frequencies, this does not necessarily mean that the field is a correspondingly good approximation to the real one.

To this we merely add that any 'predicted' frequencies are subject to similar uncertainties which it is impossible to specify numerically.

4.4.7   *Application to polymers*

We shall consider first the normal vibrational modes of an isolated perfect polymer chain in the form of an n_k helix and we shall no longer, as in chapter 3, restrict attention to the factor group modes. We define a set of coordinates $\mathbf{x}_r$ within each physical repeat unit r so that the sets are related by the screw axis and we use each set to describe the displacements of the atoms of the corresponding repeat unit. We also introduce *Born's cyclic boundary condition*, which means that we assume that any disturbance of the polymer chain repeats itself exactly in every set of $\mathcal{N}$ adjacent physical repeat units, where $\mathcal{N}$ must be a multiple of n. Sums over r will be taken over $\mathcal{N}$ terms and $\mathcal{N}$ will later be allowed to tend to infinity.

We define new sets of coordinates $\mathbf{x}_\delta$ by

$$\mathbf{x}_\delta = \sqrt{\frac{1}{\mathcal{N}}} \sum_r \mathbf{x}_r e^{ir\delta} \tag{4.68}$$

where δ takes one of the values 0 or $\pm 2\pi m/\mathcal{N}$ for m an integer between 1 and $\frac{1}{2}\mathcal{N}$ (we arbitrarily choose $\mathcal{N}$ to be even). Now consider a mode in which

$$\mathbf{x}_r = \sqrt{\frac{1}{\mathcal{N}}} \mathbf{A}_0 e^{-ir\delta'} \tag{4.69}$$

i.e. a mode in which the corresponding atoms in each repeat unit vibrate with the same frequency and amplitude but with a phase which changes progressively by δ' for each repeat unit along the chain. The possible values of δ' which give distinguishable modes are restricted by the cyclic boundary condition to be those given for δ. The column matrix of amplitudes $\mathbf{A}_0$, which may be complex to allow for differences in phase of the motions of the different components of $\mathbf{x}_r$, is assumed to include the time-dependent factor $e^{i\omega t}$, where ω is the angular frequency of the mode. Then

$$\mathbf{x}_\delta = \frac{1}{\mathscr{N}} \mathbf{A}_0 \sum_r e^{ir(\delta - \delta')}$$

$$= \mathbf{A}_0 \quad \text{for } \delta = \delta'$$

$$= 0 \qquad \text{for } \delta \neq \delta' \tag{4.70}$$

Thus $\mathbf{x}_\delta$ is a set of coordinates associated only with the set of modes for which $\mathbf{x}_r = \sqrt{(1/\mathscr{N})} \mathbf{A}_0 e^{-ir\delta}$. These are the modes of a particular symmetry species for a pseudo-factor group obtained by redefining all the operations of the line group in a similar way to that used in defining the true factor group, but considering $\mathscr{N}/n$ true translational units as the effective translational unit. This pseudo-factor group is isomorphous with the group $C_{\mathscr{N}}$, or possibly a group of higher order if the helix has other symmetry elements in addition to the screw axis. (See subsections 2.3.2 and 3.1.4 for further discussion of the normal modes of molecules which belong to these point groups.) The secular equation may therefore be separated into $\mathscr{N}$ sets of equations in the way described in subsection 4.4.5 and solved to obtain the frequencies. Each set of equations, which represents only modes corresponding to a given value of δ, has $3n'$ solutions, where n' is the number of atoms per physical repeat unit. There are only $\mathscr{N}$ sets of equations because phase differences $\delta = +\pi$ and $\delta = -\pi$ between adjacent units are equivalent.

As $\mathscr{N}$ tends to infinity the sets of $3n'$ solutions for the frequencies tend towards a set of continuous functions of δ; plots of these functions against δ are called the *frequency–phase-difference* curves or *dispersion* curves for the vibrational modes of the polymer chain. The modes for which $\delta = 0$ or $\pm 2\pi m'/n$ for m' an integer between 1 and $\frac{1}{2}n$ (or $\frac{1}{2}(n-1)$ for odd n) are the true factor group modes. The modes for other values of δ are usually inactive in infrared or Raman spectroscopy but it is sometimes possible to obtain information about them by means of inelastic neutron scattering. Modes related to them may, however, become activated under certain conditions to be discussed in the next subsection. It should be noted that although an n_k helix has been considered, a similar argument can be applied to a polymer chain with a glide plane rather than a screw axis. For the helix, n and k may both take the value 1, so that the physical and translational repeat units become identical; the results thus apply to any type of regular chain. Changing the value of δ by any integral multiple of 2π leads to a physically identical mode and the dispersion curves are therefore, in a mathematical sense, periodic in δ with period 2π.

Fig. 4.11 shows a simple model system which may be used to illustrate some of the interesting and important features of polymer dispersion curves. The system consists of a set of masses m connected together by two different kinds of springs, with force constants k_1 and k_2, which alternate along the chain. The masses are assumed to be constrained to move only parallel to the length of the chain, so that the vibrational problem is a one-dimensional one in ordinary space. The methods of subsections 4.4.2 to 4.4.5 and of the present subsection may be used to solve the vibrational problem, but it is simpler to solve this rather trivial problem completely in terms of Cartesian coordinates.

Equation (4.24) leads to the following pair of simultaneous equations:

$$m\frac{d^2 x_r}{dt^2} = (x'_r - x_r)k_1 + (x'_{r-1} - x_r)k_2 \tag{4.71a}$$

$$m\frac{d^2 x'_r}{dt^2} = (x_{r+1} - x'_r)k_2 + (x_r - x'_r)k_1 \tag{4.71b}$$

Assuming solutions of the form

$$x_r = X e^{ir\delta} e^{i\omega t} \quad \text{and} \quad x'_r = X' e^{ir\delta} e^{i\omega t} \tag{4.72}$$

leads immediately to the secular equation

$$\begin{vmatrix} k_1 + k_2 - m\omega^2 & -(k_1 + k_2 e^{-i\delta}) \\ -(k_1 + k_2 e^{i\delta}) & k_1 + k_2 - m\omega^2 \end{vmatrix} = 0 \tag{4.73}$$

which has solutions

$$m\omega^2 = (k_1 + k_2) \pm \sqrt{(k_1^2 + k_2^2 + 2k_1 k_2 \cos \delta)} \tag{4.74}$$

Fig. 4.11. Linear chain of equal masses joined by bonds with alternately different force constants, k_1 and k_2. x_r and x'_r are the displacements of the two masses in the repeat unit r, each measured from the undisplaced position of the corresponding mass.

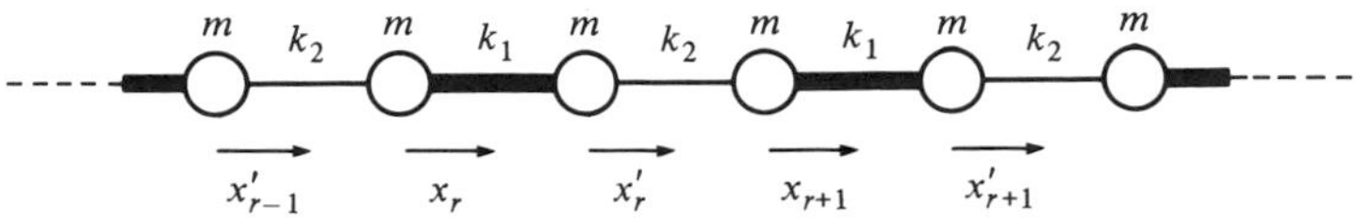

For $\delta = 0$ and π these become

$$\omega = 0 \quad \text{or} \quad \sqrt{\frac{2(k_1 + k_2)}{m}} \tag{4.75a}$$

and

$$\omega = \sqrt{\frac{2k_1}{m}} \quad \text{or} \quad \sqrt{\frac{2k_2}{m}} \tag{4.75b}$$

The dispersion curves are shown in fig. 4.12, with the assumption that $k_2 < k_1$.

The values of ω for $\delta = 0$, viz. 0 and $\sqrt{[2(k_1 + k_2)/m]}$, are the angular frequencies of the factor group modes that correspond, respectively, to translation of the chain parallel to itself and to opposite displacements of adjacent masses. The upper curve represents the *optical branch* of the dispersion curve, since it contains the factor group mode of non-zero frequency, which is the type of mode which may give rise to infrared or Raman scattering. The lower curves represent the *acoustic branch* of the dispersion curves, since for small values of δ the modes correspond to sound waves in a solid.

Fig. 4.12. Dispersion curves for the linear chain of fig. 4.11.

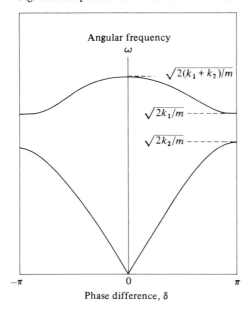

Angular frequency
ω

$\sqrt{2(k_1 + k_2)/m}$

$\sqrt{2k_1/m}$

$\sqrt{2k_2/m}$

$-\pi \qquad 0 \qquad \pi$

Phase difference, δ

If $k_2 \ll k_1$, the optical branch becomes very flat and the vibrations become essentially loosely coupled vibrations of the individual oscillators consisting of the pairs of masses which are connected internally by the springs of constant k_1. For any values of k_1 and k_2 the two masses of a pair move in antiphase in the mode for which $\delta = 0$ but all pairs move in phase, so that when one type of spring is stretched the other type is compressed. In the mode for which $\delta = \pi$ the two masses of a pair move in antiphase and alternate pairs also move in antiphase, so that alternate springs with constant k_1 are compressed and stretched, whereas springs with constant k_2 remain always at their equilibrium length. The acoustic mode with $\delta = \pi$ is similar to the second of these, but the roles of the two types of spring are reversed. Note that only the factor group mode ($\delta = 0$) with frequency $\omega = \sqrt{[2(k_1 + k_2)/m]}$ would be Raman active and no mode would be infrared active for this system. We turn now to look at the dispersion curves for one of the simplest real polymers, polyethylene.

The regular polyethylene chain takes the form of a 2_1 helix and it is important to remember that in considering the phase relationship between the motions of adjacent $—CH_2—$ units, one unit must be imagined to be brought into coincidence with the other by the screw operation. The factor group modes for an n_k helix are those for which $\delta = 0$ or $\pm 2\pi mk/n$ for m integral and $\leqslant \frac{1}{2}n$, so that for polyethylene they are those for which $\delta = 0$ or π. Dispersion curves calculated on the basis of a UBS force field and force constants determined for small hydrocarbon molecules are shown in fig. 4.13. In this figure the symmetry and nature of the factor group modes are indicated. The curves corresponding to the $v_s(CH_2)$, $v_a(CH_2)$ and $\delta(CH_2)$ modes are fairly flat; this is another way of saying that the corresponding vibrations of the individual $\diagdown CH_2$ groups are good group modes for polyethylene. All the other curves are far from flat, which indicates that the corresponding group vibrations are strongly coupled and are not good group modes.

No two dispersion curves in fig. 4.13 cross and it is important to consider whether this is an accidental or necessary feature of the dispersion curves. The (complex) eigenvectors for all modes with a particular general value of δ ($\neq 0$ or π) form a restricted set of the complete set of distortions of the molecule. In considering the symmetry species of this restricted set it is as if the molecule no longer had the full symmetry of the line group, since the only symmetry operation of a line group which may leave any member of such a set unchanged is reflection in a mirror plane, σ_c, containing the chain axis. Such a mirror plane exists for polyethylene and thus modes corresponding to a given δ ($\neq 0$

or π) will belong to one of two symmetry species, depending on whether they are symmetric or antisymmetric with respect to this plane. No two dispersion curves both of which belong to the same one of these two species can cross, because this would imply the existence of two modes of vibration of the same symmetry species and the same frequency.

Examination of table 3.1 shows that for polyethylene the dispersion curves for modes symmetric with respect to the mirror plane have as their limiting modes, when δ approaches 0 or π, factor group modes of species A_g, B_{2g}, B_{1u} and B_{3u}. Similarly, the dispersion curves for modes antisymmetric with respect to the mirror plane have as their limiting modes factor group modes of species B_{1g}, B_{3g}, A_u and B_{2u}. The B_{3g} species CH_2 rocking mode for $\delta = 0$ has a higher frequency than the A_u species twisting mode for $\delta = 0$ (see fig. 4.13)), whereas the B_{2u} species rocking mode for $\delta = \pi$ has a lower frequency than the B_{1g} species

Fig. 4.13. Calculated dispersion curves for polyethylene. (Reproduced by permission from H. Tadokoro *et al.*, in *Polymer Spectroscopy*, ed. D. O. Hummel, adapted from M. Tasumi and T. Shimanouchi, *Journal of Molecular Spectroscopy* **9** 261 (1962).)

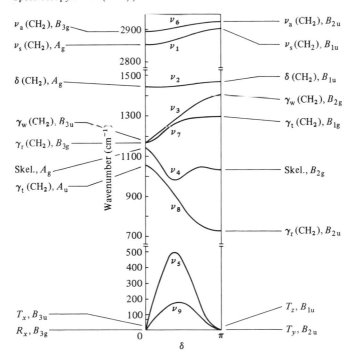

twisting mode for $\delta = \pi$. All four species go over, for $\delta \neq 0$ or π, to the same species, but instead of one curve joining the two rocking factor group modes and another joining the two twisting factor group modes, which would cross, there is in fact one curve joining the B_{3g} rocking mode to the B_{1g} twisting mode and one joining the A_u twisting mode to the B_{2u} rocking mode. It is as if there were a 'repulsion' between dispersion curves belonging to the same symmetry species.

Apart from such restrictions on crossing, it is essentially accidental that there are no crossings of dispersion curves for polyethylene. If, however, we consider a helical molecule which has a three-fold or higher screw axis, there can be no mirror plane, so that, for all helical molecules of this type, no two dispersion curves can cross. In practice, the interactions which couple different types of motion may sometimes be so weak that this non-crossing rule is of little practical consequence.

4.5   Finite chains

In the previous subsection the vibrations of infinite, perfect polymer chains were considered. In a typical crystal of polyethylene the length of chain between folds is about 125 Å or about a hundred CH_2 groups. We neglect, for the moment, the interactions between chains and consider how the vibrational modes of these finite lengths of chain are related to those of the infinite perfect chain and how the observed infrared or Raman spectra will differ from those expected for the perfect infinite chain.

It is easy to show, by a method similar to that used in discussing the diatomic chain in the previous subsection, that the normal modes of vibration of a linear chain of m coupled oscillators correspond to vibrations in which the phase difference δ between adjacent oscillators is either $k\pi/m$ or $k\pi/(m+1)$ for a chain with free or fixed ends, respectively, where k takes integral values 0 to $m-1$ for free ends and 1 to m for fixed ends. For a real polymer chain in a crystal, where the ends correspond to chain folds, or for a short-chain *normal paraffin* (normal alkane), consisting of a linear chain of CH_2 groups with CH_3 end groups, there will be conditions which do not correspond exactly to either simple free or fixed ends. As m becomes large the frequency of each of the modes will tend towards that of one of the modes of the infinite chain which has the same value of δ, but will differ increasingly from it as m becomes smaller because of the effects of the ends. The modes with the closest frequencies to those of the corresponding factor group modes for the infinite chain will be those for which δ differs by approximately π/m from 0 or π. For a polyethylene chain with a hundred CH_2 groups this difference is $\pi/100$.

A study of fig. 4.13 shows that for most optical branches of the dispersion curves the frequency of a mode with $\delta = \pi/100$ or $\delta = \pi - \pi/100$ will be virtually indistinguishable from that of the corresponding factor group mode. These modes will also be the modes of greatest intensity in the vibrational spectrum because all the translational units of the chain vibrate almost in phase. If, on the other hand, we choose a value of δ for which the frequency is significantly different from that of the corresponding factor group mode the intensity will be much lower because the interactions of the different translational units of the chain with the radiation field will tend to cancel out. In addition, the lengths of chain present in crystallites are not all the same, so that the allowed values of δ, and hence the frequencies, are not all the same. The weak peaks corresponding to modes with such values of δ are therefore spread over the whole frequency range of the particular branch of the dispersion curve involved and tend to give rise to background scattering rather than to individual peaks. For most crystalline polymers we do not, therefore, usually have to consider the finite lengths of the chains when studying the contribution of the crystalline regions to the infrared or Raman spectrum unless the crystallites are very small.

It is often useful, as explained in subsection 4.3.4, to study the spectra of small molecules as model compounds for a polymer. An important example is the study of the linear paraffins as models for polyethylene. For each different paraffin there are only a small number of widely-spaced allowed values of δ corresponding to the allowed values of k; the frequencies of the corresponding modes will approximate closely to those given by the dispersion curves for the polymer for the same value of δ, the differences being due to the effects of the end groups. Most of the modes with activity allowed by the selection rules will be observable despite their low intensity (see, e.g., fig. 5.2), because every chain is now of the same length, so that the modes are at well-defined frequencies. Cooling the sample to liquid nitrogen temperatures increases the clarity of the spectrum by removing some of the combination frequencies with acoustic modes (see section 4.6). The observed frequencies for the short-chain paraffins may be used, when corrected for the effects of end groups, to provide data for the empirical construction of dispersion curves of the polymer, since δ can be deduced provided that k is known. Fig. 4.14 shows such an empirically constructed dispersion curve for the rocking–twisting mode of polyethylene, corresponding to ν_8 of fig. 4.13. (See also subsection 5.2.1.) Note the large departures of the observed frequencies of the shorter chains from the common curve because of the greater relative importance of the end groups. As already indicated, dispersion curves can be studied more directly by means of inelastic neutron scattering, but such studies are outside the scope of this book.

4.6   Overtone and combination frequencies; Fermi resonance

So far we have been implicitly considering the *fundamental* frequencies of the various vibrational modes. In addition to these frequencies it is sometimes possible to observe infrared absorption or Raman scattering at frequencies which are approximately either integral multiples of these fundamental frequencies, i.e. *overtone* frequencies, or sums or differences of integral multiples of the fundamental frequencies of two or more different modes, i.e. *combination* frequencies. To understand the origin of these frequencies it is necessary to return to a consideration of equation (2.1) or equation (2.3).

Equation (2.1) expresses the harmonic approximation for a diatomic molecule, in which terms in $(r - r_0)$ raised to powers higher than two are neglected in the expression for the potential energy. Since in practice

Fig. 4.14. Experimental frequency–phase curve for the methylene rocking–twisting modes of the normal paraffins C_3H_8 to $C_{30}H_{62}$. Data for selected paraffins are indicated by the number of carbon atoms. The solid line indicates the dispersion curve calculated by taking end effects into account. (Reproduced by permission from H. Tadokoro *et al.*, in *Polymer Spectroscopy*, ed. D. O. Hummel, taken from an original in H. Matsuda *et al.*, *Journal of Chemical Physics* **41** 1527 (1964).)

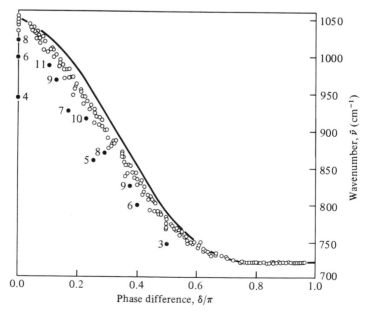

higher order terms are small, rather than zero, the vibration of the pair of atoms is not truly simple harmonic. It will, nevertheless, have a regular repetition with a period T' which is very close to the period $T = 1/v$, with v calculated from equation (2.2), which applies in the harmonic approximation. It is a well-known result that any regularly periodic function can be expressed as a Fourier series of simple harmonic functions, the frequencies of which are integral multiples of the repetition frequency. The latter is usually called the *fundamental frequency* and corresponds to $1/T' = v'$ rather than to $1/T$.

Classically, we should expect to observe a series of frequencies in the vibrational spectrum at exact multiples of the fundamental frequency v'. Quantum mechanics shows, however, that the energy levels are in fact given approximately by the expression

$$E_n/h = (n + \tfrac{1}{2})v + (n + \tfrac{1}{2})^2 x$$
$$= \tfrac{1}{2}v + \tfrac{1}{4}x + n(v + x) + n^2 x \tag{4.76}$$

where v is the frequency which would be observed in the harmonic approximation, x is a small quantity which depends on the coefficients of the cubic and quartic terms in the non-harmonic expression for the potential energy and n is an integer quantum number. This expression may be written

$$E_n/h = E_0/h + nv^0 + n(n - 1)x \tag{4.77}$$

where E_0 represents the zero point energy $h(\tfrac{1}{2}v + \tfrac{1}{4}x)$ and the observable fundamental frequency, i.e. the frequency for the transition $n = 1$ to $n = 0$, is $v^0 = (v + 2x)$, corresponding to the classical value v'.

It is not generally possible to observe transitions between all possible pairs of energy levels in the infrared or Raman spectrum but equation (4.77) shows that the frequencies for those transitions which are observable will not be simple multiples of v^0 unless $x = 0$. At normal temperatures most molecules are in the ground state ($n = 0$) before interacting with radiation, so that the transitions observed usually lead to frequencies of the form $nv^0 + n(n - 1)x$. Since x is generally negative, the observed overtone frequencies are usually slightly lower than whole multiples of the fundamental frequency v^0.

For a system such as a molecule with more than two atoms quantum mechanics shows that the energy levels are given approximately by

$$E(n_1, n_2, \ldots)/h = E_0/h + (n_1 v_1^0 + n_2 v_2^0 + \ldots)$$
$$+ [n_1(n_1 - 1)x_{11} + n_2(n_2 - 1)x_{22} + \ldots]$$
$$+ \tfrac{1}{2}[(n_1 + n_2 + 2n_1 n_2)x_{12} + \ldots] \tag{4.78}$$

where v_1^0, v_2^0 etc. are the fundamental frequencies of the normal modes and n_1, n_2 etc. are the corresponding vibrational quantum numbers. The quantities x_{ij} corresponding to x for the system with only one normal mode, are again usually negative, so that the observed overtone or summation frequencies are usually slightly lower than the corresponding integral multiples or sums of integral multiples of the fundamental frequencies. In order to predict which frequencies will be observable in the infrared or Raman spectrum it is necessary to consider the selection rules, which in turn leads to a consideration of the symmetry species to which the wave functions belong.

Consider first an energy level $E(n_1, n_2, \ldots, n_k, \ldots)$ for which all of the quantum numbers corresponding to degenerate species (if there are any) are zero. It may then be shown that the corresponding wave-function is symmetric with respect to a particular operation of the point group of the molecule if the sum $\sum n_k$, extended over all normal vibrations that are antisymmetric with respect to that symmetry operation, is even and that the wave-function ψ is antisymmetric with respect to that symmetry operation if the sum is odd. It is thus easy to deduce the symmetry species of the wave-function using character tables. If any one degenerate vibration is excited with $n_k = 1$ then it may be shown that the symmetry species of the wave-function is that of the original degenerate species or of a different species of the same degree of degeneracy. If a degenerate species has quantum number > 1 or if more than one degenerate species is involved the evaluation of the resulting symmetry species is complicated and will not be dealt with; standard tables are available showing the results for all combinations and overtones.

Quantum mechanics shows that the selection rule for the infrared activity of a transition between two states E_1 and E_2 is that the transition is active if the product $\psi_1 \psi_2$ has the same symmetry as a translation, where ψ_1 and ψ_2 are the wave-functions for the two levels. Similarly, the selection rule for Raman scattering is that the transition is active if the product of the wave-functions has the same symmetry species as a component of the polarizability tensor, i.e. as one of the quantities x^2, y^2, z^2, xy, yz, zx. If the lower state of the transition is the ground state, which is always symmetric with respect to every symmetry operation, the symmetry species of the upper state determines the spectral activity.

Transitions in which the ground state is not the lower level can take place at elevated temperatures for all modes and at ordinary temperatures for low frequency modes; such transitions can take place for any mode and temperature for which $h\nu \approx kT$. Transitions in which only one of the vibrational quantum numbers changes by 1 but which do not involve the ground state give rise to series of *hot bands* with frequencies of the form $v^0 + 2nx$, i.e. frequencies displaced by equal multiples of $2x$ from the fundamental frequency. Since x is usually negative, this gives rise to asymmetric broadening on the low frequency side of the observed peak. Another source of broadening is the occurrence of weakly active combination vibrations of optical and acoustic modes for non-zero values of δ, for which it may be shown that the appropriate selection rule is that only modes with the same value of δ can combine. Since the occurrence of difference combinations (but not sum combinations) requires the thermal excitation of the lattice mode, the intensities of these combinations will fall as the temperature is reduced, as will the intensities of hot bands.

The intensities of peaks due to overtone or combination excitations are usually low because the departures from harmonic forces are usually small. An important exception occurs through the phenomenon of *Fermi resonance*. This phenomenon, which is sometimes observed in the spectra of polymers, arises when an energy level corresponding to an overtone or combination accidentally occurs at an energy which is close to that corresponding to a fundamental of the same symmetry species. There is then a mixing of the wave-functions corresponding to the different excitations and two new energy levels appear, one on each side of the expected common level. This is analogous to the mixing of symmetry modes of the same symmetry species discussed in earlier chapters, although here cubic or higher order potential energy terms are involved rather than quadratic ones. Its particular importance lies in the fact that the two new levels usually give rise to infrared absorption or Raman scattering of similar intensity, rather than there being a strong fundamental and a weak overtone or combination.

4.7   Further reading

Most of the topics in this chapter are dealt with more fully in the books cited in section 2.6 and the books and reviews cited in section 3.3. For an excellent introduction to the use of tensors in physical science, the reader is referred to the book *Physical Properties of Crystals* by J. F. Nye, Oxford University Press, 1957.

5   The characterization of polymers

5.1   Introduction

This chapter is largely concerned with two aspects of the vibrational spectroscopy of polymers. The first of these is the interpretation of spectra, i.e. the assignment of the observed peaks in the infrared and Raman spectra at one or more of the levels listed in subsection 4.3.1, and the second is the application of a knowledge of such assignments to various aspects of the chemical analysis of polymers. Chapter 6 deals with applications aimed at understanding various aspects of the microstructure of the polymer which are generally classed as physical rather than chemical aspects, although the distinction is sometimes rather arbitrary. It should be emphasized that the division of spectral studies into separate phases of 'interpretation' and 'application' is somewhat artificial, since the two processes often take place side by side. Generally speaking, however, the aims of individual studies will be largely directed towards one or the other of these two aspects of vibrational spectroscopy.

The approach adopted for the interpretation of a particular vibrational spectrum depends upon the nature of the polymer under examination and the information being sought. If the purpose is simply the identification of the polymer, the spectrum is used on a 'finger-print' basis; strong peaks are assigned to various chemical groups and detailed assignments in terms of specific vibrational modes are unnecessary. If the objective is the quantitative analysis of a mixture, or of a copolymer, a rather less superficial approach is required. A detailed understanding of specific portions of the spectrum will be necessary for the identification of defect structures of the kinds discussed in subsection 1.2.3 and for the study of molecular orientation. A full assignment of the various vibrational modes is required for detailed studies of molecular structure.

We start by considering in section 5.2 a number of polymers for which rather detailed interpretations of the infrared and Raman spectra have been obtained and consider in section 5.3 some materials for which the group frequency approach has proved useful. In the tables which follow, π and σ indicate parallel and perpendicularly polarized infrared-active modes (see subsection 4.2.1); m indicates a mode of 'mixed' polarization, i.e. neither pure π nor pure σ; p and d indicate polarized or depolarized

162

Raman-active modes (see subsection 4.2.2); i indicates an inactive mode. The symbols v, δ, γ_r, γ_w and γ_t describe the approximate nature of eigenvectors and have the significances explained in subsection 3.1.2; v_s and v_a indicate symmetric and antisymmetric CX_2 stretching. The symbols w, s and m are used to describe weak, strong and medium absorption or scattering and v is used to mean 'very', e.g. vs = very strong.

5.2   The interpretation of polymer spectra: factor group analysis

In this section the infrared and Raman spectra of a number of polymers are interpreted in detail. The aim is to illustrate the use of the methods of assignment discussed in subsections 4.3.2 to 4.3.5 and the use of vibrational calculations where appropriate. All methods are to some extent applicable to all polymers, but the most suitable methods and the depth of interpretation possible for a given polymer depend on the degree of complexity of its structure. The polymers discussed have been chosen to illustrate this.

The discussion starts with polyethylene and illustrates the use of the full factor group analysis for the perfect isolated polymer chain. The methods used in arriving at the present interpretation and the errors made on the way illustrate well many important applications of the principles discussed in earlier chapters and we shall therefore describe in more detail than for the remaining polymers how the interpretation was arrived at.

5.2.1   *Polyethylene*

It is natural to start a discussion of the interpretation of the vibrational spectra of polymers by considering the infrared and Raman spectra of polyethylene. This polymer is one of the commonest in current use and it also has the simplest basic chemical structure. The interpretation of its spectrum is therefore of great practical and fundamental significance. Most forms of solid polyethylene are more than 50% crystalline and may be as much as 85% crystalline, so that most of the strong peaks in the infrared or Raman spectrum should be due to chains in crystallites. Since intermolecular interactions are weak in a polymer with no polar groups, it should be possible to assign most of these peaks to fundamental frequencies of the factor group modes of the isolated single chain and then to account for many other, generally weak, features as overtone or combination frequencies of these fundamentals. A much smaller number of features will then remain to be assigned to vibrations of various defect structures or end groups. The infrared and Raman spectra of polyethylene are shown in fig. 5.1.

There are only 13 spectroscopically active factor group modes (see subsections 3.1.2 and 3.2.2) and it might be anticipated that the task of assigning each of them to a peak in the appropriate spectrum would be relatively straightforward, particularly as the number of strong peaks in the observed spectra is similar to the number of factor group modes. It is therefore worth noting immediately that the first attempts at such a complete assignment were made as early as 1949, but it was not until about 1972 that the assignments now accepted could be regarded as reasonably well established for all the modes, although most of them were established much earlier. In particular, there was much controversy about the assignment of the $\gamma_w(CH_2)$ wagging modes.

The first three columns of table 5.1 show (i) the expected normal modes for the single chain, described in terms of the symmetry modes discussed in subsection 3.1.2 and illustrated in table 3.2, (ii) the symmetry species for the single chain and (iii) the spectral activity. The orientation of the reference axes is chosen as in subsection 3.1.2. It should be noted that the use of the idea of a line group was not introduced into the discussion of polymer spectra until 1955, but we shall here 'translate' earlier nomenclature into that of table 5.1.

Fig. 5.1. The infrared and Raman spectra of polyethylene: (*a*) infrared absorbance spectrum; (*b*) Raman spectrum.

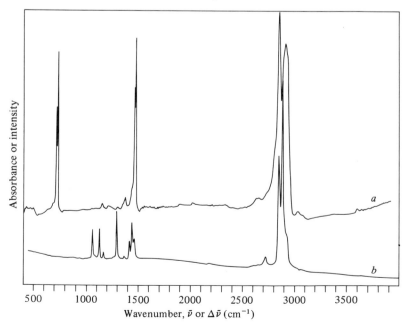

Table 5.1. *Assignment of fundamental factor group vibrations of polyethylene*

Description of mode	Single chain		Crystal		Assignments (cm⁻¹)		
	Sym. sp.	Act.	Sym. sp.	Act.	1950	1963	Current
Skeletal	A_g	R, p	A_g	R, p	1135?	1131	1133
			B_{3g}	R, d			
	B_{2g}	R, d	B_{1g}	R, d	1058 or	1061	1067
			B_{2g}	R, d	1135		
$\nu_s(CH_2)$	A_g	R, p	A_g	R, p	2846 or	2848 or	2848
			B_{3g}	R, d	2878	2865	
	B_{1u}	IR, σ	B_{1u}	IR, a	2853	2851	2851
			B_{2u}	IR, b			
$\nu_a(CH_2)$	B_{3g}	R, d	A_g	R, p	2934 or	2883 or	2883
			B_{3g}	R, d	2963	2932	
	B_{2u}	IR, σ	B_{1u}	IR, a	2925	2919	2919
			B_{2u}	IR, b			
$\delta(CH_2)$	A_g	R, p	A_g	R, p	1471	1440	1416
			B_{3g}	R, d			1440
	B_{1u}	IR, σ	B_{1u}	IR, a	1475 or	1473	1473
			B_{2u}	IR, b	1462	1463	1463
$\gamma_w(CH_2)$	B_{2g}	R, d	B_{1g}	R, d	1442 or	1415	1370
			B_{2g}	R, d	1295		
	B_{3u}	IR, π	A_u	i	1310 or		
			B_{3u}	IR, c	1375	1176	1175
$\gamma_r(CH_2)$	B_{3g}	R, d	A_g	R, p	?	1168	1170
			B_{3g}	R, d			
	B_{2u}	IR, σ	B_{1u}	IR, a	721	731	734
			B_{2u}	IR, b		720	721
$\gamma_t(CH_2)$	B_{1g}	R, d	B_{1g}	R, d	?	1295	1296
			B_{2g}	R, d			
	A_u	i	A_u	i	?		
			B_{3u}	IR, c		1050	1050

a, b or *c* indicates polarization parallel to crystal *a, b* or *c* axis. Other notation as explained in section 5.1.

Although we defer detailed consideration of the effects of crystallinity until chapter 6, it is necessary in the present discussion to refer occasionally to the factor group modes of the full crystal space group and to their spectral activities. These are shown in columns four and five of table 5.1, using the same orientation of the reference axes as in subsection 3.2.2.

Polyethylene, or 'Polythene', was discovered in 1933 and one of the earliest papers to discuss the infrared spectrum of this new material was published by Fox and Martin in 1940. They showed that in many simple compounds the $\diagdown CH_2$ group gives rise to two frequencies at about 2857 and $2927\,cm^{-1}$ which are due, respectively, to the symmetric and antisymmetric stretches of the two C—H bonds. These frequencies varied by only a few wavenumbers from compound to compound and were also observed in the spectrum of polyethylene. The specific assignments were based both on vibrational calculations and on the fact that for many small molecules containing YX_2 groups, for which complete assignments could be made by means of polarized infrared and Raman studies and the use of the selection rules, the lower frequency vibration of the YX_2 group corresponds to the symmetrical stretching mode.

During the period 1940 to 1950 attempts were made to assign the other prominent peaks in the infrared spectrum of polyethylene in terms of group frequencies by comparison with frequencies observed in normal paraffins, which consist of linear molecules of the form $CH_3(CH_2)_{n-2}CH_3$, and other small molecules containing $\diagdown CH_2$ chains. In addition to the peaks near $2900\,cm^{-1}$, assigned to C—H stretching, peaks were noted at about 1470, 1375, 1310, 1080, 890 and $725\,cm^{-1}$. Studies with a fully deuterated polyethylene showed that the peaks at 1470, 1375 and $725\,cm^{-1}$ in the spectrum of ordinary polyethylene were shifted down by a factor of approximately $\sqrt{2}$, so that they must correspond to vibrations associated primarily with the motion of hydrogen atoms. Studies with polarized radiation on oriented samples showed that the peaks at 1470 and $725\,cm^{-1}$ corresponded to perpendicular vibrations, whereas that at $1375\,cm^{-1}$ corresponded to a parallel vibration. The peaks at 725 and $1470\,cm^{-1}$ were observed as doublets in the solid state, but in the melt the lower frequency component ($721\,cm^{-1}$) of the $725\,cm^{-1}$ doublet disappeared for parallel polarization, leaving only a broad, much less intense absorption at $730\,cm^{-1}$ which was assigned to the non-crystalline material. These facts, together with studies of short-chain normal paraffins, suggested very strongly the assignments shown in the column headed '1950' in table 5.1. It was believed that the only infrared fundamental about which serious doubt remained was the $\gamma_w(CH_2)$ wagging mode, since there was another possible assignment for the $1375\,cm^{-1}$ peak, viz. to the symmetrical deformation vibration of methyl groups, which may occur at the ends of the chains or in side groups.

Assignments to the Raman fundamentals were attempted during this period, even though no actual Raman spectrum of polyethylene was available, on the basis both of a calculation, using force constants derived from infrared studies of small molecules, and on the extrapolation to infinite chain length of the frequencies of various series of peaks observed in the Raman spectra of the normal paraffins. This led to the conclusion that the most likely value of the frequency of the $\delta(CH_2)$ bending mode was about $1471\,cm^{-1}$, but there were uncertainties about the other assignments, as shown in table 5.1, where the Raman assignments made by 1950 are shown. The most serious difficulties were in the assignment of the $\gamma_w(CH_2)$ wagging mode and the two skeletal modes. The frequencies of the strongest peaks in the C—H stretching region were close to those of the corresponding infrared modes, and one group of authors assumed that these were the fundamentals whereas another group saw no reason to reject the weaker peaks as possible fundamentals. The empirical evidence from the spectra gave no clear indication of the likely positions of the B_{3g} species $\gamma_r(CH_2)$ or B_{1g} species $\gamma_t(CH_2)$ modes.

Between 1950 and 1963 a number of important developments took place. The use of the concept of the line group in discussing the spectra of polymers was introduced by Tobin in 1955, the first Raman spectrum of polyethylene was reported by Nielsen and Woollett in 1957 and more sophisticated studies were made of series of related frequencies in the infrared and Raman spectra of normal paraffins. It was also clearly recognized that the infrared doublets at $1463/1473$ and $720/731\,cm^{-1}$ observed in the spectra of highly crystalline samples were caused by correlation splitting and it was shown by polarization studies on oriented samples that in each case the higher frequency was the B_{1u} species and the lower the B_{2u} species.

Nielsen and Woollett's Raman spectrum was obtained using photographic recording and showed 12 peaks. They assigned the $\nu(CH_2)$ stretching modes to the strongest peaks in the appropriate region, viz. $\nu_s(CH_2)$ to $2848\,cm^{-1}$ and $\nu_a(CH_2)$ to $2883\,cm^{-1}$. Three other weak peaks were observed at 2912, 2932 and $2720\,cm^{-1}$ and it was suggested that the first two of these were overtones of the $1463/1473\,cm^{-1}$ infrared-active doublet in Fermi resonance with the $\nu(CH_2)$ stretching fundamentals, whereas the very weak peak at $2720\,cm^{-1}$ might correspond to the combination $1295 + 1440 = 2735\,cm^{-1}$ or to the overtone of the methyl end group. Two peaks were observed at 1440 and 1463 cm^{-1} and the first was assigned to the $\delta(CH_2)$ bending mode, because it was much the stronger. The 1463 cm^{-1} peak was assigned to either an overtone of the infrared active 720/731 cm^{-1} doublet in Fermi resonance with the $\delta(CH_2)$ fundamental or to the infrared active $\delta(CH_2)$ fundamental at 1463 cm^{-1} which may be Raman active in the amorphous regions. The lowest two observed frequencies, 1131 and 1061 cm^{-1}, were assigned to the B_{2g} and A_g skeletal modes, respectively, by comparison with the corresponding values 1137 and 1037 cm^{-1} calculated by Liang *et al.* using a very simple model in which the CH$_2$ groups were treated as point masses.

Nielsen and Woollett's assignment of the remaining three observed peaks, at 1415, 1295 and 1168 cm^{-1}, which were assumed to be the three remaining Raman-active fundamentals, illustrates an interesting use of the relationship, explained in section 4.6, between the symmetry species of a combination vibration and that of the corresponding fundamentals. Peaks observed in the infrared spectrum at 2016 and 1894 cm^{-1} were assigned, respectively, to combinations of the 1295 and 1168 cm^{-1} Raman-active fundamentals with the 720/731 cm^{-1} B_{2u} species infrared-active fundamental. Since the 2016 cm^{-1} absorption shows parallel polarization with oriented samples, and is therefore of species B_{3u}, the 1295 cm^{-1} mode must be of B_{1g} symmetry (see character table for D_{2h}, table 3.1), and was therefore assigned to the only remaining mode of that symmetry, the $\gamma_t(CH_2)$ twisting mode. Similarly, since the 1894 cm^{-1} absorption is perpendicularly polarized and is therefore of species B_{1u} or B_{2u}, the 1168 cm^{-1} mode must be B_{3g} or A_g. All A_g modes are, however, already accounted for, so that it must be the B_{3g} $\gamma_r(CH_2)$ mode, leaving as the only possibility the assignment of the 1415 cm^{-1} peak to the $\gamma_w(CH_2)$ mode.

The new studies of the normal paraffins were refinements of earlier studies, taking account of both Raman and infrared data and using selection rules to identify modes. Simple coupled oscillator theory indicates (see section 4.5) that any group vibration, such as the $\gamma_t(CH_2)$ vibration, should split into $n-2$ components for a normal paraffin with n carbon atoms and $n-2$ CH$_2$ groups. Each component corresponds to a different phase difference, δ, between adjacent oscillators and to a good approximation should have a frequency equal to that for a mode of the infinite chain which has the same value of δ. Extrapolation of the modes of highest and lowest frequency to infinite n should lead to the factor group modes for the infinite chain, for which $\delta = 0$ or π. A second method of finding these limiting modes was to assign values of k, and hence δ, to each mode and to plot dispersion curves for a given type of

Table 5.2. *Phase differences, symmetry species and selection rules for normal paraffins*

Mode	Phase difference, δ	n odd (C_{2v})		n even (C_{2h})	
		k odd sp. IR R	*k* even sp. IR R	*k* odd sp. IR R	*k* even sp. IR R
Skeletal stretch	$k\pi/n$	A_1 σ p	B_1 π d	A_g i p	B_u m i
Skeletal deform.	$k\pi/(n-1)$	A_1 σ p	B_1 π d	A_g i p	B_u m i
CH₂ bend	$k\pi/(n-1)$	A_1 σ p	B_1 π d	A_g i p	B_u m i
CH₂ wag	$k\pi/(n-1)$	B_1 π d	A_1 σ p	B_u m i	A_g i p
CH₂ rock	$k\pi/(n-1)$	B_2 σ d	A_2 i d	B_g i d	A_u σ i
CH₂ twist	$k\pi/(n-1)$	A_2 i d	B_2 σ d	A_u σ i	B_g i d

For the skeletal stretching mode k takes the values 1 to $n - 1$ and for all the other modes it takes the values 1 to $n - 2$, where n is the total number of carbon atoms in the paraffin.

Adapted by permission from Table III of M. Tasumi, *et al.*, *Journal of Molecular Spectroscopy* **9** 261 (1962).

vibration (as discussed in section 4.5), which could then be extrapolated to $\delta = 0$ and $\delta = \pi$.

Modes belonging to a given type of vibration can be picked out from the spectra using a number of guidelines, which include the selection rules for infrared and Raman activity (see table 5.2) and the expected systematic changes of frequency and intensity with δ. For the $\gamma_r(CH_2)$ modes, for instance, the selection rules for even n predict alternating absences of IR or Raman active modes as k changes in steps of one. For odd n they predict the absence of alternate modes from the infrared spectrum and the perpendicular polarization of those present. Fig. 5.2 shows the spectrum of a normal paraffin.

The first results of these studies were the confirmation that the 720/731 cm^{-1} infrared doublet was due to rocking vibrations, and the conclusion that the infrared-active B_{3u} wagging frequency was at 1176 cm^{-1}, as had recently been suggested on other grounds, rather than at either 1310 or 1375 cm^{-1} as thought earlier. The results also suggested that the Raman-active wagging fundamental was correctly assigned at 1415 cm^{-1} and that the Raman active rocking mode should occur near 1033 cm^{-1}.

In 1960 and 1961 very detailed infrared studies of polyethylene were reported by Nielsen and Holland, including polarization measurements, on 91 observed vibrations. Twenty-four of these were interpreted satisfactorily as binary combinations of a Raman-active and an infrared-active unit cell fundamental. Since several could be interpreted in more than one way, 32 of the 48 binary combinations that could be infrared active were used or accounted for in the interpretation. A further fundamental assignment was suggested, viz. that the weak peak at 1050 cm^{-1} was due to the $\gamma(CH_2)$ twisting fundamental. At this date the problem of the assignment of both the infrared and Raman fundamentals, and indeed of most features of the spectra, seemed to be essentially solved. This apparently satisfactory complete set of assignments did not, however, stand for very long against further experimental evidence and calculations.

The culmination of the work on series in the normal paraffins was in 1963 when Snyder and Schachtschneider made a very detailed set of assignments for the infrared spectra of the normal paraffins C_3H_8 to $C_{19}H_{40}$ and plotted fairly complete dispersion curves. Their most important achievement was the assignment of the skeletal modes in the region 1150–950 cm^{-1}. In the region above 1065 cm^{-1} regularity is apparent, but in the region below, no order is perceptible at first sight. This difficulty was understandable in terms of simple model calculations for the skeletal modes, of the type already referred to, which suggested that the dispersion curve for the skeletal modes would have a minimum in it. By noting this possibility they were able to discern the underlying regularity of the distribution of modes and to verify the existence of a

Fig. 5.2. The infrared spectrum of n-$C_{12}H_{26}$ at $-196°C$. (Reproduced by permission from R. G. Snyder and J. H. Schachtschneider, *Spectrochimica Acta* **19** 85 (1963).Copyright (1963), Pergamon Journals Ltd.)

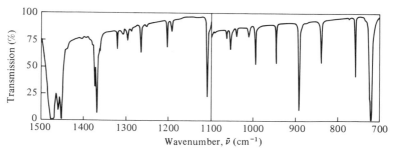

minimum in the dispersion curve. Limiting frequencies near 1135 and 1065 cm^{-1} were clearly indicated, confirming that the Raman lines at 1131 and 1061 cm^{-1} in polyethylene are indeed due to the skeletal modes. Earlier infrared studies on the normal paraffins indicated that the 1135 cm^{-1} limit was that of a series of perpendicularly polarized modes and therefore corresponded to the A_g mode, whereas the simple calculations which led to the prediction of a minimum in the dispersion curve had led Nielsen and Woollet, incorrectly, to believe that the A_g mode should have a lower frequency than the B_{2g} mode.

During the course of the work, high speed computer methods of solving the vibrational problem came into use, and once all the assignments had been made it was possible to adjust a set of force constants common to a group of molecules to give a simultaneous least squares fit of their frequencies to the observed values. A set of 35 valence force constants was fitted in this way to 270 frequencies of C_2H_6, C_3H_8, the normal paraffins C_4H_{10} to $C_{10}H_{22}$ and polyethylene. The fit was excellent, with an average deviation of only 0.25%. It was possible, in addition, to predict from the fitted force constants the frequencies of 240 modes of C_2D_6, C_3D_8 and the branched hydrocarbons isobutane, $CH_3 . CH(CH_3) . CH_3$, and neopentane, $C(CH_3)_4$. The calculations supported the reassignment of the skeletal fundamentals in polyethylene and the earlier assignment of the Raman-active CH_2 wagging mode to 1415 cm^{-1}. No other changes of assignment were indicated.

The calculations supplied normal mode eigenvectors as well as frequencies and the only surprise provided by these was the fact that the dispersion curve which starts out at $\delta = 0$ from the 720 cm^{-1} pure rocking mode in polyethylene gradually mixes with the twisting symmetry mode to become the pure twisting mode of polyethylene for $\delta = \pi$, at a predicted frequency of 1063 cm^{-1}. Since this mode is activated by the crystal field, the calculations supported the earlier assignment of the observed weak infrared mode at 1050 cm^{-1} to the $\gamma_t(CH_2)$ mode. In a similar way, the mode which is the pure B_{1g} Raman active twisting mode of polyethylene at 1295 cm^{-1} mixes with the rocking mode and eventually becomes the pure B_{3g} Raman active rocking mode at 1168 cm^{-1}. This mixing of modes occurs because rocking and twisting modes not only have similar frequencies but also have the same symmetry for the infinite chain when $\delta \neq 0$, as explained in subsection 4.4.7. Mixing of the symmetry modes therefore takes place to produce the actual normal modes and the corresponding dispersion curves, which do not cross. Dispersion curves for polyethylene, produced by a more recent calculation, have been shown already in fig. 4.13.

Snyder and Schachtschneider believed that the spectra of polyethylene were now correctly assigned and that, in particular, the question of the wagging assignments for the polymer was settled 'with finality'. Ironically, it was Snyder himself who later provided some of the most important evidence for the reassignment of the Raman active wagging mode.

The first doubts about the assignment of the wagging mode were expressed in a paper concerned with the vibrational spectra of methyl laurate, $H(CH_2)_{11}COOCH_3$, and related compounds, and an alternative assignment was suggested. Further evidence that the assignment to 1415 cm^{-1} was incorrect was provided by an analysis of the vibrational spectra of non-planar polyethylene chains by Snyder, in which the wagging modes of shorter segments are infrared activated because of the lowering of the symmetry. No modes of frequency higher than 1370 cm^{-1} were observed, although there was good reason to expect that there would be some measurable absorption very near to the frequency of the B_{3g} wagging fundamental of polyethylene. It should be noted that in the earlier work it had not been possible to trace the experimental frequency–phase curve beyond 1360 cm^{-1}, so that the value of 1415 cm^{-1} represented a fairly large extrapolation.

In order to study the CH_2 wagging modes further and to avoid overlap with symmetric deformation modes of methyl groups Snyder studied some α,ω-dichloroalkanes $(Cl(CH_2)_nCl)$. He showed that if the B_{2g} wagging fundamental of polyethylene was given the frequency 1382 cm^{-1} a set of force constants could be found which 'predicted' the dichloroalkane frequencies in the region 1310–1378 cm^{-1} without essentially changing the predicted frequencies of 185 observed modes for the normal paraffins. Snyder points out that 'although there are abundant illustrations in the literature of the dangers of "predicting" or "confirming" vibrational assignments using computer calculated frequencies, the present example is singularly spectacular'.

There remained two problems: to identify a mode near 1382 cm^{-1} in the Raman spectrum of polyethylene, and to explain the origin of the observed mode at 1415 cm^{-1}. Subsequent Raman studies by Snyder established the presence at 1370 cm^{-1} of a peak in polyethylene which could not be attributed to a methyl vibration because it appeared in the spectrum of a linear polyethylene with less than one methyl group per 1000 carbon atoms. This spectrum showed no trace of the peak at 890 cm^{-1} which occurs in the spectrum of some polyethylenes and is known from vibrational calculations to be due to an in-plane rocking mode of methyl groups. Later Raman polarization studies, including studies on a single crystal of $C_{23}H_{48}$ and biaxially oriented 'single crystal texture' polyethylene, confirmed that the peak at 1370 cm^{-1} shows the polarization properties expected of a B_{2g} species mode such as the $\delta_w(CH_2)$ mode and that all the other peaks assigned to fundamentals, including the skeletal modes, also show the expected polarization behaviour.

These studies have drawn further attention to the complex nature of the spectrum in the $\nu(CH_2)$ and $\delta(CH_2)$ regions, particularly the latter. The three peaks observed, at 1416, 1440 and 1464 cm^{-1}, are found to be of symmetry species A_g, B_{3g} and B_{3g}, respectively. On the basis of these symmetries, and the fact that the component at 1416 cm^{-1} is present in the spectra of those normal paraffins that crystallize with the same unit cell as polyethylene but not in the spectra of those that have only one chain per unit cell, it would be reasonable to conclude that the peaks at 1416 and 1440 cm^{-1} are the correlation split doublet corresponding to the $\delta(CH_2)$ mode of symmetry A_g for the single chain. Present evidence suggests that the assignment of the peak at 1416 cm^{-1} to the A_g component of the correlation doublet is almost certainly correct, but that the interpretation of the remainder of this region of the Raman spectrum, which will be discussed further in subsection 6.7.2, is quite complicated.

The extreme right-hand column of table 5.1 shows the final assignments of the crystal fundamentals of polyethylene taking account of evidence produced since 1963. The precise values quoted are those obtained near 90 K, where available.

5.2.2 *Poly(vinyl chloride)*

Poly(vinyl chloride) (PVC) provides a convenient example with which to begin the discussion of the vibrational spectra of vinyl polymers, $+CH_2 . CHX+_n$, because the substituent X is monatomic. X-ray diffraction studies suggest that the crystalline regions in the polymer consist of syndiotactic chains in which the carbon backbone is in the planar zig-zag conformation. It is therefore instructive to attempt the assignment of the various peaks in the infrared and Raman spectra of the polymer in terms of an isolated chain of this type. Most of the discussion will refer to the infrared spectrum because the assignments were originally made almost entirely on the basis of infrared spectra.

It is reasonable to suppose that the vibrational modes will be of three types: those involving principally the vibrations of C—H, C—C or C—Cl units. The discussion that follows will concentrate on experimental evidence, particularly from deuterated polymers, for assignments in terms of these three types of vibrations, but reference will also be made to the results of vibrational calculations based on force constants deduced from fitting more than 90 frequencies from the spectra of secondary chlorides such as 2-chlorobutane, $CH_3 . CH_2 . CHCl . CH_3$. The infrared and Raman spectra of PVC are shown in fig. 5.3.

The line group for the planar zig-zag syndiotactic structure has a factor group isomorphous with the point group C_{2v} (see subsection) 3.1.3). From the type of symmetry mode analysis discussed for polyethylene in subsection 3.1.2 it emerges that there are 14 infrared-active modes involving principally the motions of hydrogen atoms. If the motions of the $=$CH and $>$CH$_2$ groups are separable, which should be a fairly good approximation, the approximate normal modes (strictly they are symmetry modes) may be constructed from the in-phase and out-of-phase combinations of the various modes of vibration of the

Fig. 5.3. The infrared and Raman spectra of poly(vinyl chloride): (*a*) infrared absorbance spectrum; (*b*) Raman spectrum. The absorbance spectrum was obtained from a film cast from tetrahydrofuran onto a potassium bromide plate. The Raman spectrum shows the so-called 'fluorescence' background often found for polymers (see subsection 1.6.1).

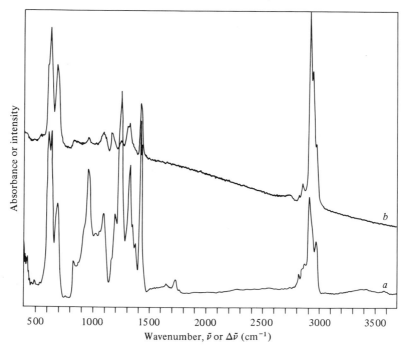

Table 5.3. *Classification of the CH and CH_2 modes of syndiotactic poly(vinyl chloride)*

IR selection rule	Species	Symmetry mode	Frequency (cm^{-1}) Exp.	Calc.	PED %
Active (strong)					
σ	A_1	$\nu(CH)_i$	2970mw	2985	
		$\delta(CH)_i$	1195w	1192	42, 31 $\gamma_t(CH_2)$
		$\nu_s(CH_2)_i$	2910	2856	
		$\delta(CH_2)_i$	1428s	1422	75
π	B_1	$\gamma_w(CH)_o$	1387w	1384	45, 39 $\nu(CC)$
		$\gamma_w(CH_2)_i$	1230mw	1226	55, 41 $\gamma_w(CH)$
σ	B_2	$\nu(CH)_o$	2970mw	2985	
		$\delta(CH)_o$	1258s	1255	79
		$\nu_a(CH_2)_i$	2930w	2928	
		$\gamma_r(CH_2)_i$	960ms	965	50, 23 $\nu(CC)$
Active (weak)					
σ	A_1	$\gamma_t(CH_2)_o$	1338ms	1320	50, 31 $\delta(CH)$
π	B_1	$\nu_a(CH_2)_o$		2928	
		$\gamma_r(CH_2)_o$	835mw	826	88, 16 $\gamma_w(CH)$
σ	B_2	$\gamma_w(CH_2)_o$	1355	1356	84
Forbidden (R active)					
	A_2	$\nu_s(CH_2)_o$		2855	
		$\delta(CH_2)_o$	1437	1431	73
		$\gamma_t(CH_2)_i$		1151	79, 19 $\gamma_w(CH)$
		$\gamma_w(CH)_i$	1316	1315	62, 59 $\gamma_t(CH_2)$

i = in phase, o = out of phase vibration of two CH_2 groups with respect to their interchange by translation parallel to the chain axis or of two CH groups with respect to their interchange by rotation about a C_2 axis. Other notation as explained in section 5.1.

$=$CH or $>$CH$_2$ groups. This approach also provides an indication of which modes are the permitted active ones, and also which of these are likely to be strong or weak, through a consideration of how the dipole moments due to each group combine (see subsection 4.1.2). The results are given in the first three columns of table 5.3.

A similar anlaysis may be undertaken for the C—Cl and C—C normal modes. In this case the two types of motion are likely to be considerably mixed with each other and it is probably not a very good approximation to consider them as separable modes. Nevertheless, despite the fact that the C—Cl modes must involve a considerable amount of carbon skeletal motion, the chlorine frequencies will be treated as separable as it will be shown in chapter 6 that specific assignments for chlorine atoms in

Table 5.4. *Classification of the skeletal and CCl modes of syndiotactic poly(vinyl chloride)*

| IR selection rule | Species | Symmetry mode | Frequency (cm^{-1}) | | PED % |
			Exp.	Calc.	
Skeletal modes					
active, strong, σ	A_1	$v(CC)$, 1	1105m	1099	36, 20 $\gamma_t(CH_2)$
weak, σ	A_1	torsion, 2		38	96
weak, π	B_1	torsion, 3	89	84	85
strong, π	B_1	$v(CC)$, 4	1122w		
			or	1117	55, 14, $\gamma_w(CH_2)$
			1090mw		
weak, σ	B_2	$v(CC)$, 5	1090mw		
			or	1079	51, 24 $\gamma_r(CH_2)$
			1030		
weak, σ	B_2	$\delta(CCC)$, 6	492vw	482	38, 19 $\delta(CH)$
inactive, (R active)	A_2	$v(CC)$, 7	1066	1065	68, 13 $\gamma_w(CH)$
	A_2	$\delta(CCC)$, 8	544	549	68, 26 $\gamma_w(CCl)$
CCl modes					
active, strong, σ	A_1	$v(CCl)_i$	640ms	638	92
		$\delta(CCl)_i$	358w	357	52, 16 $\delta(CH)$
strong, π	B_1	$\gamma_w(CCl)_o$	345w	330	83
strong, σ	B_2	$v(CCl)_o$	604s	601	89
		$\delta(CCl)_o$	315m	314	47, 30 $\delta(CCC)$
inactive, (R active)	A_2	$\gamma_w(CCl)_i$		133	68, 24 $\delta(CCC))$

Numbering of skeletal modes as in fig. 5.4.
i = in phase, o = out of phase vibrations of two CCl groups with respect to their interchange by rotation about a C_2 axis.
Other notation as explained in section 5.1.

different configurational and conformational states are possible. On this basis there should be six C—Cl frequencies, five of which are infrared active, as shown in table 5.4. Similarly, there are eight skeletal modes, six of which should be infrared active. Symmetry modes corresponding to them are shown in fig. 5.4.

Considering first the C—H modes, it might at first appear that the four expected $\geq$CH and $\geq$CH$_2$ stretching frequencies could, with confidence, be assigned to the four peaks at 2850, 2910, 2930 and 2970 cm^{-1}. By analogy with the corresponding modes at 2851 and 2919 cm^{-1} in polyethylene, it is tempting to assign the 2849 cm^{-1} peak to $v_s(CH_2)_i$, (A_1), and the 2910 cm^{-1} peak to $v_a(CH_2)_i$, (B_2), particularly as their relative intensities are also similar to those of the corresponding peaks in polyethylene. The pitfalls of this simple approach are, however, revealed when the spectra of various deuterated PVC samples are also taken into account. The 2970 cm^{-1} peak is absent from the spectrum of PVC-αd$_1$, $+$CH$_2$. CDCl$+_n$, showing that this peak must be assigned to v(CH). As no other peaks in this region are affected by α-deuteration, the one at 2970 cm^{-1} probably represents the composite peak from two overlapping peaks due to the A_1 and B_2 species modes. The corresponding peak appears at 2200 cm^{-1} in PVC-αd$_1$ and the shift ratio, 1.35, is in good agreement with the value of 1.34 which is estimated using the approximate isotope frequency rule referred to in subsection 4.3.5. PVC-d$_3$ has a peak at 2217 cm^{-1}, which is clearly the corresponding v(CD) mode. The additional peaks which occur for this fully deuterated polymer at 2110 and 2160 cm^{-1} must therefore be due to CD$_2$ modes. The predicted isotopic shift ratios 1.38 for v_s(CH$_2$) and 1.36 for v_a(CH$_2$) show that the peak at 2910 cm^{-1} in the normal polymer must be assigned to v_s(CH$_2$) and the one at 2930 cm^{-1} to v_a(CH$_2$). This latter assignment is supported by the polarization. As the intensity of the 2850 cm^{-1} peak decreases with increasing crystallinity of the sample it is to be associated with a C—H vibration of chains in amorphous regions.

Turning to the C—H deformation and skeletal modes, the strong peak at 1428 cm^{-1} is, undoubtedly, the δ(CH$_2$) mode of species A_1. The assignment of the δ(CH) modes of species A_1 and B_2 proves more difficult, partly because of the probable presence of other modes, such as twisting, at similar frequencies. Furthermore, significant mixing of symmetry modes may occur. Once again the study of deuterated polymers is a useful aid in deducing the correct assignments, but because of the mixing of different types of group or symmetry modes the results of vibrational calculations must also be taken into account.

The spectrum of PVC-$\alpha\beta$d$_2$, $+$CDH—CDCl$+_n$, shows two strong σ peaks at 1282 and 1330 cm^{-1}. This deuterated polymer almost certainly

Fig. 5.4. Skeletal symmetry modes for poly(vinyl chloride).

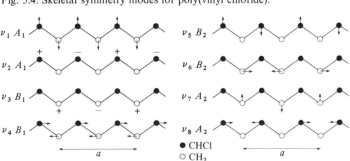

has the hydrogen atoms, as well as the chlorine atoms, syndiotactically placed, so that its planar zig-zag conformation belongs to the line group C_s and the only reasonable assignments for these two peaks are to the two $\delta(CH)$ modes expected for that structure. The fact that there is a splitting of 48 cm^{-1} between modes due to the motions of the CH groups of β-carbon atoms suggests that a large splitting may also be observed for the modes of the CH groups on the α-carbon atoms of normal PVC, so that the strong peaks observed at 1338 and 1258 cm^{-1} are probably associated with the modes of A_1 and B_2 species, respectively. The spectrum of PVC-αd_1 shows three peaks: at 1297 cm^{-1} (σ), 1353 cm^{-1} (π) and 1360 cm^{-1} (σ). Since this polymer contains methylene ($\diagdown$CH$_2$) groups but no lone CH groups, these peaks are most reasonably assigned to the A_1 species $\gamma_t(CH_2)$ and B_1 and B_2 species $\gamma_w(CH_2)$ modes, respectively, of the C_{2v} chain. In view of these frequencies it is necessary to consider the possibility that in normal PVC the corresponding modes may interact with the $\delta(CH)$ modes.

Polarization studies and the results of calculations of the normal mode frequencies lead to the assignments shown in table 5.3. The calculations show that in fact only the $\delta(CH_2)_i$, $\delta(CH)_o$ and $\gamma_w(CH_2)_o$ modes may be considered as reasonably pure modes of the forms indicated. All other CH and CH$_2$ bending or deformation modes are mixtures, several of which involve large components of C—C stretching. The last column in table 5.3 indicates the approximate nature of the mode in terms of the calculated potential energy distribution (PED), the first figure being the percentage contribution of the dominant type of motion and the second the next most important contribution if this is greater than 10%, but it should be noted that the coordinates used in the calculation were not precisely the symmetry coordinates discussed here.

Those skeletal modes which occur in the region 1050 to 1150 cm^{-1} are particularly difficult to assign even when empirical evidence and the results of vibrational calculations are combined. There is general agreement that the A_1 species mode ν_1 (see fig. 5.4) occurs at 1105 cm^{-1}, but the calculations show that it is very far from being a pure skeletal mode. Table 5.4 shows the alternative assignments which have been made for the B_1 and B_2 species modes. One of the difficulties of the assignment is that the absorption at 1090 cm^{-1} occurs as a weak shoulder on the 1105 cm^{-1} peak so that its dichroism is difficult to establish with certainty. There is no ambiguity about the bending mode of B_2 species, which is calculated to be at 482 cm^{-1} and observed at 492 cm^{-1}. Only one of the two infrared-active torsional modes has been observed with certainty, at 89 cm^{-1}.

The C—Cl stretching region (600–700 cm^{-1}) and, to a lesser extent, the C—Cl bending region (300–400 cm^{-1}) are rich in information on the configurational and conformational isomerism of the chain; the former region will be considered in detail in the following chapter. The vibrational calculations predict that the A_1 and B_1 species v(CCl) modes should occur at 638 and 601 cm^{-1}, respectively. The main components of the observed complex of peaks are at about 640 and 604 cm^{-1} and experiments with biaxially oriented samples show that these peaks have the expected dichroisms. The A_1 and B_2 species δ(CCl) modes are assigned to 358 and 315 cm^{-1}, respectively, and the B_1 species γ_w(CCl) mode to 345 cm^{-1} on the basis of vibrational calculations and polarization measurements. The assignments for the C—Cl modes are listed in table 5.4; the comments made about the PED in connection with table 5.3 also apply to this table.

There are a limited number of additional weak peaks in the spectrum that may reasonably be assigned to the vibrational modes of various conformational or configurational isomers of chains in amorphous regions. The fact that normal commercial poly(vinyl chloride) is largely disordered and yet the above analysis, based on the planar zig-zag skeletal structure, is able to account for most of the observed peaks demonstrates the value of this simplified approach to the understanding of the spectra of partially crystalline polymers. Its success is probably due partly to the fact that chains in crystalline material, because of their regular interactions with their neighbours, give sharper peaks with consequently higher peak absorption than a comparable mass of material with a single isomeric structure in the amorphous regions. Additional factors are that the amorphous material contains a number of different isomeric structures, so that the absorption due to similar group vibrations is even further smeared out, and that modes such as the CH_2 and skeletal modes are not very sensitive to configurational isomerism because they do not involve much motion of the chlorine or lone hydrogen atoms, as the potential energy distribution shows.

Further empirical justification of the use of the simplified approach emerges from a comparison of the Raman spectra of a typical commercial polymer and a highly ordered polymer prepared from monomer in the form of a urea clathrate complex. This latter polymer is known, from X-ray diffraction evidence, to contain mainly planar zig-zag chains. Although its spectrum contains a few peaks not present in that of the commercial polymer, the two spectra have a basic similarity. The various peaks are readily assignable to the symmetry species expected for an isolated chain of C_{2v} symmetry, using the type of reasoning given in detail for the infrared assignments and taking into account experimental depolarization ratios (A_1 species are polarized,

A_2, B_1 and B_2 species are depolarized). The assignments for the A_2 species, active only in the Raman effect, are shown in tables 5.3 and 5.4 with the assignments already discussed.

5.2.3   *Poly(vinylidene chloride)*

The molecule of poly(vinylidene chloride) (PVDC), with the formula $+CH_2 . CCl_2+_n$, has only one type of CH unit and it might therefore be supposed that the interpretation of its vibrational spectrum would pose few problems. In practice, there are features which make the interpretation a matter of considerable interest. The most important of these is that although PVDC is usually highly crystalline its crystal structure is not known with certainty. X-ray studies show that there are two chains per unit cell and that the physical repeat unit, which must involve two chemical repeat units, has a length of 4.68 Å in the chain axis direction. The conformation of the individual chains is, however, not known.

In early attempts to assign the observed peaks in the infrared spectrum to vibrational modes of the perfect single chain two plausible models were considered for the conformation. The first, due to Reinhardt, has a planar carbon backbone of alternating cis and trans bonds with all C—C—C bond angles equal to 120° (see fig. 5.5); the planes of the CCl_2 and CH_2 groups are not perpendicular to the fibre axis, but inclined at an angle. The second structure, proposed by Fuller, has almost the normal planar zig-zag carbon chain, but it is shortened slightly by rotations of about 34° around the C—C bonds and by a reduction of the C—CCl_2—C angle to 99°, while the C—CH_2—C bond angle remains at the tetrahedral angle; the planes of the CCl_2 groups are perpendicular to the fibre axis, neighbouring CCl_2 groups being staggered with respect to each other (see fig. 5.5). An attempt to decide between these two possible structures was made by comparing the characters of the CCl_2 stretching modes, as predicted by each of the models, with the characters observed in the infrared spectrum. The evidence from these modes unambiguously favoured the Fuller model, but there were some difficulties with the assignments of other modes. Later, particularly when Raman spectra became available, it became clear that it was necessary to consider other models. Fig. 5.6 shows the infrared and Raman spectra of PVDC.

All models must satisfy the requirement that the physical repeat unit consists of two monomer units and has a length of 4.68 Å in the chain direction. Table 5.5 lists six of the structures that have been considered more recently, all of which satisfy these requirements, at least approximately. They may all be considered to be derived from the Reinhardt model by rotations around the C—C bonds; the angles X and Y in the table are rotations from an imagined 'all-cis' conformation. If $X = 180°$, an originally cis bond becomes trans, and this is indicated by T in the table. All models except the $XY\,XY$ and $XXX'X'$ (Fuller) models assume that all the C—C—C bond angles are 120°, a value somewhat larger than the tetrahedral angle, 109.5°, but comparable to that found in the backbones of many other polymers. The models are rated 'favourable' (F) or 'unfavourable' (U) on geometric grounds by taking into account first the departures of the repeat distance from 4.68 Å, secondly the departure of the C—C—C angle from 120° and finally whether the non-bonded Cl—Cl distance is less than the estimated diameter (2.75 Å) of the chlorine atoms in the partially negatively charged state in which they occur in the polymer. Also shown in the table are the line groups (factor groups) to which the structures belong and the predicted characters of the CCl stretching modes in the infrared spectrum. The relative intensities within the CCl stretching modes were estimated from simple considerations of the way the dipole moments of the individual C—Cl bonds would add in the various vibrations. All modes are Raman active for all the models.

All four of the strong infrared peaks in the CCl stretching region (500–800 cm^{-1}) are observed to have σ polarization in the spectra of oriented

Fig. 5.5. Conformational models for poly(vinylidene chloride): (*a*) Reinhardt model, cis-planar 2_1 helix (C_{2v}); (*b*) Frevel model, distorted 'cis-plan' 2_1 helix (C_s); (*c*) Fuller model, twisted zig-zag (C_s); (*d*) $TXTX'$ and $TGTG'$ models (C_s); (*e*) uniform helix (D_2). (From M.S. Wu *et al.*, *Journal of Polymer Science Polymer Physics Edn* **18** 95 (1980). Copyright © (1980) John Wiley & Sons, Inc. Reprinted by permission of John Wiley & Sons, Inc.)

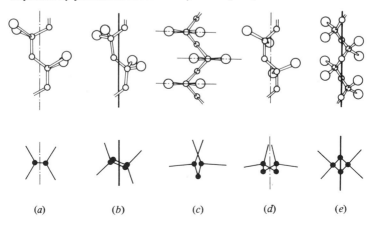

(*a*) (*b*) (*c*) (*d*) (*e*)

Table 5.5. *Predicted infrared characters of ν(CCl₂) modes for different chain models for poly(vinylidene chloride)*

Structure	Repeat length (Å)	C—C—C angle (deg.)	Rotation angles X (degrees)	Y	Line group	Non-bonded C—Cl sepn	Geom.	IR activity ν_{si}	ν_{so}	ν_{ai}	ν_{ao}
TCTC (Reinhardt)	4.62	120	180		C_{2v}	3.88	F	π	σ	σ	i
XYXY	4.56	120	141.1	20	C_2	3.55	U	π, s	σ, w	π, s	σ, w
XXX'X' (Fuller)	4.68	109.5, 99	145.7		C_s	2.45	U	σ, s	σ, w	σ, s	σ, w
TXTX'	4.68	120	32.5		C_s	3.19	F	m, s	σ, s	mσ, s	σ, w
XXXX	4.36	120	70.5		D_2	3.63	U	i	σ	π	σ

T = trans; C = cis; X' = −X.
F = favourable; U = unfavourable.
mσ = mixed but predominantly σ.
i = in phase, o = out of phase vibration of two CCl₂ groups.
Other notation as explained in section 5.1.

samples, but the polarizations of the weak peaks at 568 and 688 cm^{-1} are unknown. These results immediately rule out the $XYXY$ model, since it requires only two σ modes and also requires two stronger π modes. The $TCTC$ (Reinhardt) and the XXXX models each require two σ modes, one π mode and one inactive mode in the infrared. At first sight it would seem possible to assign the 688 cm^{-1} peak to the inactive mode and the 568 cm^{-1} peak to the π mode. This is not, however, consistent with the fact that the intensity of the latter peak increases in the spectrum of largely non-crystalline PVDC and is probably due to disordered chains. Furthermore, vibrational calculations strongly suggest that for a wide variety of conformational models, including the

Fig. 5.6. Infrared and Raman spectra of poly(vinylidene chloride): (*a*) Raman spectrum of single crystals; (*b*) FTIR spectrum of single crystals; (*c*) FTIR spectrum of quenched amorphous polymer. (Reproduced from M. M. Coleman *et al.*, *Journal of Macromolecular Science – Physics B* **15** 463 (1978), by courtesy of Marcel Dekker, Inc., NY.)

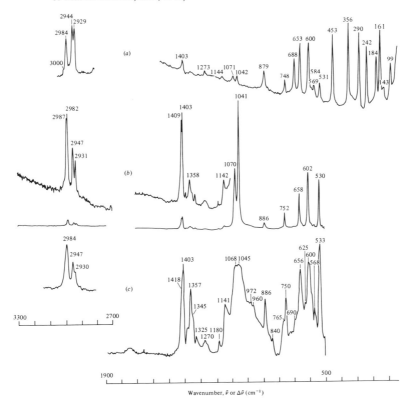

TCTC model, the four ν(CCl) modes should lie fairly close to the observed peaks at 530, 602, 658 and 688 cm^{-1}. For the Reinhardt model the π mode corresponds to the totally symmetric mode, which is expected to be relatively strong in the Raman spectrum, whereas the observed mode at 568 cm^{-1} is weak. For the *XXXX* model the mode inactive in the infrared corresponds to the totally symmetric mode and since antisymmetric stretching modes usually have higher frequencies than symmetric stretching modes it would be difficult to assign the peaks observed at 530, 602 and 658 cm^{-1} if this model were accepted. For these reasons both the Reinhardt and *XXXX* models may be eliminated; the latter is also unlikely on geometrical grounds.

The choice now lies between the *XXX'X'* (Fuller) and the *TXTX'* model. The Fuller model requires four σ modes and before Raman spectra or vibrational calculations were available it seemed reasonable to assign these to the observed peaks at 530, 602, 658 and 750 cm^{-1}, although there was some controversy about whether the 750 cm^{-1} mode was a CCl stretching mode or not. Vibrational calculations suggest that it is not, but even if this is accepted it is not possible to eliminate the Fuller model on the grounds of symmetry considerations for the ν(CCl) modes alone. The *TXTX'* model can satisfactorily account for the spectrum in the CCl stretching region, particularly when the difficulties in making accurate polarization measurements on samples which are not fully oriented are taken into account. No numerical estimates of the degree of polarization are available but a close look at the spectra suggests that although the 530 cm^{-1} peak is predominantly of σ character the 602 and 658 cm^{-1} peaks may have some π character. Since the *TXTX'* model seems more likely on geometric grounds the evidence seems to be slightly in its favour. Further supporting evidence comes from a study of the CH stretching region.

There is strong evidence that the doublet at 2984 cm^{-1} is due to correlation splitting. For oriented samples the peaks at 2931 and 2947 cm^{-1} seem to have predominantly σ character, whereas that at 2984 cm^{-1} appears to have some π character. The Fuller model predicts two strong σ peaks and two weak π peaks, whereas the *TXTX'* model predicts the same types of modes that it does for the corresponding ν(CCl$_2$) modes, viz. one strong and one weak σ mode, one strong mainly σ mode and one strong mixed mode. The absence of any definitely π modes in the observed spectrum favours the *TXTX'* model over the Fuller model.

A calculation of the frequencies of all the normal modes has been made on the basis of the *TXTX'* model, using a VFF with force constants transferred without refinement from sets determined from studies of secondary dichlorides or secondary chlorides. All the calculated frequencies could be satisfactorily associated with observed peaks in the infrared or Raman spectra. Despite this, it must not be concluded that the PVDC molecule in the crystalline regions is necessarily in the $TXTX'$ conformation, merely that this is one of the more likely structures and possibly the most likely structure. More evidence, such as Raman depolarization data and more accurate infrared polarization data, are required before the modes can be classified with certainty into symmetry classes. Even then, many structures with the correct symmetry to account for these could be envisaged, for instance, $TXTX'$ models with a different value of the angle X. The above discussion has, however, shown that symmetry arguments can be used to eliminate some otherwise plausible models.

Some of the assignments made on the basis of the calculations of the normal mode frequencies using the $TXTX'$ model were in substantial agreement with earlier assignments made on the basis of the group frequency approach and comparison with model compounds or by comparison with the spectra of similar polymers, but many were not. Even if the model or the force constants are incorrect, the fact that a consistent set of frequencies can be assigned which are not always in agreement with those deduced using such simple arguments suggests that the simple arguments may be misleading when applied to PVDC. The reason is, of course, that none of the modes are pure group modes and some are very far from pure. Fortunately the calculations do suggest that the CCl and CH stretching modes, on which the foregoing arguments are based, are fairly good group modes. The complete set of assignments made on the basis of the calculations is shown in table 5.6. It should be noted that all infrared-active modes which were judged to be π modes can be satisfactorily assigned to modes predicted to be of mixed character for the $TXTX'$ model. It should also be noted, in accordance with the comments just made, that the division of the modes into CH_2 modes, CCl_2 modes and skeletal modes of various types is only an attempt to indicate the approximate nature of the true, mixed, normal modes.

5.2.4 *Polytetrafluoroethylene*

X-ray diffraction studies show that the individual chains, $+CF_2+_n$, of polytetrafluoroethylene (PTFE) have a helical conformation and that below 19°C the translational (identity) period is 16.8 Å. The structure corresponds to a planar zig-zag chain of 13 carbon atoms that is twisted 180° about its axis in this distance. As a result, the

Table 5.6. *Assignment of fundamental vibrations of poly(vinylidene chloride)*

Symmetry mode	Symmetry species and IR character		Wavenumber (cm^{-1})		
			Raman	IR	Calculated
CH_2 modes					
$\nu_s(CH_2)_i$	A'	m	2944	2947σ	2853
$\nu_s(CH_2)_o$	A''	σ	2929	2931σ	2854
$\nu_a(CH_2)_i$	A'	m	2984	2982, 2987	2920
$\nu_a(CH_2)_o$	A''	σ	3000	3000	2918
$\delta(CH_2)_i$	A'	m		1385	1391
$\delta(CH_2)_o$	A''	σ	1403	1403, 1409	1405
$\gamma_w(CH_2)_i$	A'	m	1361?	1358π	1377
$\gamma_w(CH_2)_o$	A''	σ		1342	1373
$\gamma_r(CH_2)_i$	A'	m	1071	1070σ	1053
$\gamma_r(CH_2)_o$	A''	σ	879	886σ	871
$\gamma_t(CH_2)_i$	A'	m	1273	1265	1258
$\gamma_t(CH_2)_o$	A''	σ		1327σ	1320
CCl_2 modes					
$\nu_s(CCl_2)_i$	A'	m	600	602σ	585
$\nu_s(CCl_2)_o$	A''	σ	531	530	523
$\nu_a(CCl_2)_i$	A'	m	653	658σ	670
$\nu_a(CCl_2)_o$	A''	σ	688	688	699
$\delta(CCl_2)_i$	A'	m	290	291π	268
$\delta(CCl_2)_o$	A''	σ		307	283
$\gamma_w(CCl_2)_i$	A'	m	356	359π	388
$\gamma_w(CCl_2)_o$	A''	σ	161	162	168
$\gamma_r(CCl_2)_i$	A'	m	453	454π	451
$\gamma_r(CCl_2)_o$	A''	σ	242	245σ	254
$\gamma_t(CCl_2)_i$	A'	m	143		148
$\gamma_t(CCl_2)_o$	A''	σ		382σ	392
Skeletal modes					
$\nu_s(CC)_i$	A'	m		980	920
$\nu_s(CC)_o$	A''	σ	1042	1041σ	1075
$\nu_a(CC)_i$	A'	m	1144	1142π	1173
$\nu_a(CC)_o$	A''	σ		1180	1124
$\delta(CCC)_i$	A'	m	184	185π	172
$\delta(CCC)_o$	A''	σ	748	752σ	790
torsion, τ_i	A'	m			65
torsion, τ_o	A''	σ	99	102	90

i = in phase, o = out of phase vibration of the two chemical units of each physical repeat unit. Other notation as explained in section 5.1.

Adapted from Table III of M. S. Wu *et al.*, *Journal of Polymer Science Polymer Physics Edn* **18** 111 (1980). Copyright © (1980) John Wiley & Sons, Inc. Reprinted by permission of John Wiley & Sons, Inc.

distance between alternate carbon atoms is increased from 2.54 Å, the value for polyethylene, to 2.59 Å. Although a full twist of the chain occurs in 33.6 Å, the identity period is half of this because the 14th CF_2 group along the chain is related to the first by translation through 16.8 Å parallel to the axis of the helix. The normal to the plane of a CF_2 group makes an angle of 15–20° with this axis. Above 19°C the helix unwinds slightly to include 15 CF_2 groups per turn. The predictions of group theory for the number and character of the optically active modes are essentially the same for both forms. Since most recent spectra, particularly Raman spectra, have been obtained with samples cooled below room temperature we shall consider the low temperature form in discussing the theoretical predictions and in quoting experimental wavenumbers where these are significantly different for the two forms.

Two models may be postulated for the purpose of determining the symmetry species. Both models assume an infinitely long helical chain of CF_2 groups containing 13 groups in the repeating unit. In the first model no assumption is made about the orientation of the two-fold rotation axis of each CF_2 group with respect to the axis of the helix. The symmetry operation required to bring any CF_2 group into coincidence with an adjacent CF_2 group consists of a rotation of $(\pi - \pi/13)$ about the axis of the helix, followed by a translation along the axis of 1/13 of the identity period, 16.8 Å. This operation, $C_{13}^{(6)}$, will be denoted for simplicity by the symbol C^1. The third CF_2 group along the helix is obtained from the first chosen CF_2 group by the symmetry operation consisting of a rotation of $2(\pi - \pi/13)$ about the axis followed by a translation along the axis of 2/13 of the identity period. This symmetry operation is denoted by C^2. By similar symmetry operations, the succeeding CF_2 groups are obtained from the first. The operation C^{13} is equivalent to a translation equal to the identity period, and is denoted by E, the identity operation. The sequence of symmetry elements $E, |C^1|,$ $|C^2|, |C^3|, \ldots, |C^{12}|$, where the vertical bars have the significance explained in subsection 3.1.2, forms a cyclic group which is the factor group of the line group corresponding to the infinite helical chain. This factor group, which is isomorphous with the point group C_{13}, is sometimes denoted by $C(12\pi/13)$ because a total rotation of 12π, six turns, corresponds to one translational repeat unit; the structure is a 13_6 helix. If C^1 were defined in terms of a rotation of $\pi + \pi/13$ in the opposite direction the identical 13_7 helix would result

The X-ray diffraction results suggest that the two-fold rotation axis of each CF_2 group is perpendicular to and intersects the axis of the helix. Additional symmetry elements are then present. When these are added to the group already discussed the new factor group of the line group,

Table 5.7. *Symmetry species and activities for PTFE*

$C(12\pi/13)$				$D(12\pi/13)$			
	Activity		Active		Activity		Active
Species	IR	R	modes	Species	IR	R	modes
Fundamentals							
A	π	s, p	7	A_1	i	s, p	4
				A_2	π	i	3
E_1	σ	w, d	8 pairs	E_1	σ	w, d	8 pairs
E_2	i	w, d	9 pairs	E_2	i	w, d	9 pairs

Overtones and combinations

$A \cdot A$ or $(A)^2 = A$

$A_1 \cdot A_1 =$ or $(A_1)^2 = A_1$
$A_2 \cdot A_2 =$ or $(A_2)^2 = A_1$
$A_1 \cdot A_2 = A_2$

$A \cdot E_k = E_k$

$A_1 \cdot E_k = E_k$
$A_2 \cdot E_k = E_k$

$(E_k)^2 = A + E_{2k}$
$E_k \cdot E_k = A + E_{2k}$
$E_1 \cdot E_2 = E_1 + E_3$

$(E_k)^2 = A_1 + E_{2k}$
$E_k \cdot E_k = A_1 + A_2 + E_{2k}$
$E_1 \cdot E_2 = E_1 + E_3$

$k = 1$ or 2.
$(A_k)^2$ or $(E_k)^2$ indicates the first overtone of A_k or E_k.
A dot ($\cdot$) between two species symbols indicates a combination of the two species.

designated $D(12\pi/13)$ and isomorphous with the point group D_{13}, is obtained.

The symmetry species, irreducible representations and total number of normal modes may be obtained by the usual procedures. Table 5.7 shows the numbers of normal modes for each of the spectroscopically active symmetry species for the two models, and also the symmetry species of their overtones and combinations. The approximate vibrational patterns for the active fundamentals can be deduced from the normal modes of a CF_2 group and the symmetry of the molecule. The symmetry species correspond physically to modes of vibration of the helix in which there are various possible phase differences between the identical motions of neighbouring CF_2 groups. If this phase difference is denoted by δ, then $\delta = (12\pi/13)m$, with $m = 0$ for the non-degenerate A, A_1 and A_2 modes and $m = \pm 1$ and ± 2 for the degenerate modes E_1 and E_2, respectively.

Fig. 5.7 shows the infrared and Raman spectra of PTFE. Careful studies of the infrared spectra of samples with different degrees of crystallinity have enabled the infrared-active modes which are associated specifically with the non-crystalline regions to be identified and studies on oriented samples have provided polarization data. Limited polarization data are available for the Raman-active modes, some of which have been deduced from the spectra of model compounds such as $CF_3(CF_2)CF_3$ and $CF_3(CF_2)_4CF_3$ and some from the spectra of perfluoropropylene copolymers, since PTFE itself is rather turbid.

The positions of the observed peaks and the polarization data are summarized in table 5.8. In this table infrared peaks known to be

Fig. 5.7. Infrared and Raman spectra of polytetrafluoroethylene: (a) Infrared transmittance spectrum of sample at room temperature. For the polarized spectrum $(350-1500 \text{ cm}^{-1})$. —— electric vector perpendicular to draw direction; – – – – electric vector parallel to draw direction; (b) Raman spectrum of sample at $-50°C$. ((a) reproduced by permission from C. Y. Liang and S. Krimm, *J. Chem. Phys.* **25** 563 (1956); (b) reproduced by permission from J. L. Koenig and F. J. Boerio, *J. Chem. Phys.* **50** 2823 (1969).)

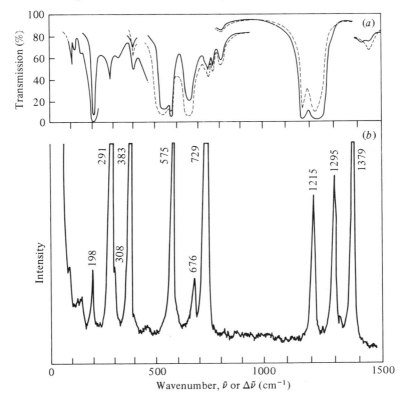

Table 5.8. *Vibrational assignments for PTFE*

IR	Raman	Assignment 1	Assignment 2
	91		?
102mw			?
124w			?
	138w		E_2
149w			?
203s, σ	201w	E_1	
	240vw		?
277m	276w	E_1	
	293vs, p	A_1]	
	308w	E_2]	
321w			E_1
	382vs, p	A_1]	
	386m	E_2]	
	404vw		?
	447vw		?
516vs, π		A_2]	
	524	E_2]	
553s, σ	(553)w	E_1	
	576m, d		defect structure
638s, π		A_2	
	676w		E_2]
	732vvs, p	A_1]	
	744w, d?	E_2]	
932w, σ		E_1	$202E_1 + 732A_1 = 934$
1152vvs, σ		E_1	
1210vvs, π?		A_2?]	
	1216m, d	E_2?]	
1242vvs, σ		E_1	
	1301m, d		E_1
	1335vw		E_2?]
	1381s, d		A_1]
1415m, σ		E_1	$202E_1 + 1210A_2 + 1412$
1450m, π		A_2	$202E_1 + 1242E_1 = 1444$

] signifies expected pairings of A and E modes. Other notation as explained in section 5.1.

associated only with non-crystalline regions have been omitted, as have data for peaks at wavenumbers greater than 1500 cm^{-1}, because all the fundamental modes should lie below this. It will be noted immediately that the symmetry $D(12\pi/13)$ is indicated by the fact that the strong or medium π polarized infrared peaks (at 516, 638 and 1450 cm^{-1}) are not observed in Raman scattering and the strong polarized Raman peaks (at 293, 382 and 732 cm^{-1}) are not observed in the infrared. Assignments to symmetry species can be made fairly readily for many of the peaks using the information given in table 5.8 together with the fact that the frequencies of E_2 modes should lie close to those of the corresponding A modes because $\delta = 332°$ for the E_2 modes. These 'obvious' assignments are shown in the column of the table labelled 'Assignment 1'.

Examination of this column shows that only three of the expected four modes of A_1 species have been assigned. The strong peak at 1381 cm^{-1} is the obvious possibility for the fourth, in spite of its apparent depolarization. Three modes which are definitely of A_2 species have been identified, and a possible fourth. Since only three A_2 fundamentals are expected, one of these modes must be an overtone or combination. There are no binary combinations or overtones which could account for the peak at 638 cm^{-1} and it seems most unlikely that the very strong modes at 516 and 1210 cm^{-1} are other than fundamentals. The medium intensity π peak at 1450 cm^{-1} could, however, be due to $202(E_1) + 1242(E_1) = 1444$ cm^{-1}. When an attempt is made to explain as many as possible of the peaks observed in the infrared spectrum above 1500 cm^{-1} as overtones or combinations it turns out that the 1210 cm^{-1} mode, if assigned to an A_2 fundamental, is considerably more helpful than the 1450 cm^{-1} mode. It is particularly difficult to explain the π mode observed at 2590 cm^{-1} except as the combination $1210(A_2) + 1381(A_1) = 2591$ cm^{-1}, which also further supports the assignment of the peak at 1381 cm^{-1} to a mode of A_1 species.

Seven E_1 modes have been identified and eight fundamentals are expected. One E_1 mode is, however, expected at a frequency lower than 100 cm^{-1}, since it corresponds to a mode at $\delta = 166.2°$ on the ν_5 dispersion curve for polyethylene (see fig. 4.13). In addition, it does not seem possible to assign the infrared mode at 321 cm^{-1} as an overtone or combination, so that it is most likely to be an E_1 fundamental, giving one E_1 species too many. The E_1 mode at 932 cm^{-1} can, however, be accounted for by several binary combinations, of which the most likely is $202(E_1) + 732(A_1) = 934$ cm^{-1}, and the mode at 1415 cm^{-1}, unlike those at 1152 and 1242 cm^{-1}, is not required for explaining any combination modes so that it seems likely that it is itself a combination mode, probably $1210(A_2) + 202(E_1) = 1412$ cm^{-1}. There are two medium depolarized peaks at 576 and 1301 cm^{-1}, either of which could be assigned to the missing E_1 species fundamental if both the 932 and 1415 cm^{-1} peaks are accepted as combination modes. Some authors have assumed that the 576 cm^{-1} mode is due to a defect structure and have assigned the 1301 cm^{-1} peak as the E_1 fundamental, whereas others have assigned the 1301 cm^{-1} peak to an E_2 species fundamental. This latter assignment seems unlikely, since there is no A species fundamental near 1301 cm^{-1}.

Finally, only five E_2 species modes have been assigned, compared with an expected nine fundamentals. Two of these modes, corresponding to the modes on the v_5 and v_9 curves for polyethylene at $\delta = 27.7°$, are expected to lie below 200 cm^{-1} and do not have a corresponding A species mode; more precisely, the A species modes have zero frequency and correspond, respectively, to translation parallel to and rotation about the chain axis. The higher of these two E_2 species modes is almost certainly to be assigned to the peak at 138 cm^{-1} and the other probably lies near 20 cm^{-1}. The only unassigned Raman peak close to the A_2 species peak at 638 cm^{-1} is that at 676 cm^{-1}, so that it is the most likely corresponding E_2 species mode. Similarly, the only unassigned Raman peak close to the A_1 species peak at 1381 cm^{-1} is the very weak peak at 1335 cm^{-1}, which may be the corresponding E_2 species. The Raman peaks at 91, 404 and 447 cm^{-1} then remain unassigned, although some authors have considered them to be E_2 species fundamentals. The infrared modes at 102, 124 and 149 cm^{-1} also remain unassigned.

The column labelled 'Assignment 2' in table 5.8 shows the further assignments made by the preceding detailed arguments. In fig. 5.8 the fundamentals are plotted against the phase angle δ and smooth continuous curves are drawn through the corresponding modes to indicate the smoothest dispersion curves which would be consistent with these fundamentals. The dispersion curves cannot cross, so that the appropriate E_1 species modes for the curves are determined simply by their order. As indicated earlier, to a first approximation each dispersion curve represents vibrations of a similar nature, differing only in the relative phases of the motions of adjacent CF_2 groups.

The totally symmetric A_1 species must correspond to symmetry modes which are of the types $v_s(CF_2)$, $\delta(CF_2)$, $\gamma_t(CF_2)$ or $C—C$ stretching, $v(CC)$. There is no doubt that the two higher frequencies correspond to $v_s(CF_2)$ and $v(CC)$ and although there is some doubt about which is which, most authors favour the assignment of the higher frequency to a mode which is predominantly $v(CF_2)$, on the basis that it is unlikely that the $v(CC)$ mode for PTFE can lie at a higher frequency than that of the corresponding mode for polyethylene (1133 cm^{-1}), bearing in mind the increased mass of the CF_2 group over the CH_2 group. Such arguments can, however, be dangerous for highly coupled systems of oscillators. There is also doubt about which of the other pair

Fig. 5.8. Approximate dispersion curves for polytetrafluoroethylene. The marked points correspond to the $A_1(\square)$, $A_2(\bigcirc)$, $E_1(\times)$ and $E_2(\triangle)$ species identified in table 5.8. The continuous curves indicate dispersion curves which would be consistent with these assignments. Except for those labelled ν_8 and ν_9, which are based on vibrational calculations, they are meant to illustrate only the general form of the curves. The dashed curves $------$ are a similar set for an alternative proposed set of assignments for the Raman- and infrared-active fundamentals. (See also text.)

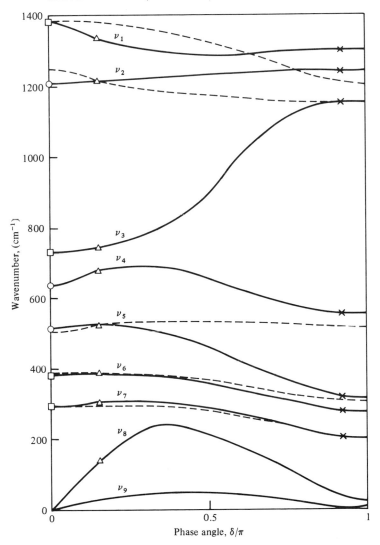

of A_1 modes is predominantly $\delta(CF_2)$ and which predominantly $\gamma_t(CF_2)$. The highest A_2 species must correspond predominantly to $\gamma_r(CF_2)$, but it is uncertain which of the lower two corresponds predominantly to $\gamma_r(CF_2)$ and which to $\gamma_w(CF_2)$. For $\delta \neq 0$, all modes with the same value of δ belong to the same symmetry species so that, in principle, all the dispersion curves represent mixtures of all the types of motion.

The dashed curves in fig. 5.8 represent smooth curves drawn through an alternative proposed set of assignments for some of the fundamental Raman- and infrared-active modes of PTFE based on a comparison of the frequencies observed for PTFE with dispersion curves deduced empirically for $CF_3(CF_2)_{12}CF_3$, which is believed to take up a helical conformation similar to that of the 13_6 helix of PTFE. For such a molecule, for which the point group is C_2, all modes are both infrared and Raman active and correspond approximately to modes of the infinite chain for which $\delta = n\pi/13$, since the only symmetry restriction is that the phase difference between the motions of the terminal groups must be 0 or π. The figure shows essential agreement with the previous assignments of the A and E_2 species fundamentals but considerable disagreements for some of the E_1 species fundamentals.

Other sets of assignments have also been suggested and because of these uncertainties we do not show examples of dispersion curves deduced from vibrational calculations performed to fit particular sets of assignments. It is, however, worth noting that all such calculations suggest considerable mixing of the simple symmetry modes. The two lowest curves shown in fig. 5.8 are based on the vibrational calculations, since their form does not depend very much on the assignments of the higher frequency fundamentals.

5.2.5   *Polystyrene*

Although polystyrene, which is a vinyl polymer, has eight carbon atoms and eight hydrogen atoms in its repeat unit,

$-CH_2 . \overset{|}{C}H-\!\!\!\langle\bigcirc\rangle$, the interpretation of its vibrational spectrum,

at least in general terms, is less difficult than might be supposed. The normal commercial type of polymer, for which the infrared and Raman spectra are shown in fig. 5.9, is atactic and completely amorphous. This implies that no regular interactions exist between monomeric units in the chain. Since there is also a variety of different possible orientations of the phenyl group around the ring-to-backbone C—C bond it should be a good first approximation to consider separately the normal vibrations of the phenyl, $\searrow CH_2$ and $\gtrless CH$ groups, and to interpret the spectrum as an essentially linear superposition of the spectra of these groups.

A further simplification can be introduced by assuming, when considering the normal modes of the phenyl group, that they are those of a monosubstituted benzene ring, ⬡—R, in which the symmetry of the substituent R need not be specifically considered. On this basis, the symmetry elements are the identity operation, E, a two-fold rotation axis passing through the substituent, a mirror plane in the plane of the ring and a mirror plane perpendicular to the ring and containing the two-fold axis. These symmetry elements define a point group C_{2v} and it follows that there will be 30 true vibrational modes, $11A_1 + 3A_2 + 10B_1 + 6B_2$, where the axes are chosen so that B_1 species are symmetric with respect to the plane of the ring. The A_1, B_1 and B_2 modes are active in both Raman scattering and infrared absorption but A_2 modes are active only in the Raman effect. Since, however, the true symmetry is not C_{2v}, because of the nature of the substituent, it is likely that these modes will be weakly active in the infrared.

Fig. 5.9. Infrared and Raman spectra of atactic polystyrene: (*a*) infrared absorbance spectrum, (*b*) Raman spectrum.

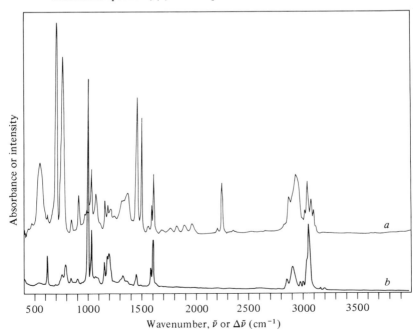

The first complete interpretation of the infrared and Raman spectra of polystyrene was attempted by Liang and Krimm at a time when the normal modes of monosubstituted benzenes had not been calculated in detail. Although this deficiency has subsequently been remedied it is instructive to follow the original treatment as, once again, it illustrates the information that can be obtained on the basis of simplifying assumptions, in this case the use of normal modes or, more precisely, symmetry modes, based on those of benzene. It is possible that modes which can be identified with some certainty in benzene may now become considerably mixed, the only restriction being that mixing can occur only between modes in the same symmetry species for the monosubstituted ring. Fig. 5.10 shows symmetry modes chosen by

Fig. 5.10. Approximate symmetry modes for monosubstituted benzenes. (From C. Y. Liang and S. Krimm, *Journal of Polymer Science* **27** 241 (1958). Copyright © (1958) John Wiley & Sons, Inc. Reprinted by permission of John Wiley & Sons, Inc.)

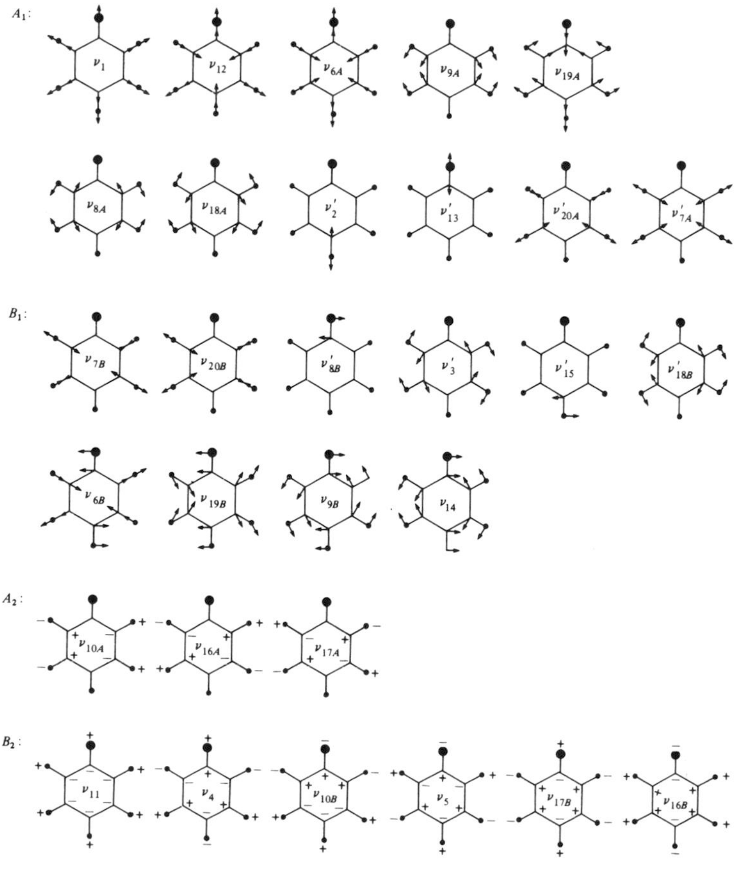

Liang and Krimm as being likely to be reasonably close to the true normal modes of the monosubstituted ring. The dashes signify symmetry modes which are mixtures of those expected for benzene. Subscripts A and B signify the A and B species components of modes which are doubly degenerate for benzene.

The mixing of modes within a symmetry species can lead to a 'borrowing' of intensity. Hence, a mode inactive for benzene may give strong absorption in the infrared spectrum or strong Raman scattering by mixing with a suitable infrared- or Raman-active mode. This seems to occur for some of the B_2 modes of the phenyl group in polystyrene. Where this effect is not present, the strong infrared peaks should be those arising from vibrations which are essentially antisymmetric with respect to inversion at the centre of the benzene ring, while Raman activity should be found mainly in modes which are essentially symmetric with respect to this operation. The study of overtone and combination peaks is also of help in assigning fundamentals. Some moderately intense combination peaks occur in the infrared spectrum of benzene and are also expected to occur for the monosubstituted ring. This proves to be the case and these combination peaks were of value for assigning some of the normal modes of species A_2 and B_2.

Table 5.9 shows the assignments suggested by Liang and Krimm, on the basis of the considerations outlined above, for those fundamentals which correspond essentially to the vibrations of the carbon framework of the ring in its own plane, to ring C—C—H angle bending in the plane of the ring or to out-of-plane modes of the ring. For each mode the symmetry species and observed activity are given and also the positions and symmetry species of the corresponding modes of benzene. It may be observed that many of the modes of polystyrene have frequencies very close to those of the corresponding modes of benzene.

The remaining modes associated with the ring are five C—H stretching vibrations, v'_2, v'_{7A}, v_{7B}, v'_{20} and v_{20B} and two modes associated with the C—C bond linking the ring to the chain, viz. the stretching of this bond, v'_{13}, and the in-plane bending of the R—C—C angles, v'_{8B}. The out-of-plane bending of the R—C bond is accounted for within the six out-of-plane ring modes of species B_2 and no mode corresponding predominantly to the bending of this bond was specifically postulated. It was not possible to assign all the individual ring C—H stretching modes with certainty for various reasons, but tentative assignments were made to specific peaks in the group of peaks lying between 3029 and 3138 cm^{-1}. Equally, it was not possible to identify the v'_{13} stretching mode, but it appeared likely that it contributed to the group of peaks near 1000 cm^{-1}. The R—C—C bending mode was tentatively assigned to the Raman peak observed at 196 cm^{-1}.

Table 5.9. *Ring vibrations of polystyrene*

Type of vibration	Mode[a]	Benzene			Polystyrene			
		Species	Act.	$\tilde{\nu}^b$ (cm^{-1})	$\tilde{\nu}$ (cm^{-1})	IR act.	R act.	Species
Ring skeleton in-plane bend or stretch								
	ν_1	A_{1g}	R, p	993	999	—	vvs, p	A_1
	ν_{12}	B_{1u}	i	1010	1011	—	vw	A_1
	ν_{14}	B_{2u}	i	1309	1310	m	—	B_1
	ν_{6A}	E_{2g}	R, d	606	558	—	w	A_1
	ν_{6B}	E_{2g}			621	vw	m, d	B_1
	ν_{9A}	E_{2g}	R, d	1599	1585	m, σ	vw	A_1
	ν_{9B}	E_{2g}			1606	s, σ	m	B_1
	ν_{19A}	E_{1u}	IR	1482	1493	vs, σ	—	A_1
	ν_{19B}	E_{1u}			1450	vs, σ	vw	B_1
Ring in-plane CCH-bending								
	ν_3	A_{2g}	R, d	1350	1328	ms, σ	vvw	B_1
	ν'_{15}	B_{2u}	i	1146	1154	mw, σ	—	B_1
	ν_{8A}	E_{2g}	R, d	1178	1200	vvw	m	A_1
	ν_{18A}	E_{1u}	IR	1037	1027	ms, σ	m, d	A_1
	ν_{18B}	E_{1u}			1070	m, σ	—	B_1
Ring out-of-plane deformation								
	ν_4	B_{2g}	i	707	540	s	—	B_2
	ν_5	B_{2g}	i	990	982	m, σ	—	B_2
	ν_{11}	A_{2u}	IR	673	700	vs	—	B_2
	ν_{10A}	E_{1g}	R, d	846	842	mw, σ	—	A_2
	ν_{10B}	E_{1g}			760	vs, π	vw	B_2
	ν_{16A}	E_{2u}			410	w	—	A_2
	ν_{16B}	E_{2u}	i	404	216	w	w	B_2
	ν_{17A}	E_{2u}			965	w, σ	—	A_2
	ν_{17B}	E_{2u}	i	967	906	m	vw	B_2

[a] Notation of Liang and Krimm. Some authors use a slightly different choice of numbering. In particular, ν_8 and ν_9 or ν_{18} and ν_{19} may be interchanged. Other notation as explained in section 5.1.
[b] Data taken from *Vibrational Spectra of Benzene Derivatives* by G. Varsanyi.

The remaining modes of the atactic polystyrene chain should all be infrared active. The CH_2 modes were readily identified by comparison with the polyethylene spectrum. On this basis, the stretching modes $v_s(CH_2)$ and $v_a(CH_2)$ were assigned to peaks at 2851 and 2923 cm^{-1}. The bending mode $\delta(CH_2)$ is found at about 1460 cm^{-1} for amorphous polyethylene and is, undoubtedly, the main component of the polystyrene peak at 1450 cm^{-1}. The fact that its intensity is greater than that of the stretching modes, contrary to what is found for polyethylene, led to the conclusion that it is a superposition of peaks, another component being the ring frequency v_{19B}. The $\gamma_w(CH_2)$ wagging mode is very weak in the spectrum of polyethylene and was not identified for polystyrene. The CH_2 rocking mode, $\gamma_r(CH_2)$, gives the well-known strong doublet at 720/731 cm^{-1} for polyethylene but is much more difficult to identify for vinyl polymers as a whole, as it moves to higher frequencies and weakens. It was suggested that for polystyrene it may contribute part of the weak peak at 965 cm^{-1}. The $\gamma_t(CH_2)$ mode was subsequently identified at 1298 cm^{-1} from studies of isotactic polystyrene.

With hydrocarbon polymers the stretching mode of the lone CH group is usually too weak to be easily identified and polystyrene proves no exception. This is also the case with the CH bending mode, $\delta(CH)$, which may contribute to the peak at 1376 cm^{-1}. The skeletal modes of the regular planar zig-zag carbon chain of polyethylene give rise to two chain stretching frequencies near 1100 cm^{-1}. For amorphous polyethylene, with a random chain, the single peak at 1080 cm^{-1} has been assigned to this mode. By analogous reasoning, a single peak is expected for atactic polystyrene and the medium intensity peak at 1070 cm^{-1} was assigned to this vibration. The peaks at 325 and 446 cm^{-1} were assigned to chain bending modes. A number of other peaks in the spectrum were assigned to combinations of the fundamental modes already assigned and, in particular, the group of peaks which appear in the infrared spectrum between 1675 and 1945 cm^{-1} were assigned in that way.

Since the original assignments were given by Liang and Krimm, studies have been made on deuterated polystyrenes and on isotactic polystyrene. The isotactic polymer is partially crystalline and the chains take up a 3_1 helical conformation in the crystallites, so that account must be taken of the symmetry of the complete chain because of the regular interactions which now take place between adjacent repeat units. The factor group of the line group is isomorphous with the point group C_3, so that there are only two symmetry species, A and E, both of which are active for both Raman scattering and infrared absorption.

Calculations of the vibrational spectrum of the isotactic form have been performed using force fields derived for hydrocarbons by Snyder and Schachtschneider (see subsection 5.2.1) and for monosubstituted alkyl benzenes. These studies have shown that the analysis of Liang and Krimm was correct in its general outlines, particularly for the ring modes. It has, however, become clear that many of the ring modes are essentially strong mixtures, within a given species, of the symmetry modes shown in fig. 5.10. As an example, the corresponding members of the two pairs of modes v_{18} and v_{19} are strongly mixed so that all four of the resulting modes involve both ring C—C—H bending and ring C—C stretching. In addition, some normal modes involve strong mixing of symmetry modes of the ring and of the backbone, for example the modes labelled v_{8A} and v'_{18B} in table 5.9. The calculations suggest that the peak assigned to v_4 in table 5.9 is predominantly a backbone vibration but includes considerable amounts of C—R out-of-plane bending, while that assigned to v_{16B} is predominantly a backbone vibration but includes a considerable amount of in-plane C—R bending. This mixing of ring and backbone vibrations leads to an explanation for a number of differences observed when comparing the spectra of atactic polystyrene and amorphous and crystalline isotactic polystyrenes which are often described as being due to 'conformationally sensitive' peaks.

5.2.6   Poly(ethylene terephthalate)

The results presented in subsection 5.2.5 show that reasonably complete and certain assignments are possible for a polymer with a repeat unit as large as that of polystyrene. The work which has been done on the interpretation of the vibrational spectrum of poly(ethylene terephthalate) (PET), shown in fig. 5.11, is of particular interest in showing what can be achieved for a still more complex molecule, $+OCH_2CH_2O—OC—\langle O \rangle—CO\xrightarrow{}_{n}$. The vibrational spectrum of PET is more complex than that of polystyrene for two reasons: the presence of an oxygen-containing group and the fact that PET is usually partially crystalline, so that the intensities of several peaks depend upon the crystalline/amorphous ratio. Nevertheless, it has been possible to make reasonably complete assignments. Two aspects of the problem will be considered briefly, first the approximate specification of the symmetry species of the repeat unit in terms of the symmetry species of two sub-units and secondly the assignment of the normal modes primarily associated with the oxygen-containing group.

X-ray diffraction studies suggest that in the crystalline regions the structure takes up a centrosymmetric trans conformation. The molecule is nearly planar and almost fully extended (see fig. 4.6). Departures from planarity arise as a result of the plane of the —CO . O— group being about 12° out of the plane of the benzene ring and from a rotation of the CH_2—CH_2 bond around the O—CH_2 axis by about 20° from the planar conformation. The only symmetry elements of the single chain are centres of inversion at the centres of the benzene rings and of the —CH_2—CH_2— groups and the true factor group of the single chain is isomorphous with the point group C_i. For this symmetry the mutual exclusion rule applies in its simplest form and exactly half of the 62 normal modes are infrared active but not Raman active (species A_u) and the other half are Raman active but not infrared active (species A_g). It is useful, because of this low symmetry, to consider the structure in terms

of the two sub-units —OC—⬡—CO— and —OCH_2CH_2O—, for each of which the number of normal modes and their symmetry species can be calculated by the methods used for small molecules.

Consider first the substituted benzene unit. Many of its normal modes will involve vibrations of the ring and of the bonds joining the ring to the adjacent carbon atoms. The symmetry of the paradisubstituted ring and

Fig. 5.11. Infrared and Raman spectra of poly(ethylene terephthalate). (*a*) Infrared absorbance spectrum; (*b*) Raman spectrum. The infrared spectrum was obtained from a free-standing film of thickness 20 μm and shows the characteristic interference fringes observed for such thin films in regions of low absorbance.

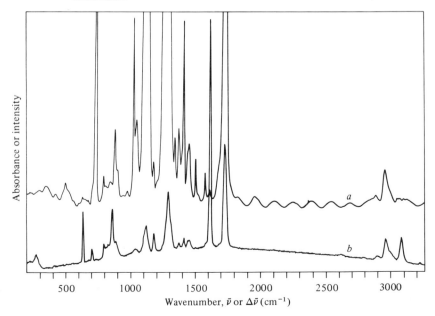

Wavenumber, $\bar{\nu}$ or $\Delta\bar{\nu}$ (cm^{-1})

its relationship to that of the full structural unit was discussed briefly in subsection 4.3.2. Although the benzene ring now has two substituents in the para positions, the principles involved in making assignments are similar to those involved for polystyrene and will not be considered in detail. It suffices to say that the assignment of a large number of the expected ring modes can be satisfactorily made with the help of studies on model compounds such as dimethyl terephthalate, $CH_3O.OC$—⬡—$CO.OCH_3$, and dihydroxyethylene terephthalate, $HO(CH_2)_2O.OC$—⬡—$CO.O(CH_2)_2OH$, and on deuterated polymers.

If there were no coupling with other groups there would be three infrared-active modes involving the $C{=}O$ group: the stretching mode, $v(C{=}O)$, the deformation mode, $\gamma_w(C{=}O)$, in which the oxygen atom moves parallel to the plane of the —C—⬡—C— group, and the rocking mode, $\gamma_r(C{=}O)$, in which it moves in the perpendicular direction. In view of the large amount of work on simple carbonyl compounds, it is evident that the strong peak at 1727 cm^{-1} must be $v(C{=}O)$. The $\gamma_r(C{=}O)$ mode should give a σ peak. On the basis of its assignment in ketones of low molecular weight to a peak at 390 cm^{-1}, the σ peak at 355 cm^{-1} in the PET spectrum is the obvious candidate. The $\gamma_w(C{=}O)$ mode has been assigned to a peak at 639 cm^{-1} in the spectrum of methyl acetate, CH_3COOCH_3, and to one at 527 cm^{-1} in the spectrum of acetylacetone, $CH_3COCH_2COCH_3$, which indicates some variability of position. It should show π polarization and, for this reason, it has been suggested that the peak in the spectrum of PET at 502 cm^{-1} may correspond to this mode.

Turning to the —OCH_2CH_2O— unit, it is necessary to remember that it adjoins the $C{=}O$ group, so that all the stretching and deformation vibrations of the —$(C{=}O)$—O—C— unit must be considered together. Simple esters have been studied in detail and it has been clearly established that there are two strong peaks associated with the stretching of the $(C{=}O)$—O and O—C bonds. In the present case the former can be assigned with confidence to the strong π peak at 1263 cm^{-1}. There are two peaks, at 1100 and 1120 cm^{-1}, that appear to be assignable to the O—C stretching mode and if this assignment is correct the doubling may reflect the influence of the conformation of the neighbouring CH_2 groups of chains in non-crystalline regions. There is some doubt as to the expected polarization of the $\delta(CCO)$ and $\delta(COC)$ modes, but they may well give π peaks. On this assumption, and the detailed analysis for methyl acetate, $\delta(CCO)$ has been assigned to the peak at $437\,\text{cm}^{-1}$ and $\delta(\text{COC})$ to the one at $383\,\text{cm}^{-1}$. Vibrational calculations have been made for a hypothetical completely planar PET molecule and the results suggest that many of the vibrations of the $-(\text{C}{=}\text{O})-\text{O}-\text{C}-$ group, including the ring to $(\text{C}{=}\text{O})$ bond, are not really assignable to simple vibrations of the constituent parts of the group, at least for the regular structure, so that some of the earlier suggested assignments may represent oversimplifications. The calculations do, however, confirm the assignment of the peak at $1727\,\text{cm}^{-1}$ and, to a lesser extent, that at $1263\,\text{cm}^{-1}$ to essentially the simple group vibrations indicated.

It is interesting to note that, in line with what is established for low molecular weight carbonyl-containing compounds, the presence of oxygen adjacent to a C—H bond leads to a shift in the C—H stretching frequencies. It is clear that, for PET, $\nu_s(\text{CH}_2)$ and $\nu_a(\text{CH}_2)$ are found at 2908 and $2970\,\text{cm}^{-1}$, respectively, shifts of some $50\,\text{cm}^{-1}$ by comparison with polyethylene. Consideration of the remaining modes of the $-\text{CH}_2-\text{CH}_2-$ group will be deferred until chapter 6 (subsection 6.2.3), as they are sensitive to conformational isomerism about the C—C bond.

5.3 The interpretation of polymer spectra: group frequencies

5.3.1 *Introduction*

In subsection 4.3.3 we considered briefly the phenomenon of group vibrations and group frequencies, i.e. the fact that a frequency or several frequencies of vibration of a particular group of atoms in a molecule may, under suitable conditions, be largely independent of the nature of adjacent groups of atoms. Good examples are the vibrations of the CH, CH_2 and phenyl groups in polystyrene, discussed in subsection 5.2.5. As this example illustrates, the simplest type of group with which a group frequency may be associated is a bonded pair of atoms, such as C—H, C=O, C—N etc., but it may also be a larger group of atoms bonded in a particular way, such as the phenyl group, which shows a number of characteristic group frequencies.

The examination of a very wide range of low molecular weight organic compounds has led to extensive frequency/structure correlations and these form the basis for the widespread use of infrared spectroscopy for the characterization of organic compounds. A group frequency may be specific not only for a particular group in a broad sense, but also for the environment of the group. This is exemplified by the stretching frequency of the carbonyl group, $(\text{C}{=}\text{O})$, which is readily identifiable and occurs within narrow ranges for specific types of carbonyl groups, such as aldehydes, acids, ketones and esters, and is also influenced by conjugation with a C=C group. When the C=O group occurs as part of the larger group —NHCO— which constitutes the amide linkage, it is no longer possible to identify a specific v(C=O) frequency because this mode becomes coupled with other modes of the group, which would individually have similar frequencies, to give a new set of frequencies for the group as a whole, the amide I, II and III modes, which are considered in subsection 5.3.5. Fig. 5.12 shows some of the more important group frequencies which occur in the spectra of polymers and table 5.10 shows, in the form of a decision tree, some of the frequencies which may be used to narrow down the possibilities when attempting to identify a polymer from its infrared spectrum.

The application of the group frequency concept to the characterization of polymers, particularly those with more complex structures to which the group-theoretical methods are less easily applied, will now be considered. The various examples selected all involve infrared spectroscopy. Analogous group frequencies exist for Raman spectra but they have not yet been so extensively tabulated. Raman spectroscopy

Fig. 5.12. Approximate locations of some useful group frequencies in the range 600–4000 cm^{-1} for some of the more important groups that occur in polymer molecules. This figure is based largely on data and much more detailed figures in *Introduction to Infrared and Raman Spectroscopy* by N. B. Colthup, L. H. Daly and S. E. Wiberley. See also fig. 4.9.

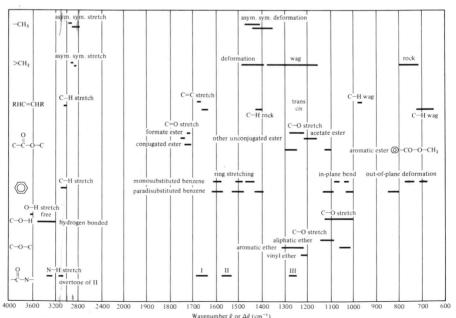

Table 5.10. Polymer classification scheme.

Carbonyl band near 1725 cm⁻¹

Present

Bands near 1590 and 1490 cm⁻¹

Present (aromatic)

cm⁻¹	
1540 1220	Polyurethanes
1300 1230 725	Isophthalate alkyds and polyesters
1230 1120 740 705	O-phthalate alkyds and polyesters
1265 1110 867 725	Terephthalate alkyds and polyesters
1235 1175 826	Bisphenol epoxy esters
813 781 700	Vinyltoluene esters
778 700	Styrenated esters

Absent (aliphatic)

cm⁻¹	
1430 1235	Poly(vinyl acetate)
1430 690 (b)	Poly(vinyl chloride-acetate) and poly(vinylidene chloride-acetate)
1265 1240 1190–1150 (s)	Polymethacrylates
1250 (b) 1190–1150 (s)	Polyacrylates
1110–1150 (b)	Cellulose esters

Absent

Bands near 1590 and 1490 cm⁻¹

Present (aromatic)

cm⁻¹	
3330 1220 910–670	Phenolics
1430 1110–1000 (s)	Phenylsiloxane
1235 1180 826	Bisphenol epoxides
814 780 700	Polyvinyltoluene
760 700	Polystyrene

Absent (aliphatic)

cm⁻¹	
3330 1430 1100 (b)	Poly(vinyl alcohol)
2940 (vs) 1470 1380 (w) 730–720 (d, s)	Polyethylene
2940 (vs) 1470 (s) 1380 (s) 1160 970	Polypropylene
2260	Polyacrylonitrile
1640 1540	Urea-formaldehyde and polyamides
1640 1280 834	Cellulose nitrate
1540 826	Benzoguanamine-formaldehyde
1540 813	Melamine-formaldehyde
1430 690 (b)	Poly(vinyl and vinylidene chlorides)
1265 1110–1000 (s)	Methylsiloxane
1220–1150 (d, vs)	Polytetrafluoroethylene
1110	Polyvinylethers and acetals, cellulose ethers
1250–1110 1000–910	Polychlorotrifluoroethylene

b = broad, s = strong, vs = very strong
w = weak, d = doublet

Reproduced by permission from 'An Infrared Spectroscopy Atlas for the Coatings Industry', published by the Federation of Societies for Coatings Technology, Philadelphia, PA, USA (1980).

should be regarded as complementary to infrared, because it provides vibrational spectra for materials that are difficult to sample for infrared spectroscopy and also because modes inactive or weakly active in the infrared spectra of highly symmetric molecules may be strongly active for Raman scattering. For instance, Raman scattering often gives strong peaks for modes involving skeletal vibrations. These may be useful for diagnostic purposes, an example being the peak in the frequency range $1000–1100$ cm^{-1} that is specific for the isocyanurate ring present in some types of polyurethanes. This ring consists of three units of the form —NR—CO—, where the group R may be different for each of the three units.

For disordered structures, modes of groups which are not themselves highly symmetric will usually be active in both Raman and infrared spectra at the same frequency, but the efficiencies may be very different. Whether Raman or infrared spectroscopy is used, it must always be remembered that any group frequency may be perturbed by the surrounding atoms in the molecule. In highly ordered structures coupling between groups may lead to significant splittings of group modes. An extreme example of this is the coupling of the rocking and twisting modes of individual $\diagdown$CH$_2$ groups in polyethylene to give very different frequencies for the Raman and infrared active rocking and twisting modes, which are almost pure, and highly mixed character to the corresponding non-factor group modes.

5.3.2   *Hydrocarbon polymers: polypropylene, polyisobutene*

We begin by considering the interpretation of the infrared spectrum shown in fig. 5.13*b*. This is a simple spectrum which has only a few more peaks than that of polyethylene (figs. 5.1 and 5.13*a*). If it were not known that the spectrum is in fact that of polypropylene, the only peak which would suggest the possibility that the polymer is unsaturated is that at 969 cm^{-1}. If, however, this peak was due to the out-of-plane deformation mode of a trans —CH=CH— group, the corresponding C=C stretching mode would have appeared, though perhaps only weakly, as a peak at about 1650 cm^{-1} (see fig. 5.12). It would therefore be reasonable to conclude that the polymer is saturated, which considerably narrows the possibilities. This illustrates an important working rule in the interpretation of vibrational spectra by the group frequency approach; if possible, identify two or more peaks that are characteristic for a particular chemical group.

By comparison with the spectrum of polyethylene, there are two regions where additional peaks appear. They are the C—H stretching and deformation regions. In the former region there are peaks at 2880 and 2960 cm^{-1}, frequencies which correspond to the symmetric and antisymmetric stretching modes of a CH_3 group, respectively. In the second region there is a strong peak at 1378 cm^{-1}, which can be assigned to the CH_3 symmetrical bending vibration. The presence of a methyl group has therefore been demonstrated unequivocally and the simplest saturated polymer which contains this group is polypropylene.

Fig. 5.13. Infrared spectra of (a) polyethylene; (b) atactic polypropylene; (c) polyisobutene. The polyethylene is a high pressure type and shows some methyl absorption at 1379 cm^{-1}. (Reproduced by permission from D. O. Hummel, 'Applied Infrared Spectroscopy' in *Polymer Spectroscopy*, ed. D. O. Hummel, Verlag Chemie, 1974.)

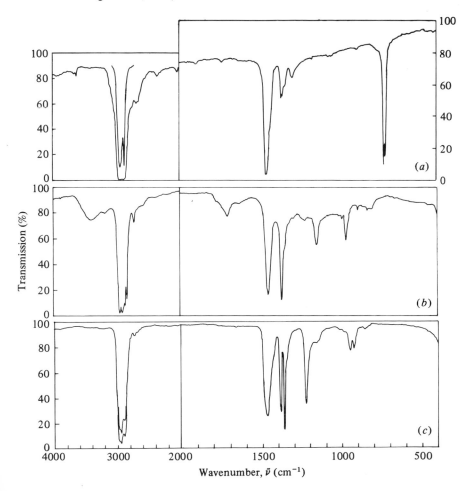

The two moderately strong peaks at 969 and 1153 cm^{-1} in the spectrum of polypropylene both arise from coupled vibrations. Normal mode calculations show that the one at 1153 cm^{-1} involves C—C stretching both of the polymer backbone and of the bond joining the methyl group to the backbone, together with C—H bending and CH$_3$ rocking vibrations. The peak at 969 cm^{-1} involves the coupling of the C—C backbone stretch with the CH$_3$ rocking mode. Coupled vibrations of this type are unreliable as group frequencies, because their positions are very sensitive to the proportions of the component vibrations. This is demonstrated very clearly by comparing the spectrum of polypropylene with that of polyisobutene, ᐧ(CH$_2$—C(CH$_3$)$_2$ᐧ), shown in fig. 5.13c. The 1153 cm^{-1} peak of the former is replaced by a peak at 1227 cm^{-1} for the latter, which normal coordinate calculations show to be due to a similar mixed mode, whereas the peak at 969 cm^{-1} splits into two components, at 922 and 950 cm^{-1}. The important diagnostic feature of the polyisobutene spectrum is the symmetrical CH$_3$ bending doublet, at 1362 and 1388 cm^{-1}, which is characteristic for two methyl groups attached to a common carbon atom.

5.3.3   *Polyesters: acetates, acrylates, methacrylates*

Polymers containing ester groups are readily recognizable as a class and it is often possible to identify an individual polyester. Poly(vinyl acetate), for which the infrared spectrum is shown in fig. 5.14, provides a particularly good example. All saturated esters have a strong, sharp absorption peak at about 1740 cm^{-1}, due to the v(C=O) stretching mode. The second oxygen atom of the ester group is involved in a C—O linkage, and has its stretching mode at a lower frequency. Furthermore, as this is a skeletal mode it is much less stable in position than v(C=O). Nevertheless, examination of the spectra of a wide range of saturated esters has shown that there are some very useful group frequencies. Acetates have v(C—O) at about 1240 cm^{-1} and this peak is readily identifiable in fig. 5.14. This frequency is not unique to acetates because a peak also occurs in this position in the spectra of all aliphatic carbonates, e.g. poly(diethylene glycol bisallyl carbonate). These latter do not, however, show the peak at 1020 cm^{-1} due to the C—O stretching of the O—CH group. Predictably, cellulose acetate also gives peaks near 1740 and 1240 cm^{-1}, but it can be distinguished from poly(vinyl acetate) by differences in the two spectra at lower frequencies.

Acrylates and methacrylates are commonly encountered commercial polymers and their characterization is typified by the best known material of this type, poly(methyl methacrylate), the spectrum of which is given in fig. 5.15. The ester group is readily identifiable by $v(C{=}O)$, at about 1735 cm^{-1}. Four peaks, at about 1150, 1190, 1235 and 1265 cm^{-1}, appear in the $v(C{-}O)$ region. These are characteristic for all methacrylates but it is possible to identify specific polymers, e.g.

Fig. 5.14. The infrared spectrum of poly(vinyl acetate). Absorbance spectrum of film cast from 1,2-dichloroethane solution on a potassium bromide plate.

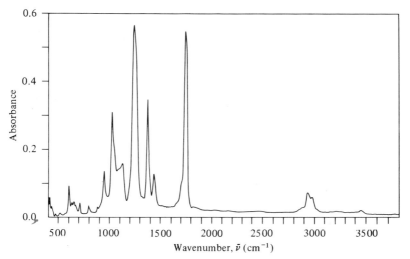

Fig. 5.15. The infrared spectrum of poly(methyl methacrylate). Absorbance spectrum of film cast from 1,2-dichloroethane solution on a potassium bromide plate.

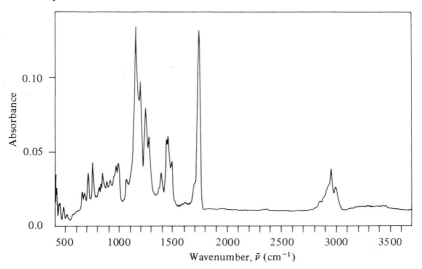

poly(ethyl methacrylate) and poly(butyl methacrylate), by comparison with reference spectra, because significant differences exist in the interval $800–1000$ cm^{-1}. Acrylates are distinguishable from methacrylates in the $v(C\!\!-\!\!O)$ region. They usually give only two peaks, one at about 1170 cm^{-1} and a broader one at 1250 cm^{-1}, the former being considerably the stronger. Poly(methyl acrylate) shows an additional peak at 1200 cm^{-1}, which, together with a highly characteristic strong peak at 830 cm^{-1}, makes this polymer uniquely identifiable.

When the ester group is attached to an aromatic system $v(C\!=\!O)$ is at a somewhat lower frequency, usually about 1720 cm^{-1}, but there is no difficulty in recognizing the aromatic moiety directly, from the ring $C\!=\!C$ stretching modes between 1500 and 1600 cm^{-1}, and one or more strong out-of-plane deformation modes in the interval $700–850$ cm^{-1}. In the specific case of terephthalates, the best known example of which is poly(ethylene terephthalate), there are two characteristic peaks in the $v(C\!\!-\!\!O)$ region, at 1110 and 1263 cm^{-1}. If the spectra shown in fig. 5.11, and already discussed in depth in relation to detailed assignments for this molecule, are considered from the viewpoint of the group frequency approach, no difficulty should be encountered in recognising them if they are found when trying to identify an unknown polymer.

5.3.4   *Other oxygen-containing polymers: phenolic resins*

The range of phenolic resins available is considerable. As noted in subsection 1.2.2, condensation using acid conditions gives structures of the type shown in fig. 5.16a, in which the aromatic rings are linked by methylene bridges; these are known as *novolak resins*. Substitution occurs mainly in the ortho and para positions, and the proportion of o–o and o–p linkages depends on the reaction conditions. A range of substituted phenols, rather than phenol, may be used as the starting material. This type of resin is usually subsequently cross-linked by reaction with hexamine. Condensation under alkaline conditions gives *resole resins*, in which the predominant structures are of the type shown in figs. 5.16b and 5.16c. Some self-condensation may occur, giving the structure shown in fig. 5.16d, which contains a methylene ether functional group, $-CH_2\!\!-\!\!O\!\!-\!\!CH_2\!\!-$. This type of reaction predominates during the self curing (cross-linking) of resole resins which takes place when they are held at $150°C$ for about two hours. As for novolaks, a range of substituted phenols may be used as starting materials for resole resins. In view of the considerable structural complexity of the phenolic resins as a whole the discussion of their spectra will be restricted to an outline of the use of the group frequency approach.

Novolak resins, as a class, are readily identifiable. They give strong peaks at about $3350 \, \text{cm}^{-1}$, from the hydroxyl stretching vibration, together with the corresponding peak near $1240 \, \text{cm}^{-1}$ due to the stretching of the ring–oxygen bond (see fig. 5.17). There are also peaks characteristic for the aromatic ring; in the $1500–1600 \, \text{cm}^{-1}$ region there are ring C=C stretching modes and out-of-plane deformation modes occur in the approximate interval $700–850 \, \text{cm}^{-1}$. These latter provide the means for distinguishing between o–o and o–p linkages: the former gives a peak at $755 \, \text{cm}^{-1}$ whereas the latter gives two, at about 760 and $820 \, \text{cm}^{-1}$. Moulding powder containing substantial amounts of free hexamine may be encountered; this curing agent gives a strong, sharp doublet centred at about $1000 \, \text{cm}^{-1}$. Novolaks prepared from substituted phenols show some spectral differences but cannot usually be identified with certainty by infrared spectroscopy. For soluble resins a sounder approach is provided by ^{1}H and ^{13}C NMR spectroscopy; for insoluble cross-linked materials pyrolysis followed by infrared examination usually leads to a characterization.

The spectra of the resole resins show distinct differences from their novolak counterparts. The various peaks of the latter appear, but the

Fig. 5.16. Some structures that may occur in phenolic resins. See text.

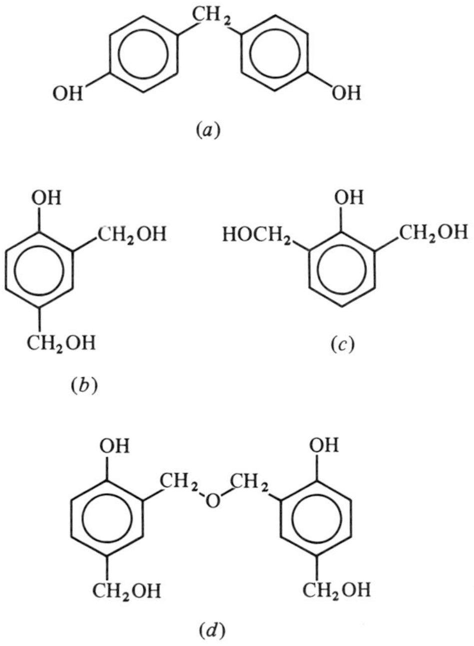

(a)

(b)

(c)

(d)

one at about 3350 cm^{-1} is considerably more intense because of the presence of the methylol (—CH$_2$OH) groups. These also give an additional peak at about 1010 cm^{-1}, the aliphatic counterpart of the phenolic 1240 cm^{-1} peak. This occurs at about the same position as the hexamine doublet but because it is a single rather broad peak resoles are readily differentiable from novolaks containing free hexamine. As with the novolaks, resole resins prepared from substituted phenols are best characterized by NMR spectroscopy.

The conversion of the methylol groups to methylene ether linkages during the cross-linking of resole resins is readily followed by infrared spectroscopy. The fall in concentration of the former leads to a reduction in intensity of the hydroxyl stretching peak at 3350 cm^{-1} and that of the 1010 cm^{-1} peak. The formation of the methylene ether groups is indicated by a peak at about 1070 cm^{-1}: this is indicative of the C—O—C stretching vibration of an aliphatic ether and is a very good characteristic group frequency. There are also changes in the aromatic out-of-plane deformation region, as the result of subsidiary reactions in which methylol groups substitute for ring hydrogen atoms.

5.3.5   *Nitrogen-containing polymers: amides, polyurethanes*

Amides provide an excellent example of a class of compounds where, although there is rather extensive vibrational coupling, it is

Fig. 5.17. The infrared spectrum of a Novolak resin. Absorbance spectrum of Novolak IM 184 cast from acetone solution on a potassium bromide plate.

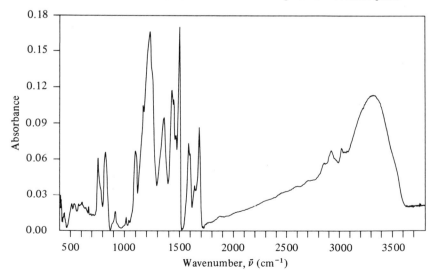

possible to use the group frequency approach successfully. Extensive studies on simple amides, which have included the examination of deuterated species and normal coordinate calculations, have shown that the two characteristic peaks, commonly referred to as amide I and amide II, that appear in the infrared spectra of both synthetic and natural polymers, e.g. polypeptides, containing the secondary amide group, —CONHR, do not originate from simple vibrations. Amide I, located at about 1640 cm^{-1} for polyamides such as Nylon 6,6 (fig 5.18) in the solid state, has the C=O stretching mode as its major component. Although the low frequency may, in part, be the result of resonance stabilization with the dipolar form —(C—O^{-})=N^{+}HR there is almost certainly some mixing with the in-plane N—H deformation vibration and, possibly, with the C—N stretching vibration. Amide II, at 1540 cm^{-1}, is largely the in-plane N—H deformation vibration but it is very probable that it also involves the v(C=O) and v(C—N) vibrations.

Despite their rather complex origin these two modes give good group frequencies, as does amide III, at about 1280 cm^{-1}, whose origin is also probably complex. Together, they provide a sure method for the recognition of the Nylons as a class when considered with the N—H stretching mode of the secondary amide group at 3310 cm^{-1} and the overtone of the amide II mode at 3070 cm^{-1}, both clearly visible in fig. 5.18. There are additional weaker peaks between 900 and 1300 cm^{-1}

Fig. 5.18. The infrared spectrum of Nylon 6,6. Absorbance spectrum of film cast from formic acid solution on a potassium bromide plate.

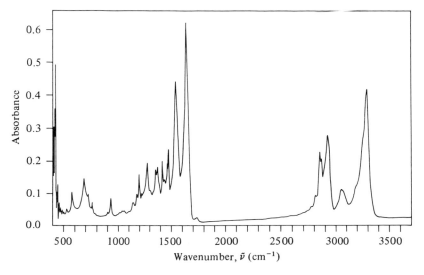

that permit differentiation between some of the Nylons but care must be exercised because some Nylons, e.g. Nylon 6, can crystallize in more than one form and the polymorphs give different spectra. An alternative approach to the characterization of a specific Nylon is to examine the acid and base products formed by hydrolysis, the latter as hydrochlorides. For example, adipic and sebacic acids, $HOOC(CH_2)_4COOH$ and $HOOC(CH_2)_8COOH$, formed from the hydrolysis of Nylon 6 and Nylon 10, are readily identified from their infrared spectra and if a mixture of these two Nylons, or a copolymer such as Nylon 6,6/Nylon 10,6, is encountered, quantitative analysis is possible using the absorption peaks at 735 and 724 cm^{-1} for adipic and sebacic acids, respectively, both due to the CH_2 rocking vibrations.

Polyurethanes provide a particularly good example of the complexity of condensation polymer reactions, as noted in subsection 1.2.2. The specificity of infrared spectroscopy is insufficient to permit a complete characterization of the wide range of materials available under the generic heading polyurethane, but the group frequency approach is, nevertheless, useful to a degree. In the condensation reaction between isocyanates and alcohols, the system most frequently encountered is the one involving an aromatic isocyanate and a polyhydric aliphatic alcohol. These mixed aromatic/aliphatic polyurethanes are recognizable on three counts: they have peaks characteristic for the urethane functional group, the aromatic moiety and the ether linkage, as the aliphatic alcohol is usually a hydroxyl-terminated polyether such as poly(ethylene oxide), $H\!+\!O(CH_2)_2\!\!\rightarrow_n\!OH$, or poly(propylene oxide), $H\!+\!O\,.\,CH_2\,.\,CH(CH_3)\!\!\rightarrow_n\!OH$. The urethane group gives peaks at 1695 and 1540 cm^{-1} which may be regarded as the analogues of peaks I and II of the polyamides. The aromatic unit is characterized particularly by the out-of-plane deformation modes. For example, if diphenyl methane-4,4'-di-isocyanate, $OCN\!-\!\!\bigcirc\!\!-\!CH_2\!-\!\!\bigcirc\!\!-\!NCO$, is used there are peaks at 725, 770 and 830 cm^{-1}. The ether linkages give the prominent $C\!-\!O\!-\!C$ stretching peak at 1110 cm^{-1}.

Polyurethanes are also prepared by reacting di-isocyanates with aliphatic polyesters that contain a terminal hydroxyl group, such as poly(ethylene adipate), $H\!+\!O(CH_2)_2O\,.\,OC(CH_2)_4CO\!\!\rightarrow_n\!OH$. These materials, known collectively as *polyester polyurethanes*, are readily distinguished from *polyether polyurethanes* by virtue of the $C\!=\!O$ stretching frequency of the ester group, located at about 1730 cm^{-1}. The $C\!-\!O$ stretching vibration, as anticipated, is also manifest as a peak at 1180 cm^{-1}. Carbamate groups, $NH_2\,.\,CO\,.\,O\!-$, are often formed and are readily detectable from peaks at about 1695 and 1540 cm^{-1}. As with the polyether polyurethanes, the spectral region $700-850$ cm^{-1} is reasonably specific for the particular aromatic isocyanate involved.

5.4 Quantitative analysis

5.4.1 *Introduction*

In subsection 1.5.1 the exponential attenuation of infrared radiation as it passes through a sample was discussed and equation (1.3) shows that the absorbance A of a sample of thickness x is given by

$$A = \log_{10}\left(\frac{I_0}{I}\right) = ax$$

Doubling the concentration of the absorbing species is found to have the same effect as doubling x, so that

$$A = \log_{10}\left(\frac{I_0}{I}\right) = kcx \tag{5.1}$$

where c is the concentration of the absorbing species and the absorption coefficient (or 'absorptivity') k is independent of x or c. This equation is known as the *Lambert–Beer law*. The quantity k is sometimes also called the 'extinction coefficient'. If, as is often the case, c is measured in units of mole litre^{-1} and x in centimetres, k is written as ε, the molar absorption coefficient. (Recent recommendations are that this term should be used only if c is measured in mole m^{-3} and x in m.) Both A and k are wavelength dependent so that for a particular wavelength λ_i we may write $A_i = k_i cx$. If k_i is obtained by measurements on the pure compound, the value A_i measured on a sample containing an unknown concentration of the material may be inserted into the Lambert–Beer equation to determine the concentration. It should be noted here that although it is desirable, when determining infrared absorbances or Raman scattering intensities, to separate out individual peaks by fitting any overlapping groups of peaks to a suitable number of components, the pseudo-baseline technique has often been used, particularly when only semiquantitative data were required. Both techniques are explained in section 1.7.

If a mixture of n absorbing species is present, the measured absorbance A_i is the sum of the absorbances of the various species present. Hence

$$A_i = k_{i1}c_1 x + k_{i2}c_2 x + \ldots + k_{in}c_n x \tag{5.2}$$

where k_{ij} denotes the absorption coefficient of component j at wavelength i. If absorbance measurements are made at n wavelengths it is possible to solve the set of simultaneous equations

$$A_i = x \sum_j k_{ij} c_j \quad i, j = 1 \text{ to } n \tag{5.3}$$

to find $c_1, c_2, c_3, \ldots, c_n$. In practice, this approach must be used circumspectly, for several reasons. The first, which is common to all types of quantitative work with infrared spectroscopy and does not relate specifically to polymers, may be illustrated by considering the simplest multi-component system, a two-component mixture. We then have

$$\begin{aligned} A_1 &= k_{11} c_1 x + k_{12} c_2 x \\ A_2 &= k_{21} c_1 x + k_{22} c_2 x \end{aligned} \tag{5.4}$$

Consider the two extreme possibilities, the first being when component 2 does not absorb at λ_1 and component 1 does not absorb at λ_2. Then $A_1 = k_{11} c_1 x$ and $A_2 = k_{22} c_2 x$ and the two components are effectively determined independently. The second extreme is when $k_{11} = k_{12}$ and $k_{21} = k_{22}$. There is then a total lack of specificity and only $c_1 + c_2$ can be determined. Clearly, it is desirable in principle to choose λ_1 and λ_2 so that the first extreme is approximated. More generally, for a multi-component mixture, the ideal condition is that $k_{ij} = 0$ when $i \neq j$, but in practice a compromise must be reached. It is evident that it becomes more difficult to maximize the specificity as the number of components increases and this factor limits the precision that may be achieved for a given error in the measurement of the values of A_i and k_{ij}.

Having selected appropriate values for λ_1 and λ_n, it is necessary to establish values for the various k_{ij} and for x in order to obtain the concentrations c_1 to c_n from the measured absorbance A_1 to A_n. It is in this respect that quantitative analyses on polymer systems often present greater problems than those for mixtures of low molecular weight organic compounds, both in the determination of k_{ij} and x. These problems will be considered in terms of examples. Raman spectroscopy can also be used for quantitative analysis but, as explained in subsection 1.6.1, it is usual to use only intensity ratios rather than absolute intensities.

The quantitative applications of infrared and Raman spectroscopy considered in this section largely relate to the determination of the compositions of copolymers and mixtures of homopolymers. For convenience the examples are grouped under types of polymer, but they should more generally be regarded as examples of methods. It is also possible to obtain a variety of quantitative information on the levels of defect structures in polymers; examples from this field will be given in the following chapter, where copolymers will also be further considered.

5.4.2 *Copolymers of vinyl chloride or ethylene with vinyl acetate*

The simplest situation, one that approximates to the analysis of mixtures of low molecular weight materials, is the examination of a mixture of polymers or of a copolymer when the pure homopolymers are available and when all the polymers are soluble in a solvent that is reasonably transparent to infrared radiation. It is then possible to work with a conventional liquid cell, for which x can be measured with adequate precision, and to measure the various k_{ij} directly. The analysis of vinyl chloride/vinyl acetate copolymers provides a particularly good example. A solution in tetrahydrofuran is placed in a cell of thickness 0.1–0.2 mm and the absorbance of the carbonyl peak at 1740 cm^{-1} is measured and compared with the corresponding value for poly(vinyl acetate) to obtain the composition. Poly(vinyl chloride) does not absorb at 1740 cm^{-1}. This method tacitly assumes that the spectrum of the poly(vinyl acetate) units in the copolymer is the same as that of the homopolymer. This is substantially so in the case of block copolymers but there may be some interaction effects with random copolymers, particularly in view of the fact that the concentration of vinyl chloride units is several times that of the vinyl acetate units in commercial copolymers of this type. An alternative approach, which may also be used to check for the presence of interactions between the different components of a copolymer, is to examine several samples by a second and more absolute method. In the case of vinyl chloride/vinyl acetate copolymers direct oxygen determination proves very satisfactory. Alternatively, ^{1}H NMR spectroscopy is a convenient and accurate method for this and many other polymer systems.

Vinyl chloride/vinyl acetate copolymers may also be examined as thin films and this may sometimes prove more convenient. Although it is possible, in principle, to measure the thicknesses of such films, which may be as low as 0.03 mm, they tend to be non-uniform and an average value must be obtained. This often leads to significant errors and it is more satisfactory to eliminate the parameter x in the Lambert–Beer equation by the use of absorbance ratios. In this particular case the ratio of the absorbance of the acetate peak at 1740 cm^{-1} to that of the vinyl chloride methylene bending peak at 1430 cm^{-1} may be used. A calibration is obtained from samples of known composition.

Copolymers of ethylene and vinyl acetate are not soluble in spectroscopically useful solvents when the ester concentration is below the level of a few per cent. Such materials, as thin films, are analysed by using the absorbance ratio of peaks at 1020 and 720 cm^{-1}, characteristic for the vinyl acetate and ethylene units, respectively.

5.4.3   *Styrene/acrylonitrile copolymers and cross-linking in polystyrene*

Thin films of styrene/acrylonitrile copolymers give infrared spectra from which the ratio of the absorbances of the C≡N nitrile stretching vibration at 2250 cm^{-1} and the polystyrene ring stretching vibration at 1600 cm^{-1} may be determined as a measure of the composition. A calibration may be obtained by the chemical analysis of several samples of varying acrylonitrile content.

Raman spectroscopy is often a particularly useful method for examining cross-linked polymers. Such materials are usually insoluble and not amenable to examination by NMR spectroscopy. Furthermore, they may be difficult to hot press into the thin films required for quantitative infrared work. Powder or pellet samples do, however, give good Raman spectra. The cross-linking of styrene by reaction with allyl alcohol, CH_2=CH—CH_2OH, provides a good example. The ratio of the intensities of $v(C$=$C)$ for the allyl group at 1645 cm^{-1} and $v(C$=$C)$ for the aromatic ring at 1600 cm^{-1} is measured for both the starting materials and the reaction product. The degree of cross-linking is then calculated from the fall in the relative intensity due to the allyl group.

5.4.4   *Polybutadienes and styrene/butadiene copolymers*

The analysis of polybutadiene provides a good example of a mixture for which it is not possible to find wavelengths at which only one component absorbs, and so resort must be made to the use of simultaneous equations. As explained in subsection 1.2.3, when butadiene polymerizes, successive units can add in the 1,2-position, to give a vinyl polymer, but 1,4-addition also occurs and the structural unit formed then can exhibit cis/trans isomerism about the double bond. In many cases, all three types of isomer are present in a sample and it is therefore necessary to analyse a three-component mixture. Although it is, in principle, possible to use the peaks originating from $v(C$=$C)$, those due to the out-of-plane deformation vibrations are stronger and more characteristic. The vinyl-1,2 unit gives a strong, sharp peak at 910 cm^{-1}, and the trans-1,4 unit a comparable one at 965 cm^{-1}. The cis-1,4 structure gives a much weaker and broader peak centred at about 730 cm^{-1}.

The analysis proves comparatively straightforward, for two reasons: polybutadienes are soluble in carbon disulphide, a useful spectroscopic solvent, and samples of pure 1,2- and cis- and trans-1,4 polymers are available, so that the nine required k_{ij} values may be determined. The values given by Silas, Yates and Thornton, who were the originators of the method, are approximately:

	1,2	cis	trans
910 cm^{-1}	184	2	2
730 cm^{-1}	5	10	1
965 cm^{-1}	7	4	133

It should be noted that these k_{ij} values are not transferable between spectrometers, because of the effect of finite resolution and other factors pertaining to the spectrometer optics that vary between instruments. They do, however, show that although the interference from the weaker components at 910 and 965 cm^{-1} is not particularly serious, it is more important at 730 cm^{-1}. The estimation of the cis-1,4 units in the commonly encountered styrene/butadiene copolymers is still more difficult because the butadiene peak at 730 cm^{-1} is overlapped by the much stronger polystyrene peak at 700 cm^{-1}, whose origin is the out-of-plane deformation vibrations of the five hydrogen atoms attached to the monosubstituted aromatic ring.

Raman spectroscopy provides an easier method for analysing such materials, using $v(C{=}C)$, which is considerably stronger and more characteristic than in the infrared spectrum. Fig. 5.19 shows the Raman spectra near 1600 cm^{-1} for ordinary polystyrene, for a piece of toughened polystyrene, which incorporates a small amount of butadiene, and for polybutadiene. In the spectrum of polybutadiene there are peaks at 1650 cm^{-1} (cis-1,4), 1655 cm^{-1} (1,2) and 1665 cm^{-1} (trans-1,4), the intensities of which are measures of the respective concentrations of the isomers indicated. These intensities are most conveniently measured as ratios against that of a convenient polystyrene peak, such as the one at 999 cm^{-1}. It is then necessary to calibrate and this is best done with samples whose compositions have already been determined by ^{13}C NMR spectroscopy. Alternatively, it is possible to make up blends if the pure 1,2 and cis- and trans-1,4 polymers are available. By a similar process, the proportion of polystyrene may also be determined.

5.5   Monomers and additives

In general, neither infrared nor Raman spectroscopy is well suited to the measurement of minor components. In many instances, the limit of detection is of the order of 1%, although this may be improved significantly in favourable cases. With the steadily increasing stringency imposed by environmental health considerations, acceptable monomer levels in polymers are only a few parts per million in some instances, and recourse must be had to other analytical methods such as gas chromatography and mass spectrometry. Similarly, the in situ characterization and estimation of additives at the levels encountered in commercial practice is often not feasible. Nevertheless, infrared spectroscopy has proved useful in several instances and with the increasing availability of Raman spectrometers this technique is now also proving advantageous for systems that give strong Raman peaks. For example, sulphur-containing materials, such as the antioxidant Santonox R, 4,4′-thiobis (2-tert.butyl-5-methyl phenol) (see fig. 5.20), give strong Raman peaks characteristic for the C—S—C group. Some illustrative examples of the study of monomers and additives are now given.

5.5.1   *Caproclactam in Nylon 6; monomer and water in poly(methyl methacrylate)*

Nylon 6 is prepared from caprolactam and may contain significant amounts of this starting material. It may be estimated by

Fig. 5.19. (*a*) The Raman spectrum in the region 1500–1700 cm^{-1} for polystyrene; (*b*) the Raman spectrum of the same region for a sample of polystyrene toughened by the incorporation of polybutadiene; (*c*) higher resolution Raman spectrum of the 1665 cm^{-1} peak of polybutadiene. See also text. ((*c*) reproduced by permission from D. L. Gerrard, *Chem. Brit.* 1984 p. 175.)

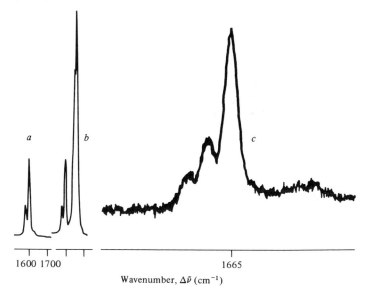

Wavenumber, $\Delta \tilde{\nu}$ (cm^{-1})

infrared spectroscopy at levels greater than 0.2 % by virtue of the mode at 870 cm^{-1}, using the Nylon in the form of a hot pressed film. At low levels of caprolactam the 870 cm^{-1} peak is weak and not readily discernible against the Nylon 6 background absorption. The technique of difference spectroscopy is therefore used to remove this background. The film under examination is placed in the reference beam of the spectrometer and a caprolactam-free film of equal thickness, obtained by precipitation from a solution in formic acid, is placed in the sample beam. Solutions of caprolactam of known concentration in carbon tetrachloride are also placed in the simple beam, to provide a calibration.

The effect of the anharmonicity of molecular vibrations involving hydrogen atoms is greater than that of other vibrations because of the low mass, and consequently larger amplitude of oscillation, of the hydrogen atom. This is manifested in terms of relatively strong absorption at overtone and combination frequencies in the near infrared spectral region. The hydrogen atoms adjacent to the double bonds present in monomers sometimes give peaks in this region which are useful for the measurement of residual monomer. Although these monomer absorptions are comparatively weak, the absence of other strong peaks in their vicinity permits substantially greater sample thicknesses to be employed and this has the added advantage that commercial sheet several millimetres in thickness may be examined directly. The best known example is the determination of methyl methacrylate monomer in poly(methyl methacrylate) (Perspex) sheet. As shown in fig. 5.21 the monomer gives a peak, characteristic for the —HC=CH$_2$ group, at 6135 cm^{-1}, of sufficient intensity to permit the determination of monomer levels down to 0.1 %.

Water gives a strong peak at 5180 cm^{-1} that is suitable for estimating it at levels down to a few hundredths of one per cent in poly(methyl methacrylate). Calibration is made by drying a sheet in a vacuum desiccator and then following its increase in weight and the increase in the absorption at 5180 cm^{-1} as the sheet takes up atmospheric water.

Fig. 5.20. The structure of Santonox R.

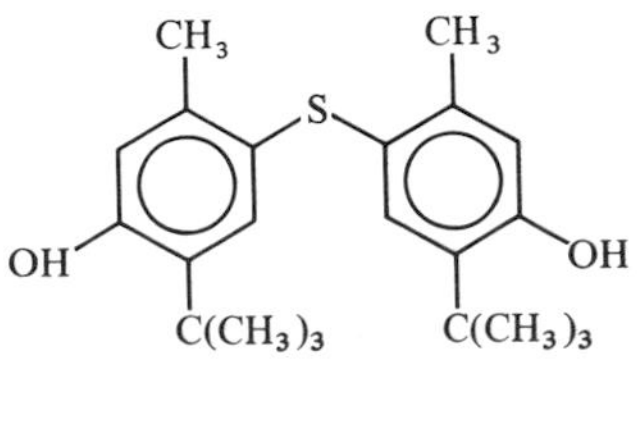

5.5.2   *Additives in polyethylene and polypropylene*

The relatively small number of strong peaks in the spectra of hydrocarbon polymers make them the most suitable type of materials for the direct determination of antioxidants and other additives. Although phenolic antioxidants usually contain a hydroxyl group which is sterically hindered so that it cannot hydrogen bond (see section 6.6) and gives a sharp $v(OH)$ peak near 3645 cm^{-1}, polyethylene also gives weak peaks, presumably combination modes, in the same region that offset the value of this antioxidant peak. Nevertheless, with the superior sensitivity and data handling facilities provided by modern infrared spectrometers, the determination of antioxidants in situ merits a reappraisal.

Fig. 5.21. The detection of monomer in poly(methyl methacrylate) using infrared spectroscopy; (*a*) shows the monomer peak (arrowed) as a small 'shoulder' on the side of the stronger peak due to saturated CH groups; (*c*) shows an accurately 'compensated' spectrum obtained by placing a monomer-free sample of the correct thickness in the reference beam of a double-beam spectrophotometer, while (*b*) and (*d*) show the effects of over- and under-compensation. (Reproduced by permission from H. A. Willis and R. G. J. Miller, *Spectrochimica Acta* **14** 119 (1959). Copyright (1959), Pergamon Journals Ltd.)

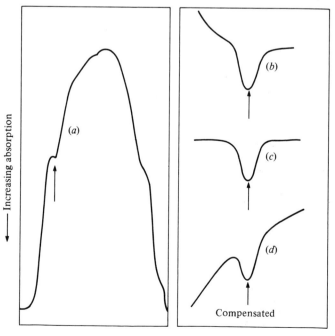

Two proven examples of the direct determination of additives are the estimation of long-chain aliphatic amides (such as oleamide), incorporated as slip agents in polyethylene, and of dilauryl thiodipropionate in polypropylene. In the former case, which is satisfactory down to levels of about 1 %, the intensity of the amide I peak at 1640 cm^{-1} is used as a measure of the long-chain amide concentration. In the latter instance the carbonyl peak at 1740 cm^{-1} is used and difference spectroscopy, with a sample free from dilauryl thiodipropionate in the spectrometer reference beam, provides enhanced sensitivity.

It is usually possible to increase the sensitivity by one or two orders of magnitude by prior extraction of the additive. Oleamide in polyethylene provides a case in point, the approach being to dissolve the sample in toluene, precipitate out the polymer by the addition of methanol, evaporate the remaining solution and dissolve the residue in carbon disulphide for infrared assay. In view of the diversity of antioxidants and other additives in commercial usage, prior extraction of unknown polymers is almost mandatory for the characterization of such materials. In some cases subsequent infrared examination may suffice for an identification.

5.6   The characterization of oxidation and degradation

The stability of polymers is a very important commercial consideration. They must not degrade appreciably at the temperatures encountered during processing operations such as extrusion and blowing, and their photo-decomposition must be minimal. Infrared and Raman spectroscopies play an important role in assessing degradation, of which oxidation is one particular form, and can sometimes provide information which is valuable for determining degradation mechanisms. Some examples are now given.

5.6.1   *Hydrocarbon polymers: polyethylene, polypropylene*

The role of infrared spectroscopy is exemplified by its use for the characterization of heat- and photo-oxidized polyethylenes. The first stage in the heat oxidation process is the formation of hydroperoxides but as these are difficult to characterize spectroscopically it is normal practice to examine their decomposition products, a range of carbonyl-containing compounds. These give a broad $v(C{=}O)$ peak centred at about 1725 cm^{-1}. It is of rather irregular shape and clearly consists of several partially overlapping peaks. When such oxidized samples are treated with alkali a shoulder at 1715 cm^{-1} disappears and is replaced by a well defined peak at about 1610 cm^{-1}. This is $v(C{=}O)$ for the

COO$^-$ ion of a salt, indicating that the shoulder at 1715 cm^{-1} is characteristic for saturated carboxylic acids. A second shoulder, at 1735 cm^{-1}, is indicative of saturated aldehydes but the major part of the overall absorption is attributable to saturated ketones. This complex carbonyl peak also appears in the spectra of photo-oxidized polyethylene specimens, which also give additional peaks at 910 and 990 cm^{-1}. These are characteristic for vinyl groups, and their presence shows that chain-terminating unsaturated groups are being formed, very probably as the result of chain scission.

Because saturated ketone groups are the major component of the carbonyl-containing compounds present it is usual to assess the extent of the oxidation in terms of the concentration of a long-chain saturated ketone, such as distearone, $[CH_3(CH_2)_{16}]_2CO$. This may be compounded into polyethylene at a known level or, alternatively, known concentrations in a paraffinic solvent may be prepared to simulate the polyethylene environment. Polypropylene gives a similar rather complex $v(C{=}O)$ peak upon oxidation and it is again customary to make quantitative measurements in terms of saturated ketones. This calls for a somewhat modified experimental approach, as polypropylene shows some absorption in the vicinity of 1725 cm^{-1}. This can be eliminated by difference spectroscopy, using carbonyl-free polypropylene in the reference beam of the spectrometer.

5.6.2   Resonance Raman studies of degradation in PVC

The degradation of poly(vinyl chloride) proceeds by a quite different route from those considered above, for both thermal and photochemical decomposition. A molecule of hydrogen chloride is eliminated at an initiation site along the chain, probably a branch point or an allylic end group, —CH=CH—CH$_2$Cl, and a double bond or additional double bond is formed. This acts as a propagation point for sequential hydrogen chloride loss by an 'unzipping' process, and conjugated polyene sequences $+CH{=}CH+_n$ of various lengths are formed. Those with about seven or more double bonds have electronic absorption spectra in or near the visible spectral region, giving the commercially unacceptable problem of sample discolouration. Although a range of stabilizers, notably tin compounds such as dibutyl tin dilaurate, is available for limiting this degradation, there has been an increasing interest in gaining a fundamental understanding of the degradation process, in order to assess the possibilities of improving the stability of the polymer.

Until recently, only two methods were available for assessing the degree of degradation. Measurement of the hydrogen chloride evolved gives the total content of double bonds and ultraviolet/visible spectroscopy is specific for sequences of up to about 10 conjugated double bonds, although quantitative work is rather difficult. The situation has been transformed by the use of a particular type of Raman spectroscopy, known as *resonance Raman spectroscopy*. If the frequency v_0 of the laser radiation used to excite a Raman spectrum approaches the frequency v_e of an allowed electronic transition in the molecule being examined, certain normal modes exhibit a pronounced enhancement of their Raman scattering intensities, sometimes by a factor as great as 10^4 or 10^5. This is the resonance Raman effect and it occurs because as v_0 approaches v_e the polarizability of the molecule increases very considerably. The scattered intensity may be expressed mathematically by the formula

$$I = \frac{I_0 A (v_0 - v)^4}{(v_e - v_0)^2 + \delta^2} \tag{5.5}$$

where I_0 represents the intensity of the incident radiation, v is the frequency of the mode and the value of A depends on the particular mode involved. For the present purposes, the denominator is the important term, for as v_0 approaches v_e it becomes increasingly smaller,

Fig. 5.22. The resonance Raman spectrum of a sample of thermally degraded poly(vinyl chloride). (Reproduced with permission from D. L. Gerrard and W. F. Maddams, *Macromolecules* **8** 54 (1975). Copyright (1975) American Chemical Society.)

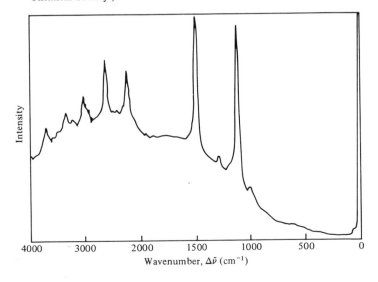

Wavenumber, $\Delta \bar{v}$ (cm^{-1})

and *I* becomes correspondingly larger. Only the presence of the damping factor δ keeps *I* at a finite value when $v_0 = v_e$. Most, but not all, of the known cases that have been studied in detail involve totally symmetric modes and intense electronic transitions. Both these conditions are fulfilled in the case of the conjugated polyene sequences present in degraded PVC specimens and two resonance peaks appear, as shown in fig. 5.22. One, in the vicinity of 1500 cm^{-1}, designated v_2, is the C=C stretching vibration and the other, at about 1120 cm^{-1}, designated v_1, arises from the stretching of C—C bonds.

These peaks are important for two reasons. Their intensities are such that degradation can be detected down to levels of one part in 10^7, and the frequencies of the peaks vary systematically with the conjugated sequence length *n*. The variation is greater for v_2 than for v_1 and it has found considerable use in estimating *n* from measured v_2 values, using a relation of the form $v_2 = a + b\mathrm{e}^{-cn}$, where the values of *a*, *b* and *c* have been established from measurements on low molecular weight conjugated polyenes and polyacetylene, which contains very long sequences. The method has the additional advantage that by varying the exciting wavelength, by using different lines from an argon ion laser, or a tunable dye laser, resonance may be excited in sequences of different lengths for $n = 9$ and longer. It is then possible to form a semiquantitative idea of the relative concentrations of conjugated polyenes of varying lengths. These studies have been very important in allowing the correct choice of plasticizers and stabilizers to be made for commercial applications of PVC.

5.7   Further reading

The books and reviews already cited in section 3.3 provide further information about the topics covered in the earlier parts of the present chapter (sections 5.1–5.3). For deeper study of the later parts the following books may be found useful: *Infrared Spectra of Complex Molecules* (2 vols.) by L. J. Bellamy, Chapman and Hall, 1975, 1980 (particularly for section 5.3); *Infrared Analysis of Polymers Resins and Additives* (2 vols.) by D. O. Hummel, Wiley, 1969 (particularly for section 5.3); *The Identification and Analysis of Plastics* by J. Haslam, H. A. Willis and D. C. M. Squirrel, Heyden, 1972 (particularly for sections 5.3–5.6).

A list of references to some of the more important papers on the topics covered in this chapter and in chapter 6 is given following chapter 6. References given in figure captions or in connection with tables may also be found useful.

6   The microstructure of polymers

6.1   The distribution of copolymerized units in the polymer chain

6.1.1   *General principles*

If a mixture is made of two or more types of monomer which polymerize by similar mechanisms, e.g. the free radical process, they may be polymerized to form chains in which the structural units of the respective homopolymers are present to varying extents and in sequences whose lengths depend on the polymerization conditions. Such copolymers are important commercially and they are also the source of a considerable amount of information on polymerization processes. Their commercial importance is that they provide the means for modifying to advantage the physical properties of the parent homopolymers. They may provide a higher tensile strength, a greater impact strength, a superior stress crack resistance, an improved resistance to thermal and photochemical degradation or a modified crystallization behaviour. At the fundamental level, the molecular structure of the copolymer can often provide information on the polymerization mechanism, such as the factors which determine the reactivity of a particular radical. It is also possible to prepare polymers with known structural defects in order to study their effects. For example, the copolymerization of small amounts of allyl chloride, $CH_2=CH—CH_2Cl$, or acetylene, $CH\equiv CH$, with vinyl chloride may be used to introduce chloromethyl branches, $—CH_2Cl$, or double bonds, respectively, into the vinyl chain in order to assess whether these types of defect are the sites of the initiation process for thermal degradation.

The overall composition of a copolymer is not determined solely by the ratio of the concentrations of the monomers used in the polymerization. This can be demonstrated by considering briefly the free radical copolymerization of monomers A and B, both of the vinyl type. When the initiator is added, radicals $A^·$ and $B^·$ are formed, where $A^·$ and $B^·$ represent chains of any length terminating in $—CH_2—C^·HA$ or $—CH_2—C^·HB$, respectively. There are four possible types of reaction, all of which usually occur simultaneously:

$$A^· + A \xrightarrow{k_{AA}} A^·; \qquad A^· + B \xrightarrow{k_{AB}} B^·; \qquad B^· + A \xrightarrow{k_{BA}} A^·; \qquad \text{and}$$

$$B^· + B \xrightarrow{k_{BB}} B^·$$

where k_{AA}, k_{AB}, k_{BA} and k_{BB} are the four rate constants for these reactions. The composition of the copolymer is determined by the probability with which A* adds to A rather than to B and the probability with which B* adds to B rather than to A. These two factors are usually expressed in terms of reactivity ratios, defined as $r_1 = k_{AA}/k_{AB}$ and $r_2 = k_{BB}/k_{BA}$. When the kinetics of the polymerization process are considered it is easy to show that

$$\frac{d[A]}{d[B]} = \frac{[A]}{[B]} \cdot \frac{r_1[A] + [B]}{[A] + r_2[B]} \tag{6.1}$$

where [A] and [B] signify the concentrations of the unreacted monomers A and B, respectively. It is evident from this equation that only if $r_1 = r_2 = 1$, so that random copolymerization occurs, does the composition of the copolymer match that of the monomer feed. Furthermore, if this condition is not satisfied the proportions of A and B in the remaining monomer will change as the polymerization proceeds, and hence the composition of the copolymer will also change. There is therefore a need for methods of determining the overall composition of a copolymer.

If random copolymerization occurs, the proportions of A and B in the copolymer are fixed by the initial values of [A] and [B] and the probability of finding a sequence of n successive A units or m successive B units is calculable. If the proportion of A is considerably in excess of that of B the probability of finding long runs of A units will be much greater than that of finding similar runs of B units. At the opposite extreme, if A and B are both present in substantial concentration, block copolymerization may occur, to give a structure in which there is a long run of A units followed by a long run of B units, but such materials are usually made by an anionic polymerization process in which only one type of monomer is present at a time. Not surprisingly, there are a number of copolymer systems for which the polymerization is largely but not completely random, so that the sequence distribution is also largely but not completely random. It is therefore necessary to have methods for determining the sequence distribution, to characterize the microstructures of these polymers in a way that cannot be done from a knowledge of their overall compositions alone.

6.1.2   *The determination of overall copolymer composition*

The principles governing the determination of overall copolymer composition by infrared and Raman spectroscopy have been discussed in section 5.4. It was then pointed out that one method in common use is to interpret the copolymer spectra in terms of a superposition of the spectra of the relevant homopolymers, but it was also noted that this approach is potentially subject to error, particularly for random copolymers containing short sequences of the individual monomer units. This problem will now be considered in rather more detail, in terms of three effects.

The first effect is that the frequency of vibration of a group mode of a unit of type A may depend on whether the unit is attached directly to other units of type A or to units of type B, and corresponding effects may occur for units of type B. The corresponding peak in the spectrum may be broadened, or become asymmetric, and may be shifted somewhat in position. Consequently, if the intensity at the position of the peak maximum in the copolymer is used as a measure of the concentration, the value so obtained will be slightly in error. Secondly, if one type of monomer unit is present as a minor component it will occur as relatively isolated units in a matrix of the major component. Such a copolymer is rather analogous to the dilute solutions used to study the infrared spectra of low molecular weight organic compounds, where the spectrum of the solute may be modified by the presence of the solvent. Hence, for the copolymer, the spectrum of the minor component may not be identical with that given by the same monomer units in the pure homopolymer. Finally, the presence of long runs of identical units in a homopolymer may lead to order, either along individual chains only or three dimensionally in crystallites. The effects of order will be considered in detail in section 6.7. For present purposes it suffices to note that peaks specific for order are often found in the spectra of polymers. Where such peaks are present in the spectrum of either of the homopolymers they will not necessarily be present in the spectrum of the copolymer, and they must therefore be recognized. Typical examples of these three effects will now be considered.

When the infrared spectra of a series of random ethylene/vinyl acetate copolymers are examined, the peak due to the carbonyl stretching mode is found to move from 1736.8 cm^{-1} for a material containing 95% of vinyl acetate units, denoted by V, to 1728.0 cm^{-1} for a copolymer with 57% of V units. Furthermore, the fact that the shape of the absorption peak changes with composition suggests the presence of more than one component peak. Examination of these copolymers by ^{13}C nuclear magnetic resonance spectroscopy shows that three types of triad units are present predominantly: VVV, VVE and EVE, where E denotes ethylene. As the concentration of this latter monomer increases, the concentration of the VVE and EVE units increases and that of the VVV units decreases. As shown in fig. 6.1, it is possible to fit the carbonyl peaks of these copolymers as the sums of three peaks, located at 1738.6,

1731.9 and 1721.5 cm^{-1} and characteristic, respectively, for the VVV, VVE and EVE triads. By assuming that the area under each peak is proportional to the concentration of the corresponding triad unit, concentration values in quite good agreement with the ^{13}C NMR results can be obtained. If, however the absorbance at 1738.6 cm^{-1}, the position of the peak for vinyl acetate homopolymer, had been used as the basis for the quantitative analysis, an erroneous result would have been obtained. In examples of this type, the use of peak areas rather than absorbance values at peak maxima or other specific positions in the spectrum largely eliminates systematic errors.

The shift in the v(C=O) frequency of these vinyl acetate/ethylene copolymers may be considered, in principle, as a combined consequence of the first two of the effects discussed above, the interaction of the V and E units at their junction points and the solvent effect that will occur when the concentration of E units increases relative to V. The former effect is likely to be the more important because isolated V units in a matrix of E will not occur frequently except at very low V concentrations. The fact that the three simple peaks in fixed positions can account for the observed complex peak shape confirms this. A very good example of the matrix effect is, however, provided by miscible polymer blends. In the spectra of blends of poly(ε-caprolactone) and poly(vinyl chloride) in the molten state the carbonyl stretching frequency of the former shifts to lower frequencies increasingly as the

Fig. 6.1. The carbonyl stretching peak for a particular ethylene/vinyl acetate copolymer, showing its analysis into peaks due to *VVV* triads (− − − −), *VVE* triads (. . . . . .) and *EVE* triads (− · − · − ·). (From J. R. Ebdon *et al.*, *Journal of Polymer Science Polymer Chemistry Edition* **17** 2783 (1979). Copyright © (1979) John Wiley & Sons, Inc. Reprinted by permission of John Wiley & Sons, Inc.)

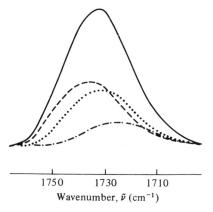

1750   1730   1710
Wavenumber, $\bar{\nu}$ (cm^{-1})

proportion of PVC is increased from 1:1, to 3:1 and finally to 5:1 molar, the shift for the last of these being about 3 cm^{-1}.

Ethylene/propylene copolymers provide a good example of a system where absoprtion peaks present in the spectra of the homopolymers may be absent from the spectra of the copolymers because of the loss of order. Polyethylene specimens vary appreciably in their crystallinity, for reasons that will be discussed in section 6.3, but all show the splitting of the $\diagdown$CH$_2$ rocking mode into components at 720 and 731 cm^{-1}, characteristic for methylene sequences in crystalline regions. In ethylene/ propylene block copolymers the long methylene sequences that are present crystallize and in the spectral region 720–731 cm^{-1} there is a basic similarity between the spectra of the copolymers and those of ethylene homopolymers. This is not so for random copolymers. If only isolated ethylene units are present the CH$_2$ rocking mode gives a single peak at about 732 cm^{-1}. When small numbers of adjacent ethylene units occur a further peak appears at 720 cm^{-1}, which is characteristic for these adjacent ethylene units in an amorphous environment. As the ethylene content increases, crystallization of the longer sequences can occur and the well-known doublet at 720/731 cm^{-1} appears in the spectrum. Similarly, in most samples of propylene homopolymer the chains take up a helical conformation and when this occurs additional peaks appear in the spectrum (see subsection 6.2.2). Such peaks are present in ethylene/propylene block copolymers but not in random copolymers.

6.1.3 *The characterization of sequence lengths*

In a copolymer of the type . . . AAAA(B)$_n$AAAA . . ., where n is rather small, the contribution to the overall spectrum from the B units may depend on n and this dependence then provides a means for determining n. Ethylene/propylene copolymers with rather low ethylene content provide an excellent example of this effect, a fact hinted at in the previous subsection.

If there is one ethylene unit sandwiched between two propylene units in a head-to-tail arrangement the structure is

$$+CH \cdot CH_3-CH_2+\ +CH_2-CH_2+\ +CH \cdot CH_3-CH_2+,$$

i.e. there are three consecutive methylene units, and if there are two adjacent ethylene units it follows that there must be five consecutive methylene units. It is well known, from the study of the infrared spectra of simple hydrocarbon systems, that the frequency of the $\diagdown$CH$_2$ rocking mode depends on the number, n, of consecutive methylene units as follows:

$$n = \quad 1 \quad 2 \quad 3 \quad 4 \quad \geqslant 5$$
$$\tilde{\nu}\,(\text{cm}^{-1}) = \quad 815 \quad 752 \quad 733 \quad 726 \quad 722$$

The presence of discrete peaks characteristic for $n = 1, 2, 3$ and $\geqslant 5$ is clearly evident in the infrared spectra of three ethylene/propylene copolymers, covering a range of compositions, shown in fig. 6.2. The concentrations of the four types of methylene sequences may be established by measuring absorbance values at the four corresponding frequencies and allowing for the overlap of contributions by setting up simultaneous equations for the absorbance of individual peaks using molar absorptivities determined from measurements on model com-

Fig. 6.2. The infrared spectra of ethylene/propylene copolymers in the region where absorption due to the $\diagdown$CH$_2$ rocking modes occurs. The compositions are: $-\cdot-\cdot-\cdot-$, 25%; ————, 50% and $------$, 75% of ethylene units, (From G. Bucci and T. Simonazzi, *Journal of Polymer Science* **C7** 203 (1964). Copyright © (1964) John Wiley & Sons, Inc. Reprinted by permission of John Wiley & Sons, Inc.)

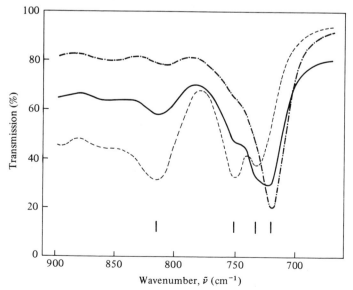

Wavenumber, $\tilde{\nu}$ (cm^{-1})

pounds. For each odd value of n there is only one corresponding sequence of propylene and ethylene (E) units, viz.

$$+CH.CH_3—CH_2+E_{(n-1)/2}+CH.CH_3—CH_2+,$$

so that the concentrations of these sequences for $n = 1$ and $n = 3$ follow directly from those of the corresponding sequences of $\diagdown CH_2$ groups. For each even value of n there are two possible sequences of propylene and ethylene units, viz.

$$+CH.CH_3—CH_2+E_{(n-2)/2}+CH_2—CH.CH_3+$$

and

$$+CH_2—CH.CH_3+E_{n/2}+CH.CH_3—CH_2+,$$

so that the concentrations of these two sequences for $n = 2$ cannot be separately determined.

Ethylene/propylene copolymers provide a striking example of the appearance of peaks characteristic for sequences of various lengths but, in general, the changes are much less marked with other polymer systems. The ethylene/vinyl acetate polymers considered in subsection 6.1.2 are more typical both in the changes that occur and the limited deductions that are possible. Propylene oxide homopolymers provide another example of effects which lead to peak broadening rather than the appearance of discrete characteristic peaks. Poly(propylene oxide) is a molecule with asymmetric carbon atoms, $—C^*H.CH_3—CH_2—O—$. When propylene oxide is polymerized in the presence of a stereospecific catalyst to give the isotactic polymer it can give rather long l or d sequences (these sequences are mirror images of each other, whereas an isotactic sequence of a vinyl polymer is identical to its own mirror image), or a more random copolymer of l and d sequences, depending on the nature of the catalyst employed. Polymers with long regular sequences give a comparatively sharp peak at $1240\ cm^{-1}$ but shorter sequences have their maxima at slightly higher frequencies and when they are present the peak broadens. On the basis of some reasonable assumptions about the dependence of the frequency of the peak maximum on the sequence length and about the individual peak shapes this broadened peak has been analysed into separate peaks due to sequence lengths of 1,2,3–9 and 10 or more units.

Raman spectra may sometimes provide more information about the presence of various sequence lengths, as they do for vinyl chloride/vinylidene chloride copolymers. A solitary vinyl chloride unit, designated VC, in a run of vinylidene chloride (VDC) units, giving the structure —VDC—VC—VDC—, has a characteristic peak at 1197 cm^{-1}. For the structural unit —VDC—VC—VC—VDC— the peak moves to 1235 cm^{-1} and it is found at 1247 cm^{-1} for vinyl chloride homopolymer. It is possible that a study of the shape of the line complex would also show the presence of —VDC—VC—VC—VC—VDC— units.

6.2   Configurational and conformational isomerism

6.2.1   *Introduction*

In subsection 1.2.3 the occurrence of configurational isomerism in vinyl polymers, $+CH_2—CHX+_n$, was discussed and the two types of stereoregular configurational isomers, syndiotactic and isotactic, were defined but the practical implications were not considered. Furthermore, it was pointed out that for a given configurational isomer there may be rotation about C—C bonds to give conformational isomers and that, in particular, when the substituent X is rather bulky such a rotation frequently occurs in the case of the isotactic configuration and leads to a non-planar conformation. Here, also, practical examples were not considered in detail.

The discovery by Ziegler of catalyst systems capable of promoting the stereoregular polymerization of vinyl monomers was exploited brilliantly by Natta and co-workers, who published their findings, from 1956, in a series of seminal papers. The degree of regularity present in these polymers is such that they are markedly more crystalline than their atactic counterparts, which enhances those physical properties, such as rigidity, that depend on the degree of crystallinity. Furthermore, the Ziegler–Natta catalysts have opened the way to the general use of such polymers, as they provide the means for their preparation on a commercial scale. Among early discoveries was a catalyst for the preparation of isotactic polypropylene and this, in turn, prompted studies on its very characteristic vibrational spectrum, which shows important features. Natta and co-workers also discovered catalysts capable of promoting the polymerization to the syndiotactic polymer, and the spectra of the two different types of polymer will now be considered as an example of the effects of configurational isomerism.

6.2.2   *Configurational isomerism: polypropylene*

Having prepared isotactic polypropylene, Natta and co-workers examined its molecular structure by X-ray diffraction studies. Their results show that it forms a regular helix with three monomer units per turn, i.e. a 3_1 helix (see fig. 3.4). This degree of chain regularity should lead to the appearance of additional peaks in the vibrational spectrum and this expectation is confirmed experimentally. Comparison of the infrared spectrum of isotactic polypropylene (fig. 6.3*d*) with that of the atactic polymer (fig. 6.3*a*) shows that whereas the former possesses sharp, medium or strong peaks at 805, 840, 898, 972, 995 and 1100 cm^{-1}, only the one at 972 cm^{-1} is present, in a rather broader form, in the spectrum of atactic polypropylene. Furthermore, when isotactic polypropylene is examined in the molten state the spectrum resembles that of the atactic polymer. This suggests that the remaining peaks are specific for various vibrations of the regular helical structure of the isotactic polymer. Substantial confirmatory evidence for this tentative conclusion is provided by factor group analysis, normal coordinate calculations, polarization studies and measurements on deuterated polymers. These will now be considered briefly.

The factor group analysis will be considered in terms of a single chain, since detailed analyses have shown that there is no evidence for appreciable inter-chain interaction. The 3_1 helical chain belongs to a line group with a factor group isomorphous with the cyclic group C_3. Analysis of this factor group indicates that the normal modes are distributed among the symmetry species as 25 totally symmetric A modes and 26 doubly degenerate E modes. Under species A there are 7 skeletal modes and 18 C—H vibrational modes, and under species E the corresponding numbers are 8 and 18, respectively.

The various C—H stretching, bending and twisting modes which lead to the appearance of peaks in the region 1200–3000 cm^{-1} will not be considered as they are not germane to the present discussion. The assignment of the peaks noted above as probably associated with the regular helical structure are shown in table 6.1. These peaks due to skeletal vibrations do not appear in the spectrum of the atactic structure because the loss of regularity causes them to be smeared out over a range of frequencies.

Syndiotactic polypropylene is capable of existing in two crystalline forms, which have appreciably different spectra, as shown in figs. 6.3*b* and 6.3*c*. In the first form the molecule takes up a 2_1 helical conformation, shown in fig. 6.4, in which the physical repeat unit consists of two chemical repeat units, with a TGGTTG′G′T conformation (T, trans; G, gauche), and in the second it takes up a simpler 2_1 helical conformation in which each physical repeat unit is a single chemical repeat unit and all backbone bonds are trans, i.e. a planar zig-zag structure. For the former, which has a factor group isomorphous with D_2, 26 optically active modes are predicted for each of the A_1, B_1, B_2 and B_3 species. The A modes are Raman active only but

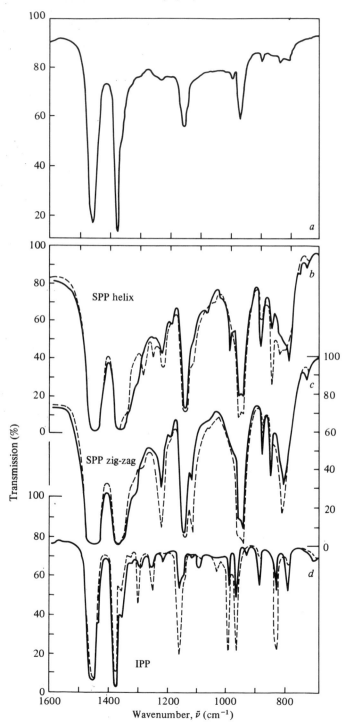

Table 6.1. *Assignments of some modes for isotactic polypropylene*

Observed IR frequency (cm^{-1})	Assignment	Main group vibrations involved
805σ, m	E	v(CC)chain, γ_r(CH$_2$), v(C—CH$_3$)
840π, s	A	v(CC)chain, γ_r(CH$_2$), v(C—CH$_3$)
898σ, w	E	γ_r(CH$_2$), δ(CH), γ_r(CH$_3$)
972π, s	A	v(CC)chain, γ_r(CH$_3$)
995π, s	A	γ_r(CH$_3$), δ(CH)
1100σ, w	E	v(C—CH$_3$), γ_r(CH$_3$), δ(CH)

Notation as explained in section 5.1.

the various B modes are both infrared and Raman active. The B_1 and B_2 modes show perpendicular dichroism whereas the B_3 modes give parallel dichroism (the Ox axis being chosen parallel to the chain axis). In the spectral region of particular interest when considering configurational order there are peaks at 776, 812, 867, 901, 906, 935, 976, 1006, 1035, 1060, 1083, 1088, 1152 and 1167 cm^{-1}, most of which differ in position from the peaks in the spectrum of the isotactic form. On the basis of normal coordinate calculations all of them may be identified as modes involving mainly mixtures of two or more of the following vibrations: C—C chain stretching, C—CH$_3$ stretching, CH$_2$ rocking, C—C—C bending, C—H bending and CH$_3$ rocking.

The spectrum of the planar zig-zag form of syndiotactic polypropylene is simpler in this spectral region, in line with the results of factor group analysis for a group isomorphous with C_{2v}. This analysis predicts 14 vibrations of A_1 species (infrared and Raman active), 11 of A_2 species (Raman active only), 11 of B_1 and 14 of B_2 species, all of which are both infrared and Raman active. The observed peaks at 828, 831, 867, 962, 972, 1095, 1130 and 1154 cm^{-1} may be assigned with reasonable confidence on the basis of normal coordinate calculations and, as with their counterparts for the more complicated helix, arise largely from various combinations of the vibrations noted in that context.

Fig. 6.3. The infrared spectra of different forms of polypropylene: (*a*) atactic; (*b*) syndiotactic, *TGGTTG'G'T* helix; (*c*) syndiotactic, planar zig-zag; (*d*) isotactic. In the polarized spectra (*b*), (*c*) and (*d*), —— indicates the perpendicular and – – – – the parallel spectra. ((*a*) reproduced by permission from D. O. Hummel, 'Applied Infrared Spectroscopy' in *Polymer Spectroscopy*, ed. D. O. Hummel, Verlag Chemie, 1974; (*b*), (*c*) and (*d*) reproduced by permission from H. Tadokoro *et al.*, *Rept. Progr. Polymer Phys. Japan* **9** 181 (1966).)

The considerable differences in the spectra of atactic, isotactic and the two forms of syndiotactic polypropylene in the region 800–1100 cm^{-1} allow the identification of the three forms. The peaks at 840 and 995 cm^{-1} are particularly useful for identifying the helical conformation of isotactic material.

6.2.3   *Conformational isomerism*

In many instances, configurational and conformational isomerism are both present in the same polymer, which complicates the task of assigning the various peaks to the two types of structure.

Fig. 6.4. The non-planar 2_1 helical structure of syndiotactic polypropylene. (Reproduced by permission from J. H. Schachtschneider and R. G. Snyder, *Spectrochimica Acta* **21** 1527 (1965). Copyright (1965), Pergamon Journals Ltd.)

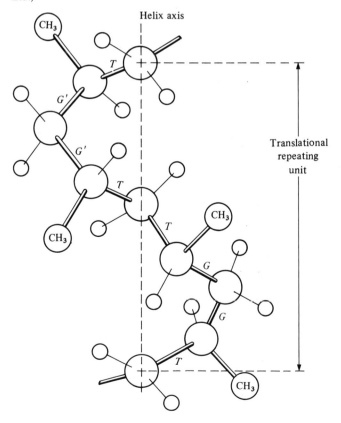

Poly(vinyl chloride), which will be considered in some detail below, provides an example of this. There are, however, polymers that exist in one configurational state only but which exhibit conformational isomerism that is clearly manifest in the vibrational spectrum; poly(ethylene terephthalate), also considered below, provides an excellent example of this type of polymer.

When two or more conformational isomers are present in a liquid the concentrations in which they are present are determined by the Gibbs free energy differences, ΔG, between them and in thermodynamic equilibrium the relative concentrations C_1 and C_2 of any two forms are given by the equation

$$C_2 = C_1 \exp(-\Delta G_{21}/kT) \tag{6.2}$$

where ΔG_{21} is the increase in Gibbs free energy per molecule in going from form 1 to form 2. This relationship applies for the low molecular weight halo-alkanes, which have been extensively investigated and some of which will be referred to below. In crystalline material one conformer may be present exclusively, because of the lowering of the Gibbs free energy by the specific intermolecular interactions which can then occur, and in the non-crystalline regions of solid polymers non-equilibrium concentrations can arise as a result of rapid quenching or of deformation in the solid state. Nevertheless, comparison of the spectra in the molten and solid states, particularly with and without quenching to vary the degree of crystallinity present, usually permits the assignment of peaks to the various conformers present.

A comparison of the infrared spectra of poly(ethylene terephthalate) in the molten form and in solid samples with different degrees of crystallinity shows that there are five pairs of peaks where the intensity of one increases as the other decreases, a clear indication for conformational isomerism. They are at 1470 and 1445 cm^{-1}; 1340 and 1370 cm^{-1}; 1120 and 1110 cm^{-1}; 975 and 1045 cm^{-1}; 850 and 896 cm^{-1}. The first pair are due to the $\diagdown$CH$_2$ bending mode and the second to the $\diagdown$CH$_2$ wagging mode, which suggests that trans/gauche isomerism about the C—C bond of the —CH$_2$—CH$_2$— group in the basic structural unit —CH$_2$CH$_2$O . OC—$\langle$O$\rangle$—CO . O— is the origin of the pairs of modes. X-ray diffraction studies show that in the crystalline state of the polymer the CH$_2$—CH$_2$ bond is in the trans conformation (see fig. 4.6) and it may be deduced that the peaks at 1470, 1340, 1120, 975 and 850 cm^{-1} are characteristic for the trans isomer and the remainder for its gauche counterpart. Further confirmation for these conclusions is provided by the changes in the spectrum of the model compound ethylene dibenzoate, $(\langle\bigcirc\rangle\!\!-\!\!CO . OCH_2)_2$, in passing from solution to the crystalline state. The absorbances in the peaks at 975 and 896 cm^{-1} have been used to follow quantitatively the conformational changes that take place in PET on orientation by drawing. For this purpose it is necessary to combine absorbance measurements for polarization directions parallel and perpendicular to the draw direction, as explained in subsection 6.10.2.

Several regions of the vibrational spectrum of poly(vinyl chloride) (PVC) are sensitive to configurational and conformational isomerism. In particular, the $\diagdown$CH$_2$ deformation mode and the C—Cl stretching and bending modes are each split into several components due to various isomeric forms. The C—Cl stretching region (600–700 cm^{-1}) is particularly complex, but detailed studies involving several approaches have proved very rewarding and the information so obtained is the basis of our current substantial knowledge of the conformational structure of PVC.

The first clear indications of the value of infrared spectroscopy for studying isomerism in PVC came from a study of the spectra of polymers prepared at temperatures between $+ 170°C$ and $- 100°C$. The spectrum in the interval 600–700 cm^{-1} changes markedly. The lower the polymerization temperature the less intense is the peak at 692 cm^{-1} by comparison with those at 615 and 638 cm^{-1}, as shown in fig. 6.5. This suggests that the peak at 692 cm^{-1} is to be associated with defects in the polymer chain. It also became evident from the spectra of polymers prepared at the lowest temperatures that there were two close peaks, at about 605 and 615 cm^{-1}. The first of these, but not the second, is present very strongly in the spectrum of the polymer prepared from the monomer in the form of a urea clathrate. X-ray diffraction studies show that this polymer is highly crystalline and syndiotactic, so that the 605 cm^{-1} vibration may be assigned, with confidence, to $v(CCl)$ for the planar syndiotactic structure, as may that at about 638 cm^{-1}, because it, too, appears strongly in the spectrum of the urea clathrate polymer.

The complete interpretation of the overlapping peaks in the interval 600–700 cm^{-1} resulted from a combination of approaches, already familiar from the detailed assignment studies for a number of polymers set out in chapter 5. They comprise studies on simple model chloroparaffins, on various deuterated PVC specimens, normal coordinate calculations and the use of curve fitting to resolve overlapping peaks (see section 1.7). Of these approaches, the examination of the spectra of a range of secondary alkyl chlorides played a very significant role.

The letters P, S and T are used to denote primary, secondary and tertiary chlorides, respectively; in the present context, only secondary chlorides are of interest, i.e. chlorides in which the carbon atom bearing the chlorine atom is itself attached to two other carbon atoms. The local isomeric structure is specified by reference to a chlorine atom. The subscript H denotes that this atom is trans to a hydrogen atom and C that it is trans to a carbon atom. The simplest secondary chloride, 2-chloropropane, $CH_3 . CHCl . CH_3$, can exist in only one conformation, corresponding to a structure S_{HH}. The higher 2-chlorohomologues can, however, exhibit three isomeric structures, S_{HH}, S'_{HH} and S_{CH}, as shown in fig. 6.6. In the S_{HH} structure the four carbon atoms are coplanar, while in S'_{HH} they are not. Isomers of higher energy can be postulated but they are unlikely to occur in practice. The study of the spectra of secondary chlorides led to the establishment of the following frequency ranges for chlorine atoms in different environments: S_{HH}, 608–615 cm^{-1}; S'_{HH}, 627–637 cm^{-1}; S_{CH}, 655–674 cm^{-1}. As already indicated, high energy conformers may be present in solid PVC, so that its spectrum may be more complex than suggested by the study of the model compounds.

In the earlier work on the polymer the presence of some isomeric structures was inferred from changes in the shapes of peaks with changes of polymerization temperature. More recently, the use of curve fitting

Fig. 6.5. The C—Cl stretching regions of the infrared spectra of two poly(vinyl chloride) samples polymerized at (a) + 50°C and (b) −78°C. (Reproduced by permission from A. R. Berens *et al.*, *Chemistry and Industry* 1959, p. 433.)

Wavenumber, $\tilde{\nu}$ (cm^{-1})

700 650 600   700 650 600

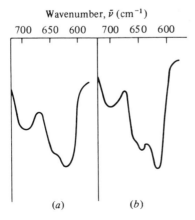

(a)   (b)

has provided a sounder approach to the analysis of the spectrum and to confirming the existence of peaks of low intensity almost hidden by stronger peaks. In particular, the presence of a peak at 647 cm^{-1} due to a non-planar syndiotactic conformer, S'_{HH}, is now established. The commercial types of polymer, of rather low syndiotacticity, have $v(CCl)$ profiles in the Raman spectrum that may be fitted in terms of nine peaks (see fig. 6.7), for which the most likely assignments are given in table 6.2, where both types of notation for defining the isomeric structures are given.

The process of fitting the spectrum in the region 600–700 cm^{-1} provides values for the areas of the component peaks and these are related to the concentrations of the respective configurational/ conformational isomers via the appropriate Raman scattering efficiencies. Unfortunately, these are not known. The areas under the component peaks may, however, be used successfully where comparative information is required. Examples include changes that occur on drawing and in the measurement of orientation by Raman spectroscopy, a topic discussed in section 6.10.

Fig. 6.6. Conformational isomers of 2-chlorobutane. Upper figures, 'ball and stick' models; lower figures, corresponding Newman projections. Full lines represent bonds to the nearer carbon atom of the two connected by the central C—C bond when the corresponding model is viewed from the right-hand end. Note that S_{HH} corresponds to the planar conformation and S'_{HH} to a non-planar or 'bent' conformation. (Upper part of the figure reproduced by permission from J. J. Shipman, V. L. Folt and S. Krimm, *Spectrochimica Acta* **18** 1603 (1962). Copyright (1962), Pergamon Journals Ltd.)

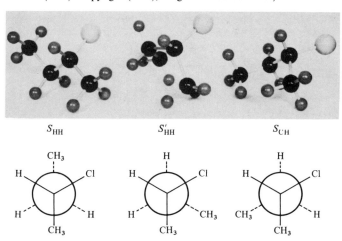

S_{HH} S'_{HH} S_{CH}

Fig. 6.7. The C—Cl stretching region of the Raman spectrum of a sample of commercial poly(vinyl chloride), fitted with nine Lorentzian peaks and a linear background. The peaks assigned to the crystalline A_g and B_{3g} species modes are shown as a double peak. The spectrum shown is actually the fitted spectrum; the difference between it and the original spectrum is shown at the top of the figure. (Reproduced from D. I. Bower *et al.*, *Journal of Macromolecular Science Physics B* **20** 305 (1981), by courtesy of Marcel Dekker, Inc., NY.)

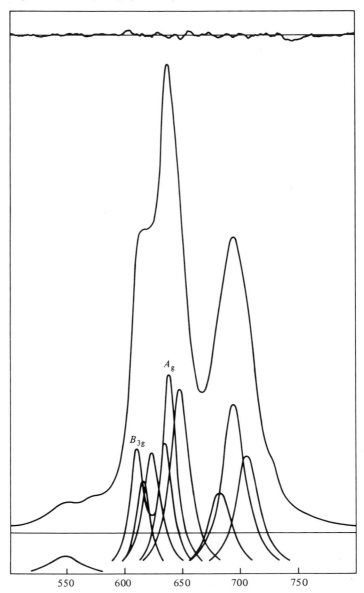

Table 6.2. *Assignment of $\nu(CCl)$ modes in the Raman spectrum of poly(vinyl chloride)*

Frequency (cm^{-1})	Configuration[a]	Type of Cl	Conformation[b]	Description of structure
608	s	S_{HH}	T | T | T | T | T	planar zig-zag, crystal
614	s	S_{HH}	T | T | T | T | T	short planar zig-zag
623	i	S_{HH}	T | G | T | T | G* | T	non-planar, high energy
634	i	S'_{HH}	T | G | T | G* | T | G	non-planar, high energy
638	s	S_{HH}	T | T | T | T | T	planar zig-zag, crystal

647	s	S'_{HH}		non-planar, low energy
680	s	S_{HC}		non-planar
692	i	S_{HC}		3-fold helix conformation
704	s	S_{HC}		non-planar

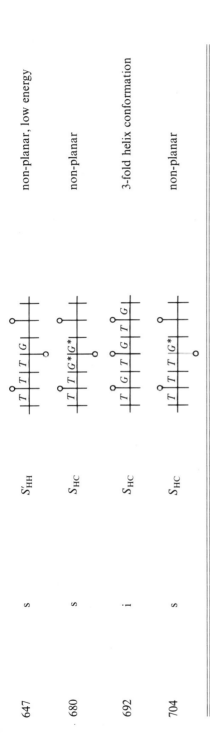

[a] Local configuration at CCl group considered: s, syndiotactic; i, isotactic.

[b] In the diagrams illustrating the structures the structures the planar zig-zag chain is imagined to be viewed so that the Cl atoms (indicated ○) are on the side of the chain nearest the eye for those structures where it is necessary to make this distinction. T stands for a trans bond, G for a right-handed gauche bond (i.e. the eighth structure represents a right-handed helix) and G^* stands for a left-handed gauche bond. (Mirror image structures would, of course, have the same frequencies.)

Adapted from Robinson et al., Polymer **19** 773 (1978).

6.3   Chain branching

6.3.1   *Introduction*

The measurement of the branching levels in polymers is of considerable importance, as various physical properties may be affected directly or indirectly by the branches present. In polyethylene, for example, the branching level strongly affects crystallinity, which may vary from about 50 % in a branched low density polyethylene to about 85 % in a linear high density polyethylene. Butyl branches are sometimes introduced deliberately, by the addition of a low level of 1-hexene, $CH_2{=}CH(CH_2)_3CH_3$, into the ethylene feed, to improve the stress crack resistance. The assessment of the types and levels of the branching present in poly(vinyl chloride) has recently attracted considerable interest for a different reason, namely that branches are one of several structural defects present in this polymer that may provide initiation sites for the thermal degradation which occurs at a significant rate at temperatures in excess of about 150°C. In the following subsections the determination of branching content by vibrational spectroscopy will be considered.

6.3.2   *Polyethylene*

If branches are present in polyethylene they will consist either of methyl groups or of short methylene sequences, each terminated by a methyl group. They should thus be detectable in a system which consists otherwise of methylene groups, apart from a very low concentration of chain terminating groups, whose nature will be discussed in section 6.5. The presence of such branches was indeed noted during the early infrared studies on the polymer.

There are basically three kinds of polyethylene produced commercially. The first to be produced, low density polyethylene, is made by a high pressure, high temperature uncatalyzed reaction involving free radicals and has about 20–30 branches per thousand carbon atoms. A variety of branches may occur, including ethyl, $-CH_2CH_3$, butyl, $-(CH_2)_3CH_3$, pentyl, $-(CH_2)_4CH_3$, hexyl, $-(CH_2)_5CH_3$, and longer. High density polymers are made by the homopolymerization of ethylene or the copolymerization of ethylene with a small amount of higher α-olefin. Two processes, the Phillips process and the Ziegler–Natta process, which differ according to the catalyst used, are of particular importance. The emergence of a new generation of catalysts has led to the appearance of linear low density polyethylenes. These have a higher level of co-monomer incorporation and have a higher level of branching, up to that of low density material, but the branches in any given polymer are of one type only, which may be ethyl, butyl, isobutyl or hexyl.

Evidence for the occurrence of methyl branches is manifest in several regions of the vibrational spectrum and the methyl deformation region proves to be the most sensitive, despite the fact that the characteristic peak appearing at 1378 cm^{-1} in the infrared spectrum is considerably overlapped by a pair of stronger peaks at 1369 and 1353 cm^{-1}, which are due to CH_2 wagging modes of sequences of methylene groups in crystalline and amorphous regions, respectively. It is possible to obtain a semiquantitative measure of the branching by taking the ratio of the absorbances at 1378 and 1369 cm^{-1}. This approach has the advantage that it avoids the need for measuring the thickness of the film used, but when the level of the branching is low, e.g. up to about five branches per thousand carbon atoms, as in most high density polyethylene specimens, the methyl absorption appears as no more than an indistinct shoulder. The use of difference spectroscopy to separate the peaks provides an excellent example of what may be achieved by this approach, as shown in fig. 6.8.

The absorption spectrum for methylene units only may be obtained from measurements on specially prepared polymethylene or a substantially linear polyethylene. Rigidex 9 is typical of several commercial materials of this type. The methylene absorption is then subtracted from the complex peak, fig. 6.8*a*, to give the required methyl absorption peak, fig. 6.8*b*. Until recently, this was done by placing a wedge specimen of polymethylene or linear polyethylene, with an appropriate thickness range, in the reference beam of the spectrometer and adjusting it manually until there was complete cancellation of the methylene absorption in the sample beam. This method is being superseded by the computer subtraction of a suitable stored spectrum. It is not difficult to judge when the subtraction conditions are approximately correct, as the 1460 cm^{-1} methylene peak is then of negligible intensity. Detailed studies have shown that the best criterion for exact subtraction is equality of the absorbance values at 1364 and 1401 cm^{-1}. A precision of 5% in the determination of the absorbance at 1378 cm^{-1} is readily achievable with a medium performance infrared spectrometer.

The calculation of the branch content from the absorbance at 1378 cm^{-1} requires an appropriate calibration factor and there are two approaches to this problem. The first is to use low molecular weight model compounds; for example, 19,28-dimethylhexatetracontane, $CH_3(CH_2)_{17}CHCH_3(CH_2)_8CHCH_3(CH_2)_{17}CH_3$, dissolved in known concentration in a normal paraffin of comparable molecular weight provides a realistic approximation to the molecular environment of the methyl branches in high density polyethylene. Alternatively, samples whose compositions have previously been established by ^{13}C NMR spectroscopy may be used as secondary calibration standards.

Studies on model compounds and on copolymers of ethylene and 1-alkenes, $CH_2=CH(CH_2)_nCH_3$, show that the molar absorption coefficients at 1378 cm^{-1} depend on the branch length. When the former is plotted against the reciprocal of the number of carbon atoms present in the branch a straight line results, as shown in fig. 6.9. It is evident that

Fig. 6.8. Subtraction of the infrared spectrum of polymethylene from the spectrum (*a*) of a branched polyethylene in the region 6–8 μm (1667–1250 cm^{-1}) to give the methyl absorption (*b*) at 7.26 μm (1378 cm^{-1}). (Reproduced by permission from *Identification and Analysis of Plastics* by J. Haslam, H. A. Willis and D. C. M. Squirrel.)

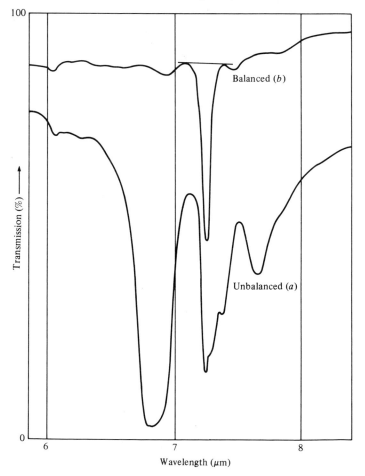

the absorption coefficient for a butyl branch is smaller by a factor of almost two than that for a methyl branch. Hence, unless there is prior information on the nature of the branches present, the infrared method may yield results that are seriously in error. Its value for the characterization of branching in polyethylenes of unknown origin is therefore somewhat limited. Likewise, as several types of branches are present in low density polymers made by the high pressure process the method is not ideal for the examination of such materials.

This drawback has prompted studies aimed at finding peaks elsewhere in the spectrum that are characteristic for specific branches. These have achieved a limited success. It has been established that methyl branches give a weak but characteristic peak at 935 cm^{-1}, which is probably due to the methyl rocking mode, whereas propyl and longer branches give

Fig. 6.9. Absorption coefficient, k, at 1378 cm^{-1} for methyl groups plotted against the reciprocal of n, the number of carbon atoms in the branch. ● methyl branches; ○ ethyl branches; □ butyl branches; (d) $C_{36}H_{74}$. N is the number of methyl groups per 1000 carbon atoms. (Reproduced by permission from C. Baker, W. F. Maddams, G. S. Park and B. Robertson, *Makromol. Chemie* **165** 321 (1973), Fig. 1, Hüthig & Wepf Verlag, Basel.)

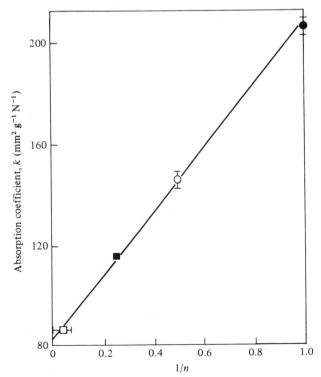

a peak at 890 cm^{-1}. The $\diagup$CH$_2$ rocking mode of ethyl branches is responsible for the weak peak that appears at 770 cm^{-1}.

The measured branching levels in near-linear polymers are seriously in error if a correction is not applied for chain-terminating methyl groups. The number average molecular weight of a typical high density polymer is such that if both ends of the chain are terminated by methyl groups, their concentration is about two per thousand carbon atoms. Furthermore, as reference to fig. 6.9 shows, the molar absorption coefficient of such chain-terminating groups is smaller than that of a methyl branch by the factor 2.2, and this must be taken into account when the correction is made. The correction for methyl end groups is relatively simple for a polymer made by the Phillips process, as one end of each chain terminates in a methyl group and the other in a vinyl group. The concentration of the vinyl groups may be estimated by infrared spectroscopy, as explained in subsection 6.5.1.

Although ^{13}C NMR spectroscopy is far more specific for the various types of branches that may be present in the range of commercially available polyethylenes, the wide availability of facilities for infrared spectroscopy and the fact that it is less time consuming go some way towards countering its relatively poor specificity. At low branching levels it has superior sensitivity. It may also be applied indirectly to poly(vinyl chloride).

6.3.3   *Poly(vinyl chloride)*

The examination of a range of PVC samples prepared in different ways, and expected to show differences in their branching levels, reveals no significant differences in their infrared spectra. If, however, these materials are chemically reduced to polyethylene the infrared method discussed above may be applied. It has been common practice recently to assume that the branches are methyl groups, an assumption that has been substantiated by ^{13}C NMR studies on the reduced samples. In the case of commercial polymers, the level is about four per thousand carbon atoms and it decreases somewhat, in line with expectations, as the polymerization temperature is lowered. These branches are, of course, present as chloromethyl groups, —CH$_2$Cl, in the original poly(vinyl chloride).

6.4   **Head-to-head placements in vinyl polymers**

The presence of head-to-head (h–h) or tail-to-tail (t–t) units in a vinyl polymer can cause a reduction in the degree of crystallinity. For

PVC it causes a lowering of thermal stability and it is therefore useful to be able to estimate the concentration of such placements. The possible occurrence of t–t units, $+CH.CH_3—CH_3)(CH_2—CH.CH_3)+$, in polypropylene and their detection by virtue of a characteristic peak at $752 \, cm^{-1}$ has already been noted in subsection 6.1.3. In this particular instance it proved possible to predict, on the basis of simple correlations among the spectra of low molecular weight model compounds, what peaks would prove specific for the $—CH_2—CH_2—$ group present in the t–t unit. In general this is difficult because suitable model compounds are not available and indirect approaches must be used to identify peaks characteristic for t–t or h–h units, particularly as the concentration of such units is usually rather low and the corresponding absorption peaks are therefore comparatively weak.

It is normally possible to predict with some certainty the resonance frequencies of h–h and t–t units in ^1H and ^{13}C NMR spectra. Hence, if two or more samples are available which can be shown by this method to have different h–h (and t–t) contents it may be possible to find peaks in the vibrational spectrum whose intensities parallel those in the NMR spectra. Alternatively, it may be possible to prepare the 'pure h h' polymer, which is more precisely described as an alternating h–h, t–t polymer, by an indirect route and thereby obtain a reference spectrum. This is then used to identify any weak h–h or t–t absorptions present in the spectrum of a normal head-to-tail (h–t) polymer containing a low concentration of h–h and t–t units. This is often the more practical of the two approaches, particularly as the infrared spectra of several pure h–h polymers, prepared by indirect routes, are available in the literature, for example those of h–h poly(vinyl chloride) and of h–h polystyrene.

The infrared spectrum of h–h poly(vinyl chloride) differs somewhat from that of the conventional h–t polymer. In particular, it shows a broad peak at about $790 \, cm^{-1}$ that is probably the corresponding $(CH_2)_2$ rocking mode to that observed at $752 \, cm^{-1}$ with t–t polypropylene. The conventional polymer shows only background absorption in the vicinity of $790 \, cm^{-1}$ and it is clear that the concentration of t–t units cannot exceed 1–2 %. In fact, the absence of a resonance specific for this type of unit in the ^1H NMR spectrum sets an upper limit of one per thousand carbon atoms.

6.5   End groups

The growth of a chain molecule during polymerization is terminated by one of a number of factors which depend on the polymerization conditions. It would be out of place in the present text to embark upon a detailed discussion of these processes; it suffices to note that they lead to a variety of chain-terminating chemical groups. These often differ significantly from the repeat unit of the polymer chain, facilitating their detection by vibrational spectroscopy, even at the rather low levels at which they occur in the many polymers whose molecular weights are measured in units of tens of thousands. One obvious application of the ability to measure the concentration of these end groups is the determination of number average molecular weight, which has an important effect on many properties of the polymer. The characterization of the end groups, both qualitatively and quantitatively, may also be important for those polymers, e.g. polyoxymethylene, where they provide the initiation sites for thermal degradation. Several examples of the determination of end group concentrations by means of vibrational spectroscopy will now be considered.

6.5.1   *Polyethylene*

Polyethylene provides a particularly simple and striking example of the ability of both infrared and Raman spectroscopy to characterize end groups. The various types of polyethylene that will be referred to have already been described briefly in subsection 6.3.2.

The infrared spectra of high density polymers made by the Phillips process show weak but sharp peaks at 910 and 990 cm^{-1} (see fig. 6.10). These are due to the out-of-plane deformation vibrations of the hydrogen atoms of the vinyl groups, $-CH=CH_2$, that terminate one end of each chain, the other terminating group being a methyl group, as noted in subsection 6.3.2. These peaks fall in a region of the polyethylene spectrum that is virtually free from other absorptions and they are therefore detectable at the low levels at which they occur with polymers whose molecular weights are 10^5 or rather greater. It is possible to convert the corresponding absorbances into vinyl group concentrations by the use of a suitable calibration compound, such as 1-pentadecene, $CH_2=CH(CH_2)_{12}CH_3$.

Low density polymers made by the high pressure polymerization route give a peak at 890 cm^{-1}. This is the $=CH_2$ out-of-plane deformation mode of the chain-terminating vinylidene group, $-CR=CH_2$, where R is usually ethyl or butyl. The use of this peak for quantitative purposes is complicated by the fact that it is virtually coincident with one arising from the methyl rocking mode of the butyl and longer branches present in this type of polymer. If, however, the polymer films used for the infrared measurements are exposed to bromine vapour for a few minutes, addition of bromine occurs across almost all the double bonds. A difference spectrum between the brominated and unbrominated films then gives the required vinylidene absorption at 890 cm^{-1} free from interference. The high pressure, low density polymers also show weakly the vinyl absorptions at 910 and 990 cm^{-1} and a weak peak at 965 cm^{-1}, characteristic for trans —CH=CH— groups within the chain ('trans-in-chain' groups). They

Fig. 6.10. The infrared transmittance spectra of different types of polyethylene in the region 800–1100 cm^{-1}: (a) Phillips polyethylene; (b) low density polyethylene; (c) Ziegler–Natta polyethylene. All samples were hot pressed film of thickness 500 μm. (Adapted by permission from *Identification and Analysis of Plastics* by J. Haslam, H. A. Willis and D. C. M. Squirrel.)

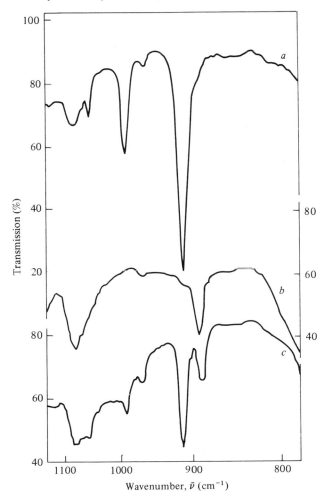

are therefore very heterogeneous with respect to unsaturation, as well as branching, and this diversity of structural defects is not unexpected in view of the various reactions, such as back-biting (see subsection 1.2.3), that occur under the rather extreme conditions used for the polymerization.

The third major type of polyethylene, high density polymer made by the Ziegler–Natta route, tends to have rather low levels of unsaturation, in which vinyl, vinylidene and 'trans-in-chain' groups may all occur. It is therefore clear that infrared spectroscopy provides a ready means for distinguishing between the three types, on the basis of the peaks appearing in the region 890–1000 cm^{-1}. These peaks are weak in the Raman spectrum but the C$=$C stretching mode is useful for identification purposes, particularly as the technique provides the means for examining samples without prior preparation. It is also valuable for the examination of very small samples, such as inclusions or inhomogeneities in polyethylene specimens. Frequently, these occur because some high molecular weight polymer that is present is incompletely dispersed. Such higher molecular weight material is readily identifiable in Phillips type polymers because v(C$=$C) for the vinyl end groups, at 1642 cm^{-1}, is of lower intensity than it is for the remainder of the polymer specimen. The corresponding stretching frequencies for the vinylidene and 'trans-in-chain' groups are 1655 and 1672 cm^{-1}, respectively.

The changes that occur upon the irradiation of polyethylene with X-rays or γ-rays involve the double bonds that are present at chain ends. In the case of the Phillips type polymers the concentration of vinyl groups decreases with increasing radiation dose. As the 910 and 990 cm^{-1} absorptions weaken, the one at 965 cm^{-1}, indicative of 'trans-in-chain' unsaturation, intensifies. Similarly, with the high pressure, low density type of polymer, the 890 cm^{-1} vinylidene absorption peak gives way to the peak at 965 cm^{-1}. These changes are associated with the formation of cross-links, as evidenced by a decrease in polymer solubility, but there are no peaks in the spectrum that are characteristic for cross-links.

6.5.2   *Vinyl polymers*

Vinylidene end groups are also present in polypropylene but are difficult to estimate directly because the absorption at 890 cm^{-1} is overlapped by the stronger polypropylene absorption at 900 cm^{-1}. The method used to overcome this problem is identical to that employed for the estimation of vinylidene groups in high pressure, low density polyethylene. A sample is brominated, to remove the double bonds, and a difference spectrum is then obtained from its spectrum and that of the starting material. Although a similar approach could, in principle, be employed for poly(vinyl chloride), it has not been pursued.

The residual unsaturation in this polymer has, however, been characterized by studies on the weak C=C stretching peak that appears at $1667\,cm^{-1}$. This work is of interest more as a demonstration of the capabilities of a Fourier transform spectrometer, both in terms of sensitivity and data processing capability, than as a practical method. This is because the absorbance of the peak at $1667\,cm^{-1}$ cannot be converted into a double-bond concentration in the absence of a calibration factor, and a series of samples whose double-bond contents have been established by a chemical method must be used as secondary standards, whereas $^1H\,NMR$ spectroscopy provides a more specific and easier method. The value of $1667\,cm^{-1}$ is at the upper end of the expected range for $\nu(C=C)$ but its assignment to chain-terminating unsaturation is demonstrated by the appearance of a peak at comparable frequencies in model compounds and by the fact that the one at $1667\,cm^{-1}$ intensifies in low molecular weight extracts. Firm proof for the precise nature of the unsaturation comes from $^1H\,NMR$ spectroscopy, which shows that the chain-terminating allylic chlorine group, $-CH=CH-CH_2Cl$, is involved. The peak at $1667\,cm^{-1}$ is very weak, both because it is due to a chain-terminating group and because $\nu(C=C)$ is rather weak in the infrared spectrum. It is therefore necessary to use thick specimens, which give substantial background absorption in the vicinity of $1667\,cm^{-1}$. This may be subtracted by commonly available spectrometer data handling routines, leaving the required unsaturation peak.

6.5.3 *Other polymers*

It is readily predictable on chemical grounds that the ends of polyoxymethylene chains will be terminated with hydroxyl groups, and these are readily detectable in the infrared spectrum by virtue of the $\nu(OH)$ mode at about $3300\,cm^{-1}$. The presence of these terminal hydroxyl groups leads to thermal instability, which makes processing difficult. This has led to the introduction of chemical modifications in which, for example, the terminal hydroxyl groups are replaced by acetate groups. These are readily identified by the $\nu(C=O)$ peak at about $1730\,cm^{-1}$. It is possible, in principle, to use the intensity of this peak to determine number average molecular weights but it is difficult to establish a precise value for the molar absorptivity of the acetate end group.

The chain-terminating group of epoxy resins is the oxirane ring, $\overline{O-CH_2}-\overset{}{C}H-$, which gives a characteristic peak at $910\,cm^{-1}$. This is useful for following the curing of such resins, as the chain-terminating epoxy group is involved in the reactions. Quantitative information may be obtained by using the ratio of the absorbances at 910 and 1610 cm^{-1}, the latter being a measure of the invariant aromatic content of the resin.

6.6   Hydrogen bonding

6.6.1   *Introduction*

So far, in the present chapter, the vibrational spectrum of a polymer has been considered in terms of the chemical groups and order present in an individual chain. It has been tacitly assumed that the spectrum of an ensemble of such chains does not differ from that of an isolated chain. It was pointed out in chapter 3, however, that interactions between chains in crystals can give rise to correlation splitting, and this will be discussed further in section 6.7. There are other types of intermolecular interactions that influence the vibrational spectrum and hydrogen bonding is by far the most important of these. Its occurrence is not restricted to crystalline systems, but it is often important in such systems.

There is a strong tendency for hydrogen bonds to form if a hydrogen atom attached to an electronegative atom, usually O or N, occupies a position between the atom to which it is attached and an adjacent atom, such as O or N, which also carries a local concentration of negative charge. The bond formed between this atom and the hydrogen atom is electrostatic in nature and weaker than a covalent bond. It is usually indicated by a broken line. Amides provide an excellent example, the hydrogen bond forming between the hydrogen atom of one ⟩NH group and the oxygen atom of a neighbouring carbonyl group (see fig. 6.11). Most hydrogen bonding in synthetic polymers is of the intermolecular type, where the hydrogen atom and the acceptor group are on different molecules. Occasionally, hydrogen bonds are able to form because of the close proximity of a hydrogen atom and an acceptor group in the same molecule. This is known as intramolecular bonding and it can survive the separation of the molecules spatially, as in a dilute solution, whereas intermolecular bonding will be destroyed. Hydrogen bonding can modify the properties of polymers appreciably. For example, the bonds act as cross-links, conferring a degree of rigidity on the system and increasing the softening temperature. The insolubility of cellulose in water is a consequence of the very strong hydrogen bonding between chains, a result of the presence of many hydroxyl groups.

The changes in the spectra of compounds that are hydrogen bonded are well understood, from the very large number of studies that have been undertaken on simple compounds during the past 30 years, and the most important effect is a lowering of the stretching frequency of the hydrogen bonded OH or NH group with respect to that of the 'free' group. For instance, when a simple alcohol, such as ethanol, is examined at a concentration of a few tenths of one per cent in solution in a substantially inert solvent such as cyclohexane, a sharp peak appears at 3635 cm^{-1}. This is the so-called 'monomeric' or free hydroxyl vibration, i.e. v(OH) for hydroxyl groups that are not hydrogen bonded to their neighbours. When the concentration of ethanol rises towards 1% a second peak appears at 3540 cm^{-1}. This is v(OH) for dimeric species, systems where the hydroxyl groups of two molecules hydrogen bond to each other. At concentrations of 1% and greater there is multiple hydrogen bonding and a broader peak appears at about 3350 cm^{-1}. For polymers in the solid state, rather than in dilute solution, it is reasonable to anticipate that in all systems containing OH or NH groups there will be substantial hydrogen bonding, so that differences may occur in the v(OH) regions of the spectra of groups of closely related materials, e.g. the various Nylons, and changes may occur in the v(OH) region of the spectrum of a given polymer as the temperature is changed. Examples of these effects will now be considered.

6.6.2 *Polyamides*

For the ⟩NH group, most frequently encountered in polymers in the form of the secondary amide group, the free v(NH) mode occurs at approximately 3450 cm^{-1}, whereas the frequency for the hydrogen-

Fig. 6.11. Hydrogen bonding in amides (schematic). The dotted line represents a hydrogen bond. $\delta-$ and $\delta+$ indicate fractional charges.

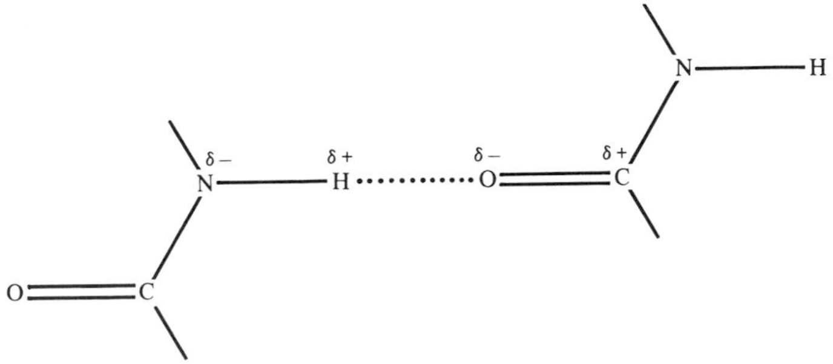

bonded group drops to about $3300 \, \text{cm}^{-1}$. The amide I peak (see subsection 5.3.5) is somewhat affected by hydrogen bonding, dropping by $30\text{–}40 \, \text{cm}^{-1}$ from the value found in dilute solutions, whereas for the amide II peak there is, on average, a rise of $20 \, \text{cm}^{-1}$ in passing from the free to the bonded form. Hence, $\nu(\text{NH})$ is of considerably greater diagnostic value than the amide I and II vibrations for hydrogen-bonding studies on amides.

At ambient temperature the spectrum of a typical polyamide, such as Nylon 6,6, shows no evidence for unbonded $\diagup$NH. As the temperature is raised $\nu(\text{NH})$ gradually broadens, a result of the general weakening of the hydrogen bonds and a concomitant broadening of the distribution of their force constants. Despite this broadening it is possible to detect a definite shoulder at about $3450 \, \text{cm}^{-1}$, indicative of free NH, at a temperature of about 200°C and at 250°C it increases in intensity significantly, indicating an appreciable concentration of NH groups of this type. At ambient temperature, however, the $3450 \, \text{cm}^{-1}$ peak is not detectable and the concentration of free NH groups cannot exceed 1%. This is also true for various other Nylons, e.g. Nylon 10.

6.6.3   Polyurethanes

Polyurethanes, as explained in subsection 1.2.2, are condensation products formed by the reaction between a di-isocyanate and a diol, and thermoplastic polyurethanes provide a particularly interesting and important example of the value of hydrogen-bonding studies by infrared spectroscopy. The most important of these materials are linear, block copolymers, consisting of alternate hard and soft blocks. They are described as segmented block copolymers because the blocks are relatively short and numerous compared with those in some other types of copolymer chain and are usefully thought of as chain segments. The hard segments contain the polyurethane linkages, whereas the soft segments are often low molecular weight polyethers or polyesters. The copolymers have a two phase microstructure in which the hard segments separate from a rubbery matrix formed by the soft segments and aggregate in glassy or partially crystalline domains, so that the materials show elastic properties typical of a composite material. They behave as elastomers at moderate temperatures but become thermoplastic at temperatures above the softening point of the hard segments.

Polyurethanes are extensively hydrogen bonded, the proton donor being the $\diagup$NH group of the urethane linkage (see subsection 1.2.2).

The hydrogen-bond acceptor may be in either the hard segment, the carbonyl of the urethane group, or the soft segment, an ester carbonyl or ether oxygen. The relative amounts of the two types of hydrogen bonds are determined by the degree of microphase separation; clearly, an increase in this separation favours the inter-urethane hydrogen bonds. The distribution of NH bonding with respect to each type of acceptor is dependent on many factors, including the electron-donating ability, relative proportions and spatial arrangement of the various proton-acceptor groups in the polymer chain. Not surprisingly, infrared spectroscopy has been widely used to study this hydrogen bonding. The two examples now to be considered are representative of the information that may be obtained. The first relates to the degree of hydrogen bonding and the nature of the acceptor sites in polyether urethanes, the particular example cited having polytetrahydrofuran, $HO-[(CH_2)_4-O-]_nH$, for the soft segment diol and p,p'-diphenyl-methane di-isocyanate (MDI), $OCN-\langle\bigcirc\rangle-CH_2-\langle\bigcirc\rangle-NCO$, chain-extended with 1,4-butanediol, $HO(CH_2)_4OH$, to give the structure

$$OCN-\langle\bigcirc\rangle-CH_2-\langle\bigcirc\rangle-NH.CO.O(CH_2)_4O.OC.$$
$$NH-\langle\bigcirc\rangle-CH_2-\langle\bigcirc\rangle-NCO$$

for the hard segment di-isocyanate.

The degree of hydrogen bonding of the NH groups was studied by examining the $v(NH)$ region of the spectrum (see fig. 6.12) and looking for the presence of a shoulder in the vicinity of $3450\ cm^{-1}$, indicative of free NH groups. This appears on the much stronger peak due to bonded NH groups at $3320\ cm^{-1}$ and its intensity is difficult to measure with precision, but it cannot amount to more than 15% of the total, so that at least 85% of the NH groups are hydrogen bonded.

The peak arising from the carbonyl stretching mode is split into two components, one at $1733\ cm^{-1}$ and the other at $1703\ cm^{-1}$. Studies on model compounds such as N-phenylurethane, $\langle\bigcirc\rangle-NH.CO.OC_2H_5$, show that the two maxima are assignable to free and bonded carbonyl groups, respectively. In order to determine the degree to which the carbonyl groups are bonded it is necessary to know the relative molar absorptivities of the 1733 and $1703\ cm^{-1}$ modes. Studies on a range of model compounds suggest that there is some increase in molar absorptivity when hydrogen bonding occurs, but generally not greater than 20%. It is therefore useful to consider two extreme possibilities: that there is no change in absorbance or that it increases by a factor of 1.2. As the ratio of the absorbances of bonded and free carbonyl groups is 1.55, it follows that 61% or 56% of the urethane carbonyl groups are involved in hydrogen bonding for the two assumptions, respectively. This then leads to the conclusion that the remaining bonded NH groups, at least 25% of the total, must be involved with a second acceptor group, and this can only be the ether

Fig. 6.12. Free $\diagdown$NH absorption at 2.90 μm (3450 cm^{-1}) for the polyurethane $\text{-}(CH_2)_8OCONH(CH_2)_6NHCOO\text{-}_n$ at (a) 22°C and (b) 100°C. (From D. S. Trifan and J. F. Terenzi, *Journal of Polymer Science* **28** 443 (1958). Copyright © (1958) John Wiley & Sons, Inc. Reprinted by permission of John Wiley & Sons, Inc.)

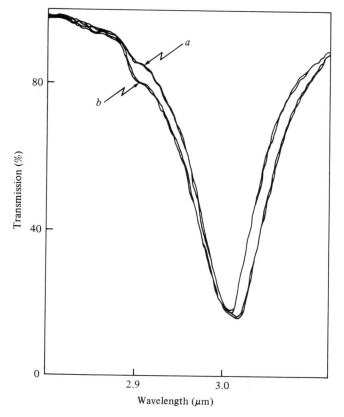

oxygen linkage. The complexity of the peaks in the $\nu(C=O)$ region for polyester urethanes prevents the straightforward use of this approach.

The second example covers a more quantitative interpretation of the infrared spectra and their temperature dependence for two types of polyurethane. For one type, a polyether polyurethane (ET), the soft segment diol is again polytetrahydrofuran, and for the other, a polyester polyurethane (ES), it is poly(tetramethylene adipate),

$$H{+}O(CH_2)_4 \cdot O \cdot CO \cdot (CH_2)_4 \cdot CO{+}_n O(CH_2)_4 OH.$$

For both types the hard block is MDI extended with butanediol, as for the first example.

The $\nu(NH)$ region of the spectrum was computer fitted with three Gaussian peaks, centred at 3420, 3320 and 3185 cm^{-1}. The first of these is due to free NH groups, the second to bonded NH groups and the third, which is rather weak, is attributed to bonded NH groups in the conformation in which the NH group is cis to the $C=O$ group rather than, as is more usual, trans to it. The area of this peak was therefore added to that of the peak at 3320 cm^{-1} in calculating the concentration of bonded NH groups. The ratio of the molar absorptivities at 3320 and 3420 cm^{-1} was assumed to be 3.46, a value based on studies on model compounds. The $\nu(C=O)$ profile for the ET series was fitted with two Gaussian peaks at 1733 and 1703 cm^{-1}, characteristic for free and bonded carbonyl groups, respectively, whose molar absorptivities were taken to be equal. It was not possible to fit the $\nu(C=O)$ profile in the ES series, because of the ester groups present in the soft segments.

The results for the ET series, covering a range of compositions in terms of MDI hard segments, were, with one exception, very uniform, showing $82 \pm 2\%$ of bonded NH with $66 \pm 2\%$ of bonding to the urethane carbonyl group. The amount of bonding to the urethane ether varied somewhat more, with upper and lower bounds of 20% and 12%. The exception was a material in which the soft segment was of appreciably higher molecular weight, and the percentage of hard segments was high. This material, which was semi-crystalline, had 78% of the NH groups bonded, 73% to the carbonyl group and only 5% to the urethane ether. This suggested a greater than average phase separation, in line with other properties of this material. The polymers of the ES series showed a comparable degree of NH bonding, ranging between 75% and 84%.

Studies of the temperature dependence of the degree of hydrogen bonding in the ET series showed, as expected, that it decreased with increasing temperature and the results were used, together with other information, to provide evidence about the influence of hard and soft segment lengths on the perfection and purity of the hard and soft segment phases.

6.6.4   Biopolymers

Hydrogen bonds are important in stabilizing biopolymers in their native conformations. In proteins, for example, many hydrogen bonds are formed between the $\diagdown$NH and C$=$O groups of the backbone peptide groups and these tend to stabilize the backbone into one or other of a variety of helical conformations in which the hydrogen bonds are aligned in a cooperative fashion. Additional hydrogen bonds are formed from N—H, C$=$O and O—H groups of the protein side chains and, in general, all potential hydrogen-bond forming groups are in fact involved in hydrogen bonds either to other groups on the protein or to the water molecules that are normally present in biological systems. Despite the stabilizing effects of hydrogen bonds and other interactions such as van der Waals interactions, the energy difference between the native state of a protein and the unfolded, or 'denatured', state is quite low, commonly of the order of 100 kJ mole^{-1}; in an aqueous environment, denaturation can be promoted at temperatures not very greatly above ambient, the favourable interactions within the protein molecule that are disrupted being replaced by only slightly less favourable interactions with water molecules surrounding the unfolded state. In fact, the stability of proteins is determined not so much by the enthalpic interactions as by the higher entropy of the bulk water when the protein is completely folded. The phenomenon of denaturation causes changes in the infrared spectrum because of the greater strength of the cooperative arrangement of hydrogen bonds in the native protein.

A particular and widely studied example of denaturation is that of the fibrous protein collagen, the material of tendon. This molecule consists of three helical chains super-coiled around each other in the form of a rope, stabilized by van der Waals interactions and hydrogen bonds between the three chains (see fig. 6.13). Denaturation, in which the fibre collapses into a gelatinous state, involves a 'melting' of the triple helical structure with a concomitant change in $v(NH)$ from 3330 to 3300 cm^{-1}

Fig. 6.13. Schematic diagram of part of a collagen molecule, showing the regular pattern of hydrogen bonds that link the three helical chains of the molecule so that they super-coil round each other. (From J. R. Holum, *Principles of Physical, Organic and Biological Chemistry*. Copyright © (1969) John Wiley & Sons, Inc. Reprinted by permission of John Wiley & Sons, Inc. After an original drawing by A. Rich.)

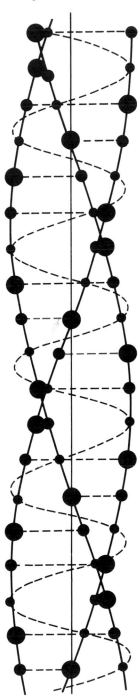

often
hydroxyproline

often
proline

glycine

as the regular pattern of inter-chain hydrogen bonds is replaced by an irregular pattern of hydrogen bonds to neighbouring water molecules.

Raman spectroscopy is often the favoured approach with biopolymers because of the ease of sampling and because of the generally high water content. This technique has been used to study the 'melting' of DNA, the genetic material, whose double helical conformation is stabilized by hydrogen bonds that provide specific interactions between complementary planar nitrogenous bases attached to the two sugar-phosphate backbone chains. In a synthetic analogue formed by one chain of polyribo-adenylic acid and one chain of polyribo-uridylic acid the Raman spectrum shows a clear peak at $1681\ cm^{-1}$ at low temperatures. This is one of the $v(C{=}O)$ vibrations from the uracil base when it is hydrogen bonded to adenine. At higher temperatures, the double helix 'melts' as the hydrogen bonds between pairs of bases are disrupted and at 59°C the $1681\ cm^{-1}$ peak is replaced by two at 1660 and $1698\ cm^{-1}$ characteristic of the free uracil moiety.

6.7   Chain order and crystallinity

6.7.1   *Introduction*

The way in which the vibrational modes of polymers are modified when the chains are present in crystalline regions has been discussed in chapter 3, primarily from the fundamental angle. The consequences will now be considered in more detail, by reference to four polymers, polyethylene, polyoxymethylene, poly(vinyl chloride) and polytetrafluoroethylene, as will the practical implications. These include the determination of crystallinity, the testing of postulated intermolecular force fields, the provision of evidence for a third phase, neither crystalline nor amorphous, in polyethylenes and the provision of evidence for or against particular crystal structures. In addition, it is now known that certain vibrations that were previously assumed to be specific for chains in crystalline regions are actually characteristic for no more than one-dimensional order along the chain length. They can, nevertheless, sometimes still prove useful for obtaining an approximate measure of crystallinity, and the general question of chain order and *regularity peaks* in vibrational spectra will be discussed after some clear examples of the effects of crystallinity have been given. By way of introduction, it will be useful to summarize the principles set out in more detail in chapter 3.

In essence, if there is a single chain passing through the crystallographic unit cell the spectrum will differ only marginally from that of a single, isolated chain or a group of chains with the same conformation in an amorphous region. Modes that were degenerate for a single chain may have different frequencies, though in general the splitting will not be very great, and modes inactive for the single chain may be activated. On the other hand, by comparison with a group of chains in the amorphous regions the peaks will tend to be somewhat sharper because of the removal of the spatial disorder which causes each vibrational frequency to be spread over a small range. If there are m chains per unit cell, correlation splitting of the modes occurs for $m > 1$. Any mode of the single chain is split into m modes which differ, to a first approximation, only in the relative phases of the motions of the m chains in each unit cell; the motions are correlated. In addition, there are $4m - 3$ low frequency lattice modes for all values of m, including $m = 1$.

Two assumptions are involved in this description of the effects of crystallinity. They are that the intermolecular forces are small compared with the intramolecular forces, so that the perturbation of the single chain vibrations is small, and that the splitting is not obscured in practice because the corresponding peaks in the spectrum are too broad or are overlaid by peaks due to non-crystalline material.

6.7.2 *Polyethylene*

The formal theoretical treatment of the factor group modes of crystalline polyethylene has already been given in subsection 3.2.2. The nature of the modes resulting from correlation splitting is very easily illustrated and a simple vectorial approach allows an easy deduction of the directions of the infrared transition moment dipoles. Consider the CH_2 rocking vibration of species B_{2u} for the single chain, shown in fig. 6.14a as it would appear if viewed along the length of the chain.

The dipole moment changes consequent upon the movements of the four hydrogen atoms are shown resolved along axes parallel and perpendicular to the projected C—C bond direction in fig. 6.14b. It is clear that the net dipole moment change is perpendicular to the C—C bond direction, as indicated by the broad arrow in fig. 6.14a. The unit cell of polyethylene is orthorhombic and contains two molecules: one molecule passes through the centre of the cell and one molecule at each corner of the cell is shared with three other unit cells. The symmetry of the factor group allows only two different phase relationships between the vibrations of the two chains in the unit cell and these are shown in fig. 6.14c. When the net dipole moments of the chains are resolved parallel to the a and b axes of the unit cell and summed, bearing in mind that the contribution per unit cell from the central chain will be four times that from each of the corner chains, the overall dipole moment is seen to lie parallel to the b axis in the upper diagram and parallel to the a axis in the lower one.

Measurements on the 720/731 cm^{-1} infrared doublet, assigned to the rocking mode, using plane polarized radiation and a biaxially oriented sample, show that the peak at 720 cm^{-1} originates from the dipole moment change parallel to the *b* axis (B_{2u} species for the crystal) and that the peak at 731 cm^{-1} originates from the dipole moment change parallel to the *a* axis (B_{1u} species). A similar conclusion has been obtained from studies on single crystals of high molecular weight normal paraffins. A similar splitting occurs for the B_{1u} species bending mode, $\delta(CH_2)$, giving absorption peaks at 1462 cm^{-1} (B_{2u}) and 1473 cm^{-1} (B_{1u}). The splitting of the other infrared-active modes is too small to be observed for two reasons. First, the size of the splitting

Fig. 6.14. The nature of the B_{1u}/B_{2u} rocking doublet modes in polyethylene: (*a*) the rocking vibration of a single chain viewed along the chain axis. For simplicity the motions of the carbon atoms are not shown. The broad arrow indicates the direction of the instantaneous dipole moment, which is the vector sum of those due to the four C—H bonds; (*b*) the bond dipole changes resolved parallel and perpendicular to the projected C—C bond direction; (*c*) the two allowed phase relationships of the vibrations of the two chains in the unit cell for factor group modes and the corresponding net dipole changes.

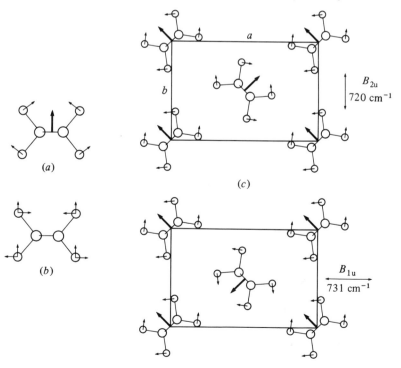

depends on the type of vibration involved and is greater for those modes that involve motions predominantly perpendicular to the chain direction because the inter-chain interaction is stronger. Secondly, the width of the peaks is a relevant factor, as two peaks separated by less than their width will not be seen as resolved.

One method for reducing this limitation on the observation of splitting is to cool the sample. This sharpens the peaks (see section 4.6) and also leads to a thermal contraction of the crystal lattice, bringing the chains closer together and so increasing the inter-chain interaction and the splitting. This method has proved of considerable value for resolving correlation splitting doublets for several modes in the Raman spectrum of crystalline polyethylene. The B_{2g} skeletal stretching mode, a single peak at $1066 \, \text{cm}^{-1}$ at ambient temperature, appears as a doublet at $1065 \, \text{cm}^{-1}$ (B_{1g}) and $1068 \, \text{cm}^{-1}$ (B_{2g}) at $-196°C$ and at this temperature there are peaks at $1295 \, \text{cm}^{-1}$ (B_{2g}) and $1297 \, \text{cm}^{-1}$ (B_{1g}) corresponding to the twisting mode $\gamma_t(CH_2)$ of species B_{1g} for the single chain, which is observed as a single peak at $1296 \, \text{cm}^{-1}$ at ambient temperature, as shown in fig. 6.15. Furthermore, splitting of the B_{2g} species wagging mode, $\gamma_w(CH_2)$, of perdeutero-polyethylene gives components at $827 \, \text{cm}^{-1}$ (B_{1g}) and $830 \, \text{cm}^{-1}$ (B_{2g}) at $-160°C$.

It is possible to calculate values for the splittings by assuming values for the intermolecular and intramolecular force constants and the values so obtained tend to be larger than those observed, particularly for small splittings. No doublets have been observed corresponding to calculated splittings less than $5 \, \text{cm}^{-1}$ and the failure to observe them is simply a consequence of the small splittings and not a failure of the predictions of group theory.

Although correlation splitting of the $\delta(CH_2)$ mode was established very early in the study of the infrared spectrum of polyethylene, an understanding of the corresponding region of the Raman spectrum has been more difficult to obtain. As shown in fig. 5.1, three peaks are observed at 1416, 1440 and $1464 \, \text{cm}^{-1}$. Polarization studies show that they are of symmetry species A_g, B_{3g} and B_{3g}, respectively. In one of the earliest studies of correlation splitting in the Raman spectrum it was suggested that the peaks at 1416 and $1440 \, \text{cm}^{-1}$ were the correlation split doublet corresponding to the $\delta(CH_2)$ mode of symmetry A_g for the single chain. Other authors have, however, suggested that the three observed peaks are simply the result of the superposition of a large group of peaks resulting from the Fermi resonance of the components of the correlation split doublet due to the $\delta(CH_2)$ mode with the overtones and combinations of the correlation split doublet due to the $\gamma_r(CH_2)$ modes at $720/731 \, \text{cm}^{-1}$.

The fact that the intensity of the peak at $1416\,cm^{-1}$, relative to the rest of the spectrum, is insensitive to temperature change between $-180°C$ and $+100°C$ even though its frequency changes by about $4\,cm^{-1}$ suggests that it cannot be involved in Fermi resonance, since the intensities of the components of such resonance split peaks show a very strong dependence on the degree of interaction, which in turn depends on the precise frequencies of the interacting vibrations. This peak is thus assignable to the A_g component of the correlation doublet. The assignment of the other two peaks is less simple and may involve Fermi

Fig. 6.15. The effect of temperature on the splitting of the B_{1g}/B_{2g} species twisting doublet observed at $1296\,cm^{-1}$ in the Raman spectrum of polyethylene: (*a*) sample at room temperature; (*b*) sample at $-196°C$; (*c*) sample at $-196°C$ with increased spectral resolution and expanded wavenumber scale ($\longmapsto$ represents $5\,cm^{-1}$ in each spectrum); (*d*) forms of the vibrations. The $+$ and $-$ signs represent displacements up and down with respect to the plane of the diagram. Only two differently oriented chains are shown for simplicity.

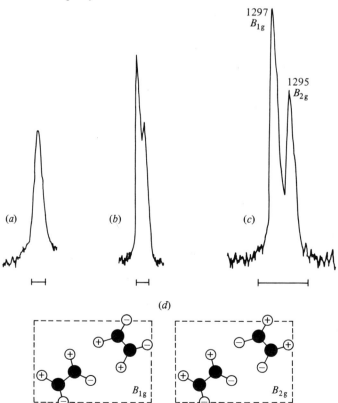

resonance. In addition, as discussed below, the peak at $1440 \, cm^{-1}$ is partially due to non-crystalline chains.

Although the occurrence of correlation doublets would appear to offer a method for the determination of crystallinity, quantitative work is not straightforward. This is true even for the $720/731 \, cm^{-1}$ doublet, which corresponds to the single peak at $720 \, cm^{-1}$ in the amorphous material, so that measurement of the ratio of the absorbances in the two peaks would in principle lead to a determination of crystallinity. The reason is that it is difficult to produce totally unoriented samples, and since the transition moments for the two components are at $90°$ to each other, correction for the polarization introduced by the spectrometer must be made. Although this is not difficult in principle, other methods have now come to be preferred.

The first of these utilizes the absorption peak at $1894 \, cm^{-1}$, which arises from the combination vibration of the Raman-active CH_2 rocking fundamental at $1170 \, cm^{-1}$ and the infrared-active CH_2 rocking fundamental at $731 \, cm^{-1}$. This combination vibration is specific for chains in crystalline regions and is less sensitive to sample orientation than the CH_2 rocking doublet. It is often convenient to determine the ratio of the intensity of the peak at $1894 \, cm^{-1}$ to that of a convenient internal standard peak, such as the peak at $910 \, cm^{-1}$ due to the out-of-plane deformation vibration of the chain-terminating vinyl group, as this eliminates the need for the measurement of sample thickness. The method requires calibration and this may be done in terms of crystallinity values measured by X-ray diffraction or density determination.

The alternative approach is to determine the proportion of chains in amorphous regions, using a peak at $1303 \, cm^{-1}$, due to CH_2 wagging in amorphous regions. This has two significant advantages. Direct calibration is possible using a paraffinic hydrocarbon that is liquid at ambient temperature, such as normal hexadecane, $CH_3(CH_2)_{14}CH_3$, or an amorphous polyethylene obtained by light cross-linking by γ-irradiation in the molten state. Secondly, although it is difficult to avoid some orientation in amorphous regions, the intensity of the peak at $1303 \, cm^{-1}$ is substantially unaffected. The method is well proven, as plots of amorphous content determined by X-ray diffraction against amorphous content based on the absorbance at $1303 \, cm^{-1}$ give straight lines over the approximate concentration ratio 20% to 100%, with a slope of unity, as shown in fig. 6.16.

Both of these approaches to the measurement of crystallinity tacitly assume that the polymer is a two phase system, consisting of wholly crystalline and wholly amorphous material. As explained in subsection

1.2.4, the crystalline regions consist of lamellae, typically a few hundred Ångströms in thickness. Originally it was assumed that a sharp change occurs at the boundary between a lamellar crystallite and the adjacent amorphous region. More recently, however, several studies have suggested that there is a layer of intermediate order at the interface. Some evidence for this model has come from the Raman spectra of various polyethylene samples.

The spectrum in the CH_2 bending region consists of three peaks, superimposed upon a much broader peak characteristic for chains in amorphous regions. The peaks at 1416 and $1440\,cm^{-1}$, both assigned in these studies to the correlation doublet, are of approximately equal intensity for highly crystalline samples. For polymers of lower crystallinity, however, their intensities differ markedly and it is not possible to account for the observed spectrum in terms of partially crystalline and partially amorphous material. The intensity at

Fig. 6.16. Correlation of infrared and X-ray diffraction estimates of amorphous content in high molecular weight hydrocarbons. (From R. G. J. Miller and H. A. Willis, *Journal of Polymer Science* **19** 485 (1956). Copyright © (1956) John Wiley & Sons, Inc. Reprinted by permission of John Wiley & Sons, Inc.)

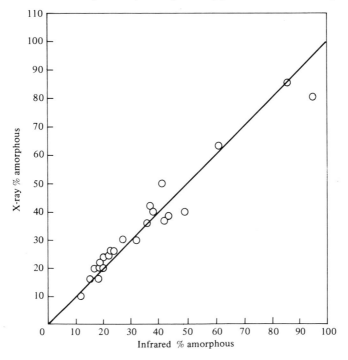

1440 cm^{-1} is too high, suggesting the presence of an additional peak, which has been assigned to regions, presumed to be at the crystal/amorphous interface, where the chains are in the all-trans conformation but do not show regular packing in the lateral direction. This assignment is based on the Raman spectra of normal alkane crystals near to, but below, their melting points, which show a single peak at 1438 cm^{-1}. Estimates of the concentration of the interfacial material have been made by using the areas under the peaks at 1416 and 1303 cm^{-1} as measures of the concentration of truly crystalline and truly amorphous material, respectively. The area under the peak at 1296 cm^{-1}, due to CH_2 twisting and insensitive to crystallinity, was used as an internal standard, with calibration factors for crystalline or amorphous material being determined from measurements on extremely highly crystalline or molten polymer, respectively.

6.7.3 *Polyoxymethylene*

Polyoxymethylene, $+CH_2-O+_n$, prepared by conventional routes crystallizes in a hexagonal unit cell with only one chain passing through it. The conformation of the chain is helical, with nine monomer units in five turns, a 9_5 helix (see fig. 6.17b and d). The vibrational spectrum of crystalline samples differs from that of almost amorphous samples mainly in having rather sharper peaks. Slow polymerization in alkaline conditions yields a polymer which crystallizes to give an orthorhombic unit cell with two molecules passing through it. The conformation of each of these chains is a 2_1 helix, illustrated in fig. 6.17 a and c, in which the bond angles and the angles of rotation around bonds are not very different from those in the 9_5 helix; a 2_1 helix may equivalently be thought of as a 10_5 helix. It is therefore not surprising that the vibrational spectrum of the orthorhombic form is quite similar to that of the hexagonal form. The most important difference is that the spectrum of the orthorhombic form shows correlation splitting of some of the modes.

The factor group of the line group for the 9_5 helix is isomorphous with the group D_9 and the spectroscopically active modes are distributed among the symmetry species as follows: $5A_1 + 5A_2 + 11E_1 + 12E_2$. The factor group for the 2_1 helix is isomorphous with the group D_2 and the 20 true vibrational normal modes are distributed equally among the four symmetry species A, B_1, B_2 and B_3. The spectroscopic activities and the correspondence between the symmetry species for the two helices are shown in table 6.3. The A modes of the 2_1 helix correspond to the A_1 modes for the 9_5 helix, the B_1 modes correspond to the A_2 modes (taking the helix axis as Oz) and the B_2 and B_3 modes correspond to the E_1

modes, which lose their degeneracy. The E_2 modes of the 9_5 helix also lose their degeneracy and become identical to A or B_1 modes of the 2_1 helix. The factor group of the space group for the orthorhombic structure is also isomorphous with the group D_2 and the relationship between the factor groups of the site and space groups is such that each of the line group modes of species A or B_1 splits into two components of species A and B_1 and each of the modes of species B_2 or B_3 splits into two components of species B_2 and B_3. The relationships between the symmetry species and spectral activities of the single 9_5 and 2_1 helices and the orthorhombic crystal are also shown in table 6.3.

The final predictions for the infrared spectrum are that active A_2 modes of the 9_5 helix should give rise to active B_1 modes of the

Fig. 6.17. The chain conformations of polyoxymethylene: (a) and (b) show the projections of the carbon and oxygen atoms of two chemical repeat units for the 2_1 and 9_5 helices, respectively, onto planes normal to the helix axes; (c) and (d) show projections which are normal to the chain axes and to the lines AB for the 2_1 and 9_5 helices, respectively.

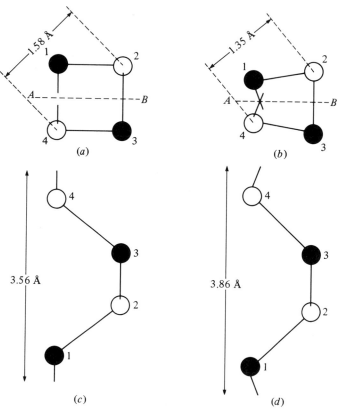

Table 6.3. *Relationships between symmetry species and spectroscopic activities for the vibrational modes of different forms of polyoxymethylene*

9_5 helix			Single 2_1 helix			Orthorhombic crystal		
species	R	IR	species	R	IR	species	R	IR
						A	p	i
A_1	p	i	A	p	i	B_1	d	π
						A	p	i
A_2	i	π	B_1	d	π	B_1	d	π
			B_2	d	σ	B_2, B_3	d	σ
E_1	d	σ	B_3	d	σ	B_2, B_3	d	σ
E_2	d	i						

Notation as explained in section 5.1.

orthorhombic crystal, active E_1 modes should give rise to active B_2, B_3 doublets and inactive A_1 or E_2 modes should give rise to active B_1 species modes. Some of the correlation split doublets, those corresponding to E_1 modes of the 9_5 helix, should thus be observable directly in the infrared spectrum, whereas others, corresponding to A_1, A_2 or E_2 modes, require the observation of Raman spectra.

The task of assigning the modes in the infrared spectrum and of seeking the B_2 and B_3 doublets, assuming that the intermolecular coupling constants are large enough, is complicated by the fact that the orthorhombic form is unstable and reverts to the hexagonal form at 70°C or on stretching. It is therefore necessary to examine the sample in powder form and polarization information cannot be obtained. Such information can, however, be obtained for the hexagonal form, for which A_2 species should be π modes and E_1 species should be σ modes. A comparison of the infrared spectra of the hexagonal and orthorhombic forms then suffices to establish which modes are doubled in the orthorhombic form and to identify which line group modes they correspond to. The results are shown in table 6.4 for those modes which can be clearly established as doubled. The labelling in terms of group modes is, as usual, only approximate.

Table 6.4. *Vibrational modes of polyoxymethylene which show correlation splitting*

Mode	Frequency (cm^{-1})		Symmetry species for single D_2 chain
	Hexagonal form	Orthorhombic form	
C—H stretching	2984	3000 2980 2965	B_1, B_3
CH$_2$ bending	1471	1488 1466	B_3
CH$_2$ rocking + COC bending	1238	1237 1220	B_2
C—O stretching	1120	1120 1110	B_3
COC bending	456	434 428	B_2

Adapted from Table III of V. Zamboni and G. Zerbi, *Journal of Polymer Science* **C7** 153 (1964). Copyright © (1964) John Wiley & Sons, Inc. Reprinted with permission of John Wiley & Sons, Inc.

Some of the splittings are quite appreciable, which shows that the intermolecular forces are comparatively large. As expected, the C—H stretching region is rather complex. Theory predicts one peak from the B_1 mode and doublets from the B_2 and B_3 modes, but the fact that only four peaks are clearly resolved is not surprising, as this spectral region is likely to be overlaid by first harmonics of CH$_2$ bending modes, which will blurr some of the detail. The doublet at 428/434 cm^{-1} is particularly clear and may be suitable for the determination of crystallinity.

6.7.4   *Poly(vinyl chloride)*

Correlation splitting in poly(vinyl chloride) has been discussed briefly and formally in subsection 3.2.2 and it will now be considered in rather more detail because both the theoretical and experimental aspects of the topic present interesting features. In the earlier discussion it was shown that the presence of two chains in the orthorhombic unit cell of the syndiotactic polymer leads to correlation splitting but that, unlike the examples considered so far, only one component of each doublet is infrared active while the other is only Raman active. Hence the proof for the existence of the splitting requires the careful comparison of peak positions in the two types of spectrum. Furthermore, because the commercial type of polymer is only marginally syndiotactic and contains only about 10% of crystalline material, the spectrum is complicated by the presence of peaks due to many different conformational and configurational isomers, as explained in subsection 6.2.3.

This latter problem can be overcome by the use of the highly syndiotactic polymer made by the urea clathrate method. Nevertheless, only one measurement has been reported in which it was clearly established that the peak positions were determined with sufficient accuracy to prove the existence of the splitting. This study gave frequencies of 601 ± 1 and 608 ± 1 cm^{-1} for the antisymmetric C—Cl stretching modes in the infrared and Raman spectra, respectively, and the frequency 639 ± 1 cm^{-1} for the symmetric stretching mode for both spectra. The C—Cl stretching regions of the infrared and Raman spectra are shown in fig. 3.7. The corresponding calculated values for the antisymmetric stretching mode are 604 and 602 cm^{-1}, and for the symmetric stretching mode they are 637 and 642 cm^{-1}. The differences between the observed and calculated values and the fact that the values are reversed in order for the antisymmetric modes shows that the interchain force field used for the calculations is incorrect. This is confirmed by the fact that the observed splittings for the C—C—Cl bending vibrations of A_1 and B_2 species, at about 358 and 315 cm^{-1}, respectively, are only about 3 cm^{-1} or less, whereas the corresponding calculated splittings are 16 and 6 cm^{-1}.

Despite the difficulties involved in working with commercial polymers, which have a low degree of syndiotacticity, the values of 608 and 601 cm^{-1} obtained for the antisymmetric C—Cl stretching mode from spectra of the urea clathrate polymer have been confirmed. It has been possible to fit the observed Raman spectrum of the commercial polymer with nine peaks over the interval 600–700 cm^{-1}, including peaks at 608 and 638 cm^{-1} (see fig. 6.7). Second and fourth derivative infrared spectra, which have sharper peaks than the normal spectrum (see section 1.7), also reveal peaks at about 601 and 638 cm^{-1}. Furthermore, it has proved possible to use the sum of the areas of the peaks at 608 and 638 cm^{-1} in the Raman spectrum of a powder sample, expressed as a fraction of the total area under all the ν(CCl) peaks, as an approximate measure of crystallinity, although the complexity of this approach precludes it as a routine practical method.

6.7.5   *Lattice modes in polyethylene and polytetrafluoroethylene*

As explained in subsection 3.2.1, the three-dimensional order present in polymer crystallites leads to the occurrence of lattice modes. These modes are vibrations in which the distortion of the crystal corresponds, to a first approximation, to the bodily movement of each chain with respect to its neighbours. Vibrations of this type have been most extensively studied for polymers containing long methylene sequences, and for polyethylene in particular. The far infrared spectrum of polyethylene at low temperatures has two absorption peaks at about 80 and 108 cm^{-1}. The assignments of these two peaks, based partly on the results of normal coordinate calculations, are well established. The one at 80 cm^{-1} results from a translational vibration in which the two chains in the unit cell move in opposite directions parallel to the b axis of the cell, whereas that at 108 cm^{-1} has its origin in a similar translational vibration in which the chains move parallel to the a axis.

The interpretation of the low frequency spectrum of polytetrafluoroethylene (PTFE) has proved more challenging and the problem has been attacked by a combination of normal coordinate calculations and careful experimental measurements, including studies of the polarized infrared spectra of oriented samples. The early experimental studies showed that several new peaks appear in the far infrared region when the temperature of the sample is below the transition temperature of 19°C (see subsection 5.2.4) and that splittings also appear in several peaks in the Raman spectrum. These observations were difficult to reconcile with the assumption that the change in structure at the transition temperature is purely intramolecular and involves only a transformation from a 15_7 to a 13_6 helix, since the most natural explanation would be that the low temperature form has at least two chains per unit cell. Early X-ray diffraction studies suggested, however, that both forms have a unit cell containing only one molecule. In an attempt to resolve this paradox it was suggested that the changes in the vibrational spectrum for samples at low temperature were a consequence of a relaxation of the helical selection rules, due to the effect of neighbouring chains, or to conformational disorder. The two possible interpretations of the spectroscopic data stimulated further low temperature studies.

The original interpretation of the splitting observed in the Raman spectrum, as correlation splitting, led to the conclusion that the crystal structure was monoclinic, with space group $P2_1$. This structure has two chains passing through the unit cell in such a way that they are related by a two-fold screw axis perpendicular to the chain axes and the factor group of the space group is isomorphous with the point group C_2. For this factor group all five lattice modes are infrared active and it follows that three of them are A modes polarized parallel to the C_2 axis, i.e. perpendicular to the chain axis, and as a consequence they should be observed as σ modes in oriented samples. These are the libration of the two chains in the same direction around the chain axis, the translational vibration of the two chains with respect to each other parallel to the chain axis and the translational vibration of the two chains with respect to each other perpendicular to the C_2 axis. The other two modes are B species modes and may have their oscillating dipoles anywhere in the plane perpendicular to the C_2 axis, so that they are neither σ nor π modes. These correspond to the translational vibration parallel to the C_2 axis and to the librational motion of the two chains in opposite directions.

The experimental results show that there are at least seven absorption peaks in the region 30–90 cm^{-1}, four of which, at 45, 54, 58 and 86 cm^{-1}, are σ polarized and the other three of which, at 32, 71 and 76 cm^{-1}, are without strong dichroism. Several slightly different interpretations of these peaks have been given, but it is generally accepted that the peak at 32 cm^{-1} is probably to be assigned to the internal E_1 mode on the dispersion curve labelled v_8 in fig. 5.8 and that those at 45, 54, 58 and 71 cm^{-1} are probably to be assigned to lattice modes. The latter four show appropriate polarization for lattice modes associated with C_2 symmetry and are found to decrease in intensity when the samples are irradiated with γ-rays, which leads to a lowering of the crystallinity. If they were due to defects they would be expected to increase in intensity, and if they were due to changes in the selection rules for the low temperature phase they would be expected to maintain essentially constant intensities, since the crystal structure is not further changed by irradiation. In addition, lowering the temperature causes the peaks at 45, 54 and 58 cm^{-1} to move towards higher frequencies, as would be expected for lattice modes, since thermal contraction of the lattice leads to increased interaction between the molecules. The observations on the low frequency vibrational spectrum thus clearly support the idea that the low temperature form of PTFE has a structure with two chains per unit cell.

The results of recent X-ray and electron diffraction studies of the low temperature form have been interpreted in terms of a unit cell with two chains passing through it. The two chains have helical conformations of opposite chirality which correspond closely, but not exactly, to 13_6 helices. No definite space group symmetry could be assigned to the structure. The gradual elucidation of the structure and spectral assignments of the low temperature form of PTFE, which is not yet complete, provides a good example of the complementary role that vibrational spectroscopy can play to conventional crystallographic studies of polymers.

6.7.6   *Differentiation between crystallinity and chain order*

It is clear from the results considered in the present chapter that the major factor which differentiates the vibrational spectrum of polymer chains in crystalline regions from the spectrum of those in amorphous regions is the presence of order, i.e. translational symmetry, along the chain direction. The additional changes that result from lateral order, due to crystalline packing of the chains, are small. Not surprisingly, therefore, a careful distinction has not always been drawn between the two types of order and values of crystallinity have been reported which were actually based on peaks characteristic for order along the chain direction only. It will therefore be useful to summarize how the two types of order manifest themselves and to consider to what extent peaks characteristic for the latter type of order may be used, albeit empirically, as a measure of crystallinity.

Over 20 years ago Zerbi, Ciampelli and Zamboni divided so-called *regularity peaks* into three types and as their system is logical, and has stood the test of time, it will be followed here. The first group of peaks characterize conformational regularity or irregularity, of which the most obvious example is trans/gauche rotational isomerism. For example, the peaks at 1450 and 1435 cm^{-1} in the spectrum of trans-1,4-polybutadiene are characteristic for the trans and gauche conformers, respectively, of the CH_2—CH_2 bond. Such peaks may occur both in the solid state and in the melt or solution, although the ratio of the concentrations of the two conformers may be different for the two states. The trans isomer is often of lower energy than the gauche isomer and is then found in the crystalline state. It is unwise, however, to assume that the intensities of peaks characteristic for trans conformers are a direct measure of crystallinity. This has often been done for poly(ethylene terephthalate), for which peaks at 975, 1340 and 1470 cm^{-1} in the infrared spectrum are characteristic for the trans conformer. It has, however, been shown that trans units occur in amorphous regions in concentrations which may depend on annealing temperature and time or on the degree of molecular orientation. Crystallinity values based on the intensity of a peak due to trans conformers may therefore be too high.

The second class of regularity peaks, often erroneously called 'crystallinity peaks', consists of peaks associated with long stereoregular chain segments. Some examples have been discussed in section 6.2, of which the best known is isotactic polypropylene. This forms a regular helix with three monomer units per turn and it was noted that peaks at 805, 840, 898, 972, 995 and 1100 cm^{-1} are characteristic for the helical structure. This type of chain regularity peak is usually absent from the spectrum of the molten polymer, but occasionally it is replaced by a much broader peak. Of the six peaks that are characteristic for the polypropylene 3_1 helix, all are absent from the spectrum of the molten polymer except the one at 972 cm^{-1}, which is considerably broadened. A detailed analysis of the profile suggests that short helical units are still present in the melt. For this reason, although regularity of individual chains is a necessary condition for the occurrence of crystallinity, the intensity of a particular peak characteristic for such regularity may not be a direct measure of the crystallinity.

The third class of regularity peaks comprises the true *crystallinity peaks*. Their occurrence is a consequence of inter-chain interaction and is predictable by group theory although, as is evident from the examples considered in subsections 6.7.2 to 6.7.4, the magnitude of the correlation splitting cannot, at present, be calculated with any degree of confidence. These true crystallinity peaks disappear when the sample is melted, but this may not prove a useful diagnostic test because of the instability of a number of polymers, e.g. polyoxymethylene and poly(vinyl chloride), at temperatures well below their melting points. Hence, there are often real practical difficulties, particularly with new or unusual polymers for which detailed vibrational analyses are not available, in diagnosing with certainty which, if any, of the observed peaks are specific for crystallinity and whose intensities may therefore be used as a measure of crystallinity, provided that allowance is made for complications such as those arising from preferred orientation.

In these circumstances it is necessary to look for alternative approaches to the measurement of crystallinity. As noted above, modes specific for chains in amorphous regions are sometimes more suitable for quantitative work although here also it may not always be easy to identify such peaks with certainty. In addition, it must be recognized that the definition of the degree of crystallinity for a polymer is not straightforward. States of order intermediate between the truly crystalline state, with three-dimensional order over regions large compared with a monomer unit, and the truly amorphous state, with no order extending for distances more than a few monomer units, may exist. Each different method of measuring crystallinity may then lead to a different result.

If, for a given polymer, samples are available covering a range of crystallinities, as determined by a technique such as X-ray diffraction or density measurement, it may be possible to deduce an empirical correlation between this measure of crystallinity and the intensity of one or more peaks in the vibrational spectrum. Such correlations can be useful, but should be treated with caution if the reason for the correlation is not understood. For example, although the peak at 1141 cm^{-1} in the infrared spectrum of poly(vinyl alcohol) has not been shown to be a crystallinity peak its intensity correlates with values of crystallinity obtained by X-ray diffraction methods. Similarly, for those ethylene/propylene/dicyclopentadiene terpolymers which contain more than 65% of ethylene and are partially crystalline, correlations have been found between the intensity of the C—C stretching mode, which gives a peak at 1065 cm^{-1} in the Raman spectrum, and the degree of crystallinity determined by X-ray diffraction. The virtue of this approach is that once a calibration in terms of a second technique has been established, vibrational spectroscopy usually provides a simple and rapid method for the measurement of crystallinity.

6.8   Chain folding in polyethylene

In the preceding section, lateral order between polymer chains was considered in the important but specific context of interactions between chains in the crystalline unit cell and the dimensions of the crystallites were merely assumed to be much greater than those of the unit cell, so that surface effects could be neglected. As already pointed out in subsection 1.2.4, the crystals of polyethylene and various other polymers obtained from solution consist of lamellae a few hundred Ångströms in thickness, but with considerably larger lateral dimensions, and it is now well established by a variety of methods that the polymer chains fold back and forth so that their axes are approximately normal to the lamellar plane. This may be represented as shown schematically in fig. 1.7.

Two assumptions have been made in drawing this figure: that the majority of the chains re-enter the lamella in a position adjacent to where they emerge, by tight chain folds, and that the folding takes place along a particular crystallographic plane. An alternative extreme assumption would be that the chains re-enter in positions randomly related to those from which they emerge, the so-called 'switchboard' model. The identification by electron microscopy of the growth face of the diamond-shaped lamellar crystal of polyethylene (see fig. 6.18a) as a (110) crystallographic plane has led to the conclusion that folding occurs in this plane, but it is possible to grow truncated crystals (see fig. 6.18b) using different conditions, suggesting that chain folding may occur in the (200) plane. The morphological characteristics of polyethylene crystallized from the melt are more difficult to observe, but the presence of lamellae has been established. Several experimental techniques have been used in attempts to decide whether or not adjacent re-entry predominates in chain-folded crystals and, if so, which are the predominant fold planes. Among these methods, infrared spectroscopy has been, and continues to be, important. Its contribution originated in 1968 with the work of Krimm and his colleagues. Their method is based on infrared studies of co-crystallized mixtures of polyethylene (PEH) and perdeutero-polyethylene (PED), and the principles may be understood by reference to fig. 6.19.

Fig. 6.19a shows the arrangement, in a cross-section perpendicular to the chain direction, for a regular 1:1 mixed crystal of PEH and PED in which folding is assumed to occur in the (110) planes. It is evident that the unit cell contains two chain segments of PEH and two of PED, and there will therefore be correlation splitting of the various modes of both species. Since each chain now interacts with only two chains of the same kind but different orientation instead of the four such chains that it interacts with in the normal crystal, the splittings will be halved. They will, nevertheless, still be readily observable for some modes. Additionally, it should be possible to distinguish between adjacent and random re-entry, because if there is a significant amount of the latter there will be an appreciable number of isolated chains of either species, giving peaks at frequencies between the two components of the

Fig. 6.18. Electron micrographs of single crystals of polyethylene crystallized from dilute solution in xylene: (a) diamond-shaped crystals; (b) truncated crystals. (From (a) V. F. Holland and P. H. Lindenmyer, *Journal of Polymer Science* **57** 589 (1962); (b) D. J. Blundell *et al.*, *Journal of Polymer Science B* **4** 481 (1966). Copyright © (a) (1962) (b) (1966) John Wiley & Sons, Inc. Reprinted by permission of John Wiley & Sons, Inc.)

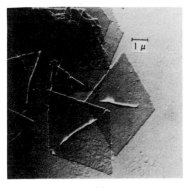

(a) (b)

Fig. 6.19. Possible regular structures for 1:1 mixed crystals of normal and fully deuterated polyethylene: (a) regular folding in the (110) planes; (b) regular folding in the (200) planes. ○ indicates $(CH_2)_n$ and ● indicates $(CD_2)_n$. (From M. Tasumi and S. Krimm, *Journal of Polymer Science A2* 995 (1968). Copyright © (1968) John Wiley & Sons, Inc. Reprinted by permission of John Wiley & Sons, Inc.)

(a)

(b)

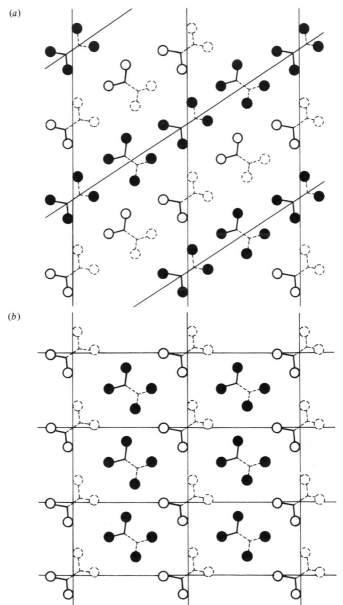

correlation split doublets. Reference to fig. 6.19*b* shows that for regular (200) folding the unit cell contains only one chain segment of each species and correlation splitting will therefore not occur. The absence of correlation splitting cannot, however, be regarded on its own as concrete proof for (200) folding because if there is extensive random re-entry with (110) folding there will also be very little splitting.

While these conclusions have been derived for regular alternating arrangements of PEH and PED chain segments, which lead to specific lattice structures, they should be valid for any crystal in which a (110) plane containing one species has nearest neighbour planes containing the other species, by the following line of reasoning. Consider a mixed crystal in which the concentration of PEH chains is very high, thus ensuring that the planes of PED chain segments are isolated from each other. A plane of PED chain segments may then be considered as a one-dimensional lattice which, for (110) folding, has two equivalent but differently oriented oscillators per repeat period, whereas for (200) folding it has only one oscillator per repeat period. The former will give split modes, whereas the latter will not.

Three modes of vibration were used for the experimental studies: the CH_2 bending and rocking modes observed at 1463/1473 and 720/731 cm^{-1}, respectively, in pure PEH and the CD_2 bending mode observed at 1084/1091 cm^{-1} in pure PED. The results of the original measurements of the splittings on a series of solution crystallized and melt crystallized (cast film) mixtures of PEH and PED of various compositions are given in table 6.5, which also shows calculated values of the splittings for pure crystals of either type. The calculated values for the pure PEH and PED are in good agreement with the experimental values. The limited number of calculated values for mixtures were somewhat lower than those observed, but the difference is not unexpected, as any aggregation of chains will increase the observed splitting. The fact that for cast films splitting of the modes of the minor component is observed even when this forms only 10 % of the mixture shows that chain folding occurs predominantly in the (110) plane with some adjacent re-entry. Furthermore, the absence of an obvious third peak between the components of the doublet suggests that the folding in fact occurs mainly by adjacent re-entry.

The spectra of melt-crystallized samples showed a broad single peak for the species at low concentration, with some residual doublet structure occurring as shoulders, and this suggested that either (200) folding or random re-entry predominates for samples of this type. Mixed crystals of PEH and PED with random re-entry were simulated by preparing and examining mixtures, of varying composition, of the

Table 6.5. *Splittings of bending and rocking modes in mixed crystals of PEH and PED*

PEH:PED	$\Delta\tilde{\nu}_b(CH_2)$ (cm^{-1})		$\Delta\tilde{\nu}_b(CD_2)$ (cm^{-1})		$\Delta\tilde{\nu}_r(CH_2)$ (cm^{-1})	
	Cast film	Melt crystallized	Cast film	Melt crystallized	Cast film	Melt crystallized
1:0 (calc.)	9.9				10.5	
1:0	10.3	10.5	—	—	10.8	11.3
10:1	9.8	10.4	5.5	0	10.7	11.0
4:1	10.1	9.9	5.9	0	10.4	10.5
2:1	9.8	9.6	6.6	0	10.3	9.9
1:1	9.3	8.8	6.7	4.8	9.5	9.0
1:2	8.3	0	7.1	6.8	8.7	0
1:4	7.4	0	7.5	7.3	8.4	0
0:1	—	—	7.7	8.0	—	—
0:1 (calc.)			8.1			

Adapted from Table IV of M. I. Bank and S. Krimm, *Journal of Polymer Science A2* **7** 1785 (1969). Copyright © (1969) John Wiley & Sons, Inc. Reprinted by permission of John Wiley & Sons, Inc.

normal paraffin $C_{36}H_{74}$ and its fully deuterated analogue. Since there is no chain folding in paraffin crystals, the distribution of hydrogenated and deuterated chains in these mixed crystals will be random. The observed splittings for the species at high concentration in the paraffin mixtures were found to be appreciably less than for the corresponding melt crystallized polymer mixture. It was therefore deduced that there must be a substantial amount of adjacent re-entry in the polymer mixes and, as a corollary, that the chain folding is largely in the (200) plane. It must be noted, however, that the lattice dimensions of melt-crystallized polyethylene and normal paraffins are different at room temperature, and this conclusion is not supported by more recent work.

The interpretation of the results in the way indicated depends upon the assumption that there is no segregation of the PEH and PED chains, since otherwise it does not follow that adjacent chain segments of the species with low concentration necessarily belong to the same molecule. It has been shown that although segregation can take place, because of a difference in the melting points of the two species, the conditions used in the original work were close to those required to eliminate this effect.

Fig. 6.20. The CD_2 bending region of the infrared spectrum of a PED/PEH mixed crystal grown from dilute solution in xylene and containing 0.5% PED of MW 216 000. The spectrum was obtained with the sample at the temperature of liquid nitrogen and is shown fitted with a singlet and two doublets; the sum of these is shown by the dashed line. Further increasing the number of doublets improves the fit. (Reproduced from S. J. Spells *et al.*, *Polymer* **25** 749 (1984), by permission of the publishers, Butterworth & Co. (Publishers) Ltd. ©)

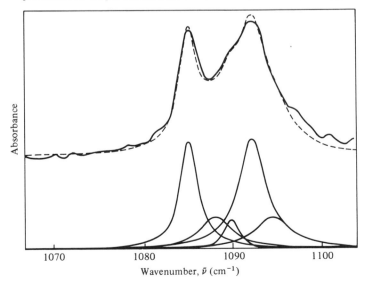

$\bar{\nu}$ (cm^{-1})

More recent studies have been made with samples cooled to liquid nitrogen temperature and it has become clear that the structures of the CH_2 and CD_2 bending regions of the spectrum are more complex than originally thought. A detailed analysis of the regions into sets of overlapping peaks (see fig. 6.20) has shown that for solution crystallized samples they can be understood in terms of a model in which a molecule folds in the (110) plane to give a sheet containing 10 to 20 parallel chain segments. The folding is largely, but not completely, accompanied by adjacent re-entry, so that the chain segments occur within the sheet as blocks separated by blocks of chain segments from another molecule. For high molecular weights the molecule 'superfolds' to give multilayer sheets of this type. The model is also consistent with the results of neutron scattering experiments and was in fact originally developed to explain those results. For melt-grown crystals the more recent infrared studies, in which the broadened single peak observed for the CD_2 bending mode has been analysed into several components, suggest that random re-entry predominates but that there is possibly a small amount of (110), rather than (200), adjacent re-entry. The results of neutron scattering studies suggest a similar structure, in which groups of a few chain segments related by adjacent or nearly adjacent re-entry occur among chain segments for which re-entry is largely random.

6.9   Longitudinal acoustic modes

As long ago as 1949, in a general study of the Raman spectra of solid normal paraffins (n-paraffins), it was noted that there was a peak in the low frequency region whose frequency appeared to be inversely proportional to the chain length. This observation was confirmed and extended nearly 20 years later, as was the original suggestion that this peak has its origin in an accordian-like motion of the all-trans chain present in the crystals. There is maximum displacement of the atoms at the surface of the crystal and a node of zero displacement at the centre of the crystal, as shown in fig. 6.21b. This motion will change the polarizability of the crystal and the vibration will thus be Raman active. It is also evident that this vibration is the fundamental of a series of longitudinal acoustic modes (LA modes or LAMs). The higher members of the series will have two or more nodes within the crystal, as shown in fig. 6.21c for three nodes. It is clear that those with an even number of nodes will not produce a net change of polarizability for the crystal as a whole and so will not be Raman active, whereas those with an odd number of nodes will give peaks in the Raman spectrum at frequencies which, to a first approximation, will be the appropriate multiple of the fundamental frequency. Such higher order LAMs are readily observable in the spectra of the n-paraffins.

To first approximation, a LAM of any order should correspond to a point on the dispersion curve v_5 for polyethylene (see fig. 4.13 and section 4.5), but if the corresponding wavelength is large enough it is simpler to consider the vibrating chain as an elastic rod. Elementary theory shows that the vibrational frequency, v, of such a rod vibrating in a mode with m nodes is given by the equation:

$$v = \frac{m}{2l} \sqrt{\frac{E}{\rho}} \tag{6.3}$$

where l is the length of the rod, E is the Young's modulus and ρ is the density. This relation fits the experimental results for solid n-paraffins well in the limit where l/m is much greater than the translational repeat unit and it is possible to obtain a 'spectroscopic' value for the Young's

Fig. 6.2.1. The displacements of the repeat units and the forms taken by the backbone of the molecule in the longitudinal acoustic modes (LAMs) of polyethylene (schematic); (a) shows the equilibrium state of the chain backbone and (b) and (c) show the first and third order LAMs. In (b) and (c) the upper and lower diagrams show the forms taken by the backbone in opposite phases of the motion and the central diagram shows the corresponding displacements, with displacement to the right shown above the straight line representing the undisplaced positions. The full curves refer to the upper diagrams and the dashed curves to the lower diagrams. The magnitudes of the distortions of the backbone are greatly exaggerated.

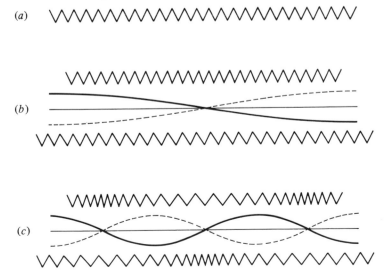

modulus of polyethylene from the limiting slope, at $v = 0$, of a plot of v against m/l, since the density of crystalline polyethylene is known.

At the time that the later work on n-paraffins was in progress the lamellar nature of the single crystals of polyethylene obtained from solution (see subsection 1.2.4) was well established, as was the fact that the lamellar thickness is, typically, 200 Å. It was therefore possible to deduce that the lowest LAM frequency for the chain segments in the lamellae occurs in the range 10–$20\,\mathrm{cm}^{-1}$, and the corresponding peak will be difficult to observe as it will lie on the 'wing' of the intense line due to Rayleigh scattering. For this reason, the LA mode was not observed in polyethylene until 1971. In this initial study it was clearly established, as anticipated, that the frequency decreased as single crystals were annealed and so became thicker. The change in lamellar thickness with annealing temperature, as assessed by this approach using the value of Young's modulus obtained from the measurements on n-paraffins, proved to be very similar to that deduced from small angle X-ray diffraction measurements. Fig. 6.22 shows the peaks due to the LA mode in the Raman spectra of two polyethylene samples prepared in different ways.

In the last decade, with the emergence of the holographic diffraction grating which gives a high discrimination against the Rayleigh line, it has proved comparatively easy to work within about $6\,\mathrm{cm}^{-1}$ of this line, and there have been numerous studies on the LA mode of polyethylene and, to a lesser degree, on that of other polymers which crystallize in the lamellar form. Higher order LA modes have been observed, but most effort has been concentrated on the lowest order LAM because of its greater intensity. There have also been a number of theoretical treatments of the LA mode, prompted largely by anomalies which appeared in the experimental results. This work has proved interesting and has provoked several questions. For example, is it possible to determine lamellar thicknesses with reasonable accuracy from the positions of the corresponding LAM peaks by using the value of Young's modulus derived from measurements on solid n-paraffins? In view of the all-trans conformation required for the LAM vibration, what will be the effect of an occasional defect structure in the crystalline region, such as a gauche unit or a chain branch? Is it correct to assume that the LAM frequency is determined only by the length of the straight chain segments ('stems') in the crystal lamellae so that it is unaffected by the presence of disordered 'amorphous' material between lammelae or chain folds at their surfaces? Since a range of lamellar thicknesses will occur in a given sample, the observed LAM peaks will be the concentration-weighted sum of the peaks for each particular stem length, and so a broader composite peak will be obtained; is it possible to obtain information on the distribution of lamellar thicknesses from the shape of this peak? These topics will now be considered briefly.

The early work on n-paraffins gave a value of 3.6×10^{11} N m^{-2} for Young's modulus. This value was derived by assuming that the

Fig. 6.22. The low wavenumber regions of the Raman spectra of two samples of linear polyethylene, showing the fundamental LAM. Each sample was compression moulded at 160°C. Spectrum (a) was obtained from a sample quenched directly into cold water from 160° and spectrum (b) was obtained from a sample which was slowly cooled to 110°C before being quenched.

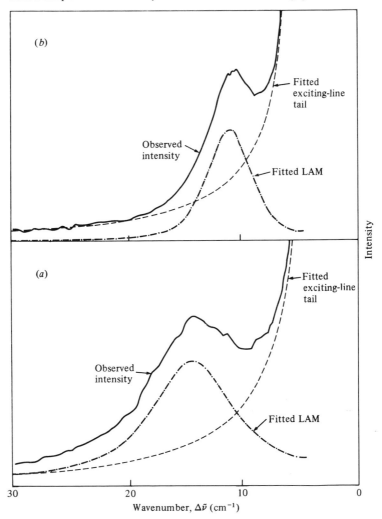

frequency depends only on the length of the sequence of CH_2 groups in an individual stem but later work indicated that weak interlamellar forces affect the peak position and an estimate of the magnitude of this effect led to a new value of 2.9×10^{11} N m^{-2}. This is in quite good agreement with the value of 3.3×10^{11} N m^{-2} obtained from inelastic neutron scattering data for the v_5 dispersion curve of polyethylene. It is also very close to the value obtained by direct stress/strain measurements on highly oriented samples of polyethylene and since these are not perfect single crystals the true crystal modulus must be somewhat higher, probably close to that obtained from the neutron scattering data.

Even if the correct crystal modulus and the observed LAM peak frequency are substituted into equation (6.3) the value of l obtained will not be the mean thickness of the lamellae in the sample, as might be expected, since it has been shown that the peak position depends on the temperature of the sample and the half-width of the peak. These factors will not be discussed in detail and it will suffice to note that even in the favourable circumstances of a narrow peak with a half-width of about 6 cm^{-1}, corresponding to a narrow distribution of sequence lengths, the correction is about 1 cm^{-1} for a peak at 20 cm^{-1}; for a half-width three times larger the corresponding correction is about 7 cm^{-1}. The correction, which reduces the observed frequency and so leads to a higher value for the mean lamellar thickness, rises rapidly with decreasing wavenumber of the peak position and may be comparable with the measured value. In these circumstances it is not surprising that the absolute measurement of lamellar thicknesses from observed LAM frequencies poses considerable problems and the method is of much greater value for following the changes that occur during treatments such as annealing. It should also be noted that if the chains are inclined at an angle to the normal to the lamellar surface, as is often observed, the angle of tilt must also be established. Nevertheless, by combining the results of Raman measurements with those from small angle X-ray diffraction studies, which give the repeat distance of the 'sandwich' structure of lamellar crystalline and amorphous layers which usually occurs, it has been possible to deduce reasonably reliable values for the thickness of the amorphous layers.

The evidence for the effect of defect structures comes primarily from normal mode calculations for models in which one or more successive gauche units are introduced near to the centre or to the end of a long straight chain segment. They show that such a structure near to the chain end has little influence on the LAM vibration but when it is near to the chain centre it reduces the intensity of the peak considerably without affecting its position perceptibly. Several pieces of evidence suggest that the presence of chain folds and amorphous material between lamellae does not affect the LAM frequency corresponding to a given stem length. One important piece of experimental evidence is provided by the results of measurements following treatment with nitric acid. This material selectively attacks the amorphous material and the chain folds, leaving isolated stems in the lamellae, but the LAM frequency does not change significantly after this reaction. This evidence is supported by the results of calculations based on an extension of the elastic rod model in which a composite rod is considered, which has one length, density and modulus for that part of the molecule contained within a lamella, and two ends of different lengths, densities and moduli for the parts of a particular chain lying within amorphous regions. Using this model it is possible to obtain a modified expression for the LAM frequency which, when applied to the specific case of polyethylene, shows that the frequency is only modified slightly by the fact that the stems traversing the lamellae are parts of chains that also cross the interlamellar non-crystalline regions. A model of a similar nature, but which treats the chain more correctly at the microscopic level, shows similarly that the dispersion curves obtained by neutron scattering should be largely unaffected by the interlamellar disordered material. These conclusions probably apply only for melt-crystallized samples, since there is some evidence that the LAM frequencies of 'mats' of sedimented solution-grown single crystals may be influenced by interactions between the lamellar surfaces, which consist of more regular chain folds than do those of melt-grown crystals.

As noted above, if there is a range of lamellar thicknesses the LAM peak will be broadened because of the superposition of the peaks, somewhat displaced with respect to each other, characteristic for each stem length. Not surprisingly, there has been considerable interest in the possibility of deducing information about the distribution of lengths from the width of the LAM peak. Unfortunately, the half-width of the chain-length distribution function is not obtained from the observed intensity distribution simply by converting the frequencies at the half-intensity points to chain lengths using equation (6.3). Just as the peak position depends on temperature and the distribution of stem lengths, as discussed above, so does the shape of the peak. The following equation relating the distribution of stem lengths, $f(l)$, and the line shape, $I(\tilde{v})$, has been derived:

$$f(l) = C\tilde{v}^2 B(\tilde{v})I(\tilde{v}) \tag{6.4}$$

where C is a constant, $B(\tilde{v})$ is the factor $1 - e^{-h\tilde{v}c/kT}$ and l is the stem length which gives rise to the LAM frequency $\tilde{\nu}$. An additional fact which must be taken into account is that the LAM peak has a finite width even for a sample consisting of stems of identical length; measurements on crystalline n-paraffins lead to half-width values of 2.5–3.5 cm^{-1} for peak positions ranging from 6 to 30 cm^{-1}. The effect of these corrections, which have not been applied in several published studies, is to reduce the spread of the calculated stem lengths. For a particular sample of solution-crystallized polyethylene, the corrected spread was found to be as low as 15 Å, in line with the results of small angle X-ray measurements, whereas an earlier study in which the appropriate corrections were not applied gave a spread of 40 Å.

Although LAM studies on polyethylene have attracted the most attention, a range of other polymers has been examined. These fall into two types: those that have $(CH_2)_n$ sequences as part of the overall structure and those that do not. The former category includes materials such as poly(decamethylene sebacate), poly(alkylene oxide)s and poly(alkylene sulphide)s. The second category comprises a range of ordered polymers, such as polytetrahydrofuran, cis-1,4-polybutadiene, polytetrafluoroethylene, polyoxymethylene, poly(ethylene oxide) and isotactic polystyrene. Suitable short-chain molecules related to these polymers as the n-paraffins are to polyethylene are not available, nor generally are neutron scattering data for constructing dispersion curves, so that spectroscopic values of Young's modulus cannot always be obtained. It has also been shown that values determined in other ways are usually less reliable. The use of LA mode frequencies in morphological studies of these polymers is therefore more difficult than it is for polyethylene and will not be discussed here.

6.10   Molecular orientation

6.10.1   *Introduction*

In earlier sections it has been explained how infrared spectra obtained from samples in which the molecular chains have been preferentially oriented, by drawing or stretching the sample, can be used as an aid to the assignment of vibrational modes. For this purpose it is usually desirable for the chains to be as highly oriented as possible, and in the ideal sample all chain axes would be parallel to the draw or stretching direction. Many polymers in practical use are oriented to some degree, either intentionally, to improve the mechanical properties, or incidentally, during injection moulding, cold forming or other processes. This latter type of orientation may lead to undesirable distortions of the products when they are heated, because of relaxation of the orientation. Both intentional and incidental orientation are often very far from complete and it is important to be able to characterize the degree of orientation so that the properties of the products may be more fully understood and usefully modified.

A number of methods have been devised for studying orientation in polymers. X-ray diffraction is often the preferred method for studying the orientations of crystallites, particularly if the degree of crystallinity is high, but it is quite difficult to obtain reliable information about the orientation of the non-crystalline chains from X-ray diffraction studies. Among the other available methods, infrared and Raman spectroscopies using polarized radiation are particularly important because they can potentially give a great deal of information, about both crystalline and non-crystalline chains, since measurements on any group mode can give information specifically about the distribution of orientations of the corresponding groups in the polymer.

Infrared spectroscopy provides the simpler method, but it cannot give as much information about the distribution of orientations as Raman spectroscopy because it involves only one beam of polarized radiation and the fraction of the radiation absorbed by a single molecule depends only on the orientation of the molecule with respect to the polarization vector of the radiation. Raman scattering, in contrast, involves two beams of radiation and the intensity scattered by a molecule depends on the orientation of the molecule with respect to the polarization vectors of both beams, which may be different. Suitable measurements of scattered intensities can thus give more information about molecular orientation. In addition, infrared absorption measurements are usually restricted to samples of thickness less than about 50 μm, because thicker samples are opaque to the infrared radiation, whereas Raman scattering can sometimes give good results on samples as thick as 1 cm. In the next subsection the theoretical background to the methods will be given and in subsection 6.10.3 some examples of their use.

6.10.2 *Theoretical background*

We shall assume that the symmetry of the distribution of orientations in the sample is such that a set of three mutually perpendicular axes $OXYZ$ may be chosen within it so that rotating it through 180° about any one of them does not change the distribution of orientations with respect to a fixed observer. If the distribution is unchanged by rotation through any arbitrary angle around one of these axes, the sample is said to be *uniaxially oriented*, or *transversely isotropic*, with respect to that axis; otherwise it is said to be *biaxially oriented*. These terms originate from crystal optics and are used because the optical properties of the samples are similar to those of uniaxial and biaxial crystals, respectively. It is usual to label the unique direction of a uniaxial sample OZ. We shall concentrate in detail on the simplest type of uniaxial orientation and the information that can be obtained by means of infrared spectroscopy. A brief indication will then be given of the additional information which can be obtained if the samples are biaxial or if Raman spectroscopy is used, or both.

We assume that the sample is made up of only one type, or possibly a small number of different types, of *structural unit* and that the units of any one type, such as a crystallite or a non-crystalline chain segment, are identical. Each unit is imagined to have embedded within it a set of axes $Oxyz$. In the simplest form of uniaxial orientation each type of unit has a special axis with the property that if this axis lies in a particular direction for a typical unit, all orientations of units obtained by rotation around that axis are equally represented in the distribution. If Oz is chosen to coincide with this special axis we can then say that there is no preferred orientation of the structural units around their Oz axes. For this type of uniaxial sample the distribution of orientations of a particular type of unit can be described fully by a function $N(\theta)$ such that $N(\theta)\,d\omega$ measures the fraction of units for which the Oz axes lie within any infinitesimal solid angle $d\omega$ at angle θ to the OZ axis and $2\pi \int_0^\pi N(\theta)\sin\theta\,d\theta = 1$. The aim of orientation studies is to determine as much as possible about the form of $N(\theta)$.

It is often convenient to expand the function $N(\theta)$ in terms of Legendre polynomials, as follows:

$$N(\theta) = \frac{1}{2\pi} \sum_l \left(\frac{2l+1}{2}\right) a_l P_l(\cos\theta) \qquad (6.5)$$

where, because of the orthogonality of the Legendre functions, it follows that the coefficients a_l are the averages, $\langle P_l(\cos\theta)\rangle$, of the corresponding polynomials $P_l(\cos\theta)$ taken over the distribution of orientations. For distributions in which equal numbers of chains point in opposite directions only even values of a_l are non-zero. The lowest values of a_l are then

$$a_0 = 1 \qquad (6.6a)$$

$$a_2 = \langle P_2(\cos\theta)\rangle = \tfrac{1}{2}(3\langle\cos^2\theta\rangle - 1) \qquad (6.6b)$$

$$a_4 = \langle P_4(\cos\theta)\rangle = \tfrac{1}{8}(3 - 30\langle\cos^2\theta\rangle + 35\langle\cos^4\theta\rangle) \qquad (6.6c)$$

Consider an infrared-active vibrational mode for which the oscillating dipole moment μ of a particular unit makes the angle θ_0 with Oz and θ_μ with OZ, as shown in fig. 6.23. If infrared radiation, of the appropriate frequency, polarized parallel to OZ falls normally on a sample in the form of a thin sheet with its plane containing the OZ axis, the intensity $I_\parallel$ transmitted will be given by (see subsection 5.4.1)

$$\log\left(\frac{I_0}{I_\parallel}\right) = k_\parallel nt \qquad (6.7a)$$

and similarly the intensity transmitted if the polarization is perpendicular to OZ will be

$$\log\left(\frac{I_0}{I_\perp}\right) = k_\perp nt \qquad (6.7b)$$

where I_0 is the incident intensity, $k_\parallel$ and $k_\perp$ are the mean absorptivities for a single structural unit for radiation of the corresponding polarization, n is the number of units per unit volume and t is the thickness of the sample. Since the energy absorbed by an individual unit is proportional to the square of the component of the electric vector

Fig. 6.23. The orientation of a structural unit and its attached infrared dipole moment μ with respect to axes $OXYZ$ fixed in a uniaxially oriented sample, with OZ parallel to the symmetry axis (draw direction) of the sample. Oz is the special axis of the unit around which it has no preferred orientation. The points P and Q lie in the OXY plane.

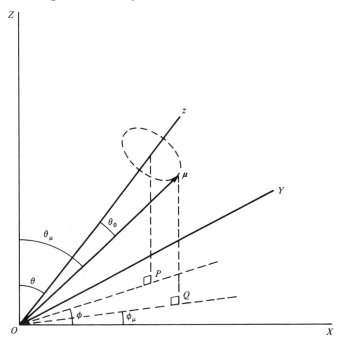

parallel to the oscillating dipole moment, it follows that

$$k_{\parallel} = 3k\langle\cos^2\theta_\mu\rangle \tag{6.8a}$$

$$k_{\perp} = 3k\langle\sin^2\theta_\mu\cos^2\phi_\mu\rangle \tag{6.8b}$$

where k is the mean absorptivity per structural unit for a randomly oriented sample, for which $\langle\cos^2\theta_\mu\rangle = \frac{1}{3}$, and depends on the vibrational mode but not on the distribution of orientations. Because of the uniaxial nature of the sample, ϕ_μ takes random values, independently of θ_μ, and thus

$$k_{\perp} = \frac{3}{2}k\langle\sin^2\theta_\mu\rangle \tag{6.9}$$

Equations (6.8a) and (6.9) show immediately that

$$3k = k_{\parallel} + 2k_{\perp} \tag{6.10}$$

and

$$\langle\cos^2\theta_\mu\rangle = \frac{k_{\parallel}}{k_{\parallel} + 2k_{\perp}} \tag{6.11}$$

It follows from equations (6.7) that

$$\log\left(\frac{I_0}{I_{\parallel}}\right) + 2\log\left(\frac{I_0}{I_{\perp}}\right) = A_{\parallel} + 2A_{\perp} = nt(k_{\parallel} + 2k_{\perp})$$

$$= 3ntk = 3A \tag{6.12}$$

$$\langle\cos^2\theta_\mu\rangle = \frac{\log\left(\dfrac{I_0}{I_{\parallel}}\right)}{\log\left(\dfrac{I_0}{I_{\parallel}}\right) + 2\log\left(\dfrac{I_0}{I_{\perp}}\right)}$$

$$= \frac{A_{\parallel}}{A_{\parallel} + 2A_{\perp}} = \frac{D}{D+2} \tag{6.13}$$

where $A_{\parallel}$ and $A_{\perp}$ are the absorbances of the sample for radiation polarized parallel and perpendicular, respectively, to the draw direction, OZ, and D is the dichroic ratio defined in section 4.2.1. The quantity A is the absorbance of a randomly oriented sample with the same value of nt.

The average value of $\cos^2\theta_\mu$ or, equivalently, $\langle P_2(\cos\theta_\mu)\rangle$ is the only quantity related to molecular orientation which may be determined directly by means of infrared spectroscopy. It may be shown, however, that

$$\langle P_l(\cos^2\theta_\mu)\rangle = \langle P_l(\cos\theta)\rangle P_l(\cos\theta_0) \tag{6.14}$$

for all l, so that if the angle θ_0 is known, a_2 or $\langle \cos^2 \theta \rangle$ may be determined. Equation (6.12) shows that the quantity $A_\parallel + 2A_\perp$ does not depend on the distribution of orientations, but only on the nature of the vibrational mode and the number of structural units per unit area of the sheet, and can thus be used to follow changes in conformational content with orientation if the absorption mode is a conformationally sensitive one.

For samples with more general distributions of orientations than the simplest uniaxial type just considered it is possible to define a function $N(\theta, \phi, \psi)$ so that $N(\theta, \phi, \psi)\, \mathrm{d}\omega'$ measures the fraction of units oriented within the infinitesimal generalized unit solid angle $\mathrm{d}\omega'$ at θ, ϕ, ψ, where the angle ψ measures rotation around the Oz axis of the unit. This function may be expanded in terms of a set of generalized spherical harmonics, of which the Legendre functions are the simplest type. Equations (6.8) are then replaced by three equations which relate the absorptivities for radiation polarized parallel to OX, OY or OZ linearly to the averages of four second order generalized spherical harmonics. The coefficients in these equations depend on the angles θ_0 and ϕ_0 which specify the orientation of the oscillatory dipole with respect to $Oxyz$. Equations (6.8) could have been written in a similar way by the use of equations (6.6) and (6.14). It is therefore possible in principle, by making absorption measurements on more than one absorption peak for which θ_0 and ϕ_0 are known, to solve the resulting equations for the four orientation averages.

Similarly, the intensities of Raman scattering from a sample with the most general type of biaxial orientation usually considered may be expressed, for all possible combinations of polarization vectors of incident and scattered radiation, as linear functions of the averages over the distribution of orientations of the four second order and an additional nine fourth order generalized spherical harmonic functions. The coefficients in these expressions are second order functions of the Raman tensor components referred to the symmetry axes of the unit. For the simplest type of uniaxial orientation the only non-zero averages are $\langle P_2 \langle \cos \theta \rangle \rangle$ and $\langle P_4 (\cos \theta) \rangle$ and if the ratios of the principle Raman tensor components and the orientation of the principal axes within the structural unit are known it is possible to determine these two averages, or $\langle \cos_2 \theta \rangle$ and $\langle \cos_4 \theta \rangle$. One of the difficulties of the Raman method is that this information about the Raman tensor is not always available. In addition, for biaxially oriented samples it is not possible to determine all the relevant averages from measurements on one mode and although, in principle, measurements on several Raman-active modes could be combined, this has not yet been satisfactorily done.

In the treatment given above, corrections for reflection at the surfaces of the sample and for certain 'internal field' effects have been neglected, but they must be taken into account in order to obtain accurate results from either the infrared or Raman method. Studies of orientation using vibrational spectroscopy have been made on a number of polymers and some of the more detailed of these will now be briefly described.

6.10.3   *Orientation in polyethylene and poly(ethylene terephthalate)*

Infrared studies of uniaxially oriented polyethylene have been made by a number of authors. In one study of linear polyethylene two sets of samples of very different molecular weight distribution were drawn at 60°C to draw ratios between 5 and 25. Some of the samples were prepared before drawing by slow cooling from the melt and some were prepared by quenching from the melt. Dichroic ratios were determined for the 1894 cm^{-1} crystal mode and the 1078, 1303 and 1368 cm^{-1} amorphous modes. The 1894 cm^{-1} mode is assigned to a combination of the Raman active A_g methylene rocking mode at 1170 cm^{-1} and the infrared-active B_{2u}/B_{1u} methylene rocking modes at 720/731 cm^{-1} and its dipole moment is perpendicular to the crystal c-axis. The amorphous mode at 1078 cm^{-1} is assigned to gauche C—C stretching with some CH_2 wagging and the 1303 and 1368 cm^{-1} modes are assigned to asymmetric and symmetric CH_2 wagging modes for the CH_2—CH_2 group where the C—C bond is at the centre of a gauche–trans–gauche sequence.

Fig. 6.24 shows a plot of dichroic ratio, D, against draw ratio for the four modes. D_{1894} rapidly approaches zero with increasing draw ratio, which indicates full orientation of the c-axes of the crystallites parallel to the draw direction. The D values for the three amorphous modes approach saturation values which do not indicate full alignment either parallel or perpendicular to the draw direction and this is clearly in accord with their assignments, since the corresponding groups are unlikely to become very highly oriented. In addition, the approach to saturation, particularly for the 1368 cm^{-1} mode, is much more gradual than that for the crystal mode and this has been interpreted as evidence for a clear segregation of the crystalline and amorphous phases.

An extensive series of studies has been performed on oriented samples of poly(ethylene terephthalate) by several spectroscopic techniques, including infrared and Raman. The first quantitative Raman studies were performed on a series of tapes drawn at 80°C to draw ratios up to 5.9. They were assumed to have the simplest type of uniaxial orientation and the Raman-active mode used was that at 1616 cm^{-1}, assigned to a good group vibration of the benzene ring with the symmetry species A_g with respect to the D_{2h} symmetry of the paradisubstituted ring (see fig. 6.25). The Raman tensor was assumed to have cylindrical symmetry around Oz_0 and the ratio α_x/α_z was determined from the linear depolarization ratio for a random sample. This assumption has been shown to be reasonable by a study of the Raman scattering from single crystals of the model compound bis(2-hydroxy ethyl) terephthalate,

$HO(CH_2)_2O \cdot OC$—$\langle O \rangle$—$CO \cdot O(CH_2)_2OH$. The values of $\langle \cos^2 \theta \rangle$

and $\langle \cos^4 \theta \rangle$ for the chain axes were calculated from the observed scattering intensities by assuming that the molecular conformation of all

Fig. 6.24. Dichroic ratio, D, plotted against draw ratio for various infrared absorption peaks of samples of linear polyethylene drawn at 60°C. The wavenumbers of the peaks are indicated for each curve and the assignments are given in the text. ○, slowly cooled Marlex 6050; ●, quenched Marlex 6050; △ slowly cooled Marlex 6002; ▲ quenched Marlex 6002. (Reproduced by permission from B. E. Read in *Structure and Properties of Oriented Polymers*, ed. I. M. Ward, Applied Science Publishers (1975), adapted from W. Glenz and A. Peterlin, *Journal of Macromolecular Science Physics* **B4** 473 (1970), by courtesy of Marcel Dekker, Inc.)

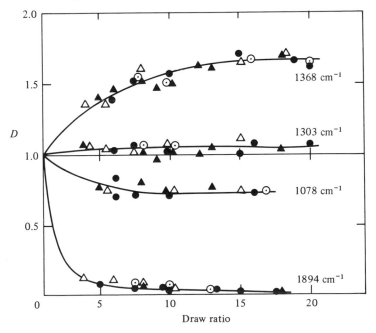

Fig. 6.25. Approximate form of the vibration assigned to the peak observed at $1616 \, \text{cm}^{-1}$ in the Raman spectrum of poly(ethylene terephthalate). The set of axes used in discussing the orientation of the ring is shown.

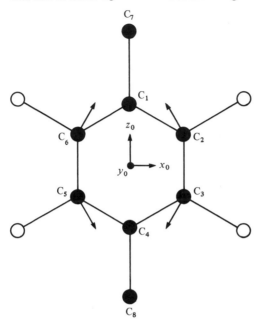

Fig. 6.26. A plot of $\langle \cos^2 \theta \rangle$ against the birefringence of the sample for a series of samples of poly(ethylene terephthalate). The values of $\langle \cos^2 \theta \rangle$ were determined by Raman spectroscopy as explained in the text. (Reproduced from J. Purvis *et al.*, *Polymer* **14** 398 (1973), by permission of the publisher, Butterworth & Co. (Publishers) Ltd. ©)

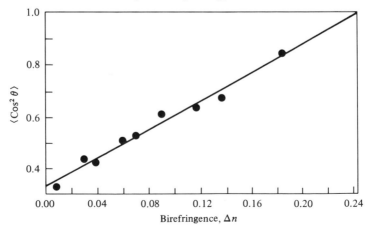

chains in the polymer was similar to that in the crystalline regions except for the possibility of gauche bonds in the —CH_2—CH_2— groups. The results were compared with the birefringences of the samples, i.e. the differences between the refractive indices for light polarized parallel and perpendicular to the draw direction, which may be shown to be approximately equal to $\langle P_2(\cos\theta)\rangle\Delta n_{max}$, where Δn_{max} is the birefringence of a fully oriented sample. A straight line resulted (see fig. 6.26) which extrapolated to the value $\Delta n = 0.24$ at $\langle\cos^2\theta\rangle = 1$, in good agreement with the value for Δn_{max} calculated for a fully oriented sample from bond polarizability data. The points representing pairs of values of $\langle\cos^2\theta\rangle$ and $\langle\cos^4\theta\rangle$ were also shown to lie close to curves relating them predicted by simple theoretical models for the development of molecular orientation in drawn polymers.

Later work showed that modes at 632, 857, 1286 and 1732 cm^{-1} could give information about molecular orientation in poly(ethylene terephthalate) but that the interpretation of the results was not straightforward. Subsequent Raman work has relied almost entirely on intensity measurements on the 1616 cm^{-1} mode and in one study they were combined with infrared measurements on the 875 cm^{-1} mode of the paradisubstituted benzene ring, which is the out-of-plane bending vibration of the C—H bonds in which all the H atoms move in phase. It was possible in this way to determine seven of the orientation averages for the benzene rings in a biaxially oriented sheet and the results showed clearly that the ring planes were preferentially oriented parallel to the plane of the sheet although the molecular chain axes were almost uniaxially oriented with respect to the draw direction. These results confirm X-ray data which also show similar preferential orientation of the ring planes and they have been used in an attempt to understand the high degree of mechanical anisotropy of such biaxially oriented sheets.

6.11 Further reading

The following books and reviews, in addition to those referred to in section 5.7, should be found particularly useful for further reading on the topics of this chapter: *The Chemical Microstructure of Polymer Chains* by J. L. Koenig, Wiley, 1980; *Infrared and Raman Spectroscopy of Polymers* by H. W. Siesler and K. Holland-Moritz, Dekker, 1980; 'Polymer Characterisation by Raman Spectroscopy' by D. L. Gerrard and W. F. Maddams in *Applied Spectroscopy Reviews*, **22**(2–3) 251 (1986); *Biological Spectroscopy* by I. D. Campbell and R. A. Dwek, Benjamin-Cummings, 1984; 'Determination of Molecular Orientation by Spectroscopic Techniques' by I. M. Ward in *Advances in Polymer Science* **66** 81 (1985).

A list of references to some of the more important papers on the topics covered in this chapter and in chapter 5 is given on the following pages. References given in figure captions or in connection with tables may also be found useful.

References for chapters 5 and 6

The following are some specific references to topics in chapters 5 and 6. References given in figure captions or in connection with tables will also be found useful but they are not repeated here.

5.2

J. J. Fox and A. E. Martin, 'Investigations of infrared spectra. Determination of C—H frequencies ($\sim 3000\,cm^{-1}$) in paraffins and olefins, with some observations on "polythenes"', *Proc. Roy. Soc.* **A175** 208–33 (1940).

M. C. Tobin, 'Selection rules for normal modes of chain molecules', *J. Chem. Phys.* **23** 891–6 (1955).

J. R. Nielsen and A. H. Woollett, 'Vibrational spectra of polyethylene and related substances', *J. Chem. Phys.* **26** 1391–400 (1957).

C. Y. Liang, G. B. B. M. Sutherland and S. Krimm, 'Selection rules and frequencies of the skeletal vibrations of long chain polymers in the crystalline state', *J. Chem. Phys.* **22** 1468–9 (1954).

J. R. Nielsen and R. F. Holland, 'Dichroism and interpretation of the infrared bands of oriented crystalline polyethylene', *J. Mol. Spectroscopy* **6** 394–418 (1961).

R. G. Snyder and J. H. Schachtschneider, 'Vibrational analysis of the n-paraffins – I. Assignments of infrared bands in the spectra of C_3H_8 through n-$C_{19}H_{40}$'. '...–II. Normal coordinate calculations', *Spectrochim. Acta* **19** 85–116; 117–68 (1963).

R. G. Snyder, 'A revised assignment of the B_{2g} methylene wagging fundamental of the planar polyethylene chain', *J. Mol. Spectroscopy* **23** 224–8 (1967).

J. Barnes and B. Fanconi, 'Critical review of vibrational data and force field constants for polyethylene', *J. Phys. Chem. Ref. Data* **7** 1309–21 (1978).

S. Krimm and C. Y. Liang, 'Infrared spectra of high polymers. IV. Polyvinyl chloride, polyvinylidene chloride, and copolymers', *J. Polymer Sci.* **22** 95–112 (1956).

W. H. Moore and W. Krimm, 'The vibrational spectrum of crystalline syndiotactic poly(vinyl chloride), *Makromol. Chemie Suppl.* **1** 491–506 (1975).

J. L. Koenig and F. J. Boerio, 'Raman scattering and band assignments in polytetrafluoroethylene', *J. Chem. Phys.* **50** 2823–9 (1969); 'Refinement of a valence force field for PTFE by a damped least squares method', *J. Chem. Phys.* **52** 4826–9 (1970).

C. Y. Liang and S. Krimm, 'Infrared spectra of high polymers. VI. Polystyrene', *J. Polymer Sci.* **27** 241–54 (1958).

R. W. Snyder and P. C. Painter, 'Normal coordinate analysis of isotactic polystyrene: I. A force field for monosubstituted alkyl benzenes'; '...II. The normal modes and dispersion curves of the polymer', *Polymer* **22** 1629–32, 1633–41 (1981).

F. J. Boerio, S. K. Bahl and G. E. McGraw, 'Vibrational analysis of polyethylene terephthalate and its deuterated derivatives', *J. Polym. Sci.: Polym. Phys.* **14** 1092–46 (1976).

303

5.4

J. G. Grasselli, M. A. S. Hazle, J. R. Mooney and M. Mehicic, 'Raman spectroscopy: an update on industrial applications', *Proc. 21st Colloq. Spectrosc. Int.*, Heyden, London, 1979, pp. 86–105.

R. S. Silas, J. Yates and V. Thornton, 'Determination of unsaturation distribution in polybutadienes by infrared spectrometry', *Anal. Chem.* **31** 529–32 (1959).

5.6

G. Martinez, C. Mijangos, J. L. Millan, D. L. Gerrard and W. F. Maddams, 'Polyene sequence distribution in degraded poly(vinyl chloride) as a function of the tacticity', *Makromol. Chemie* **180** 2937–45 (1979).

D. L. Gerrard, W. F. Maddams and J. S. Shapiro, 'Influence of annealing on the formation of long polyenes during PVC degradation', *Makromol. Chemie* **185** 1843–54 (1984).

6.1

M. M. Coleman and J. Zarian, 'Fourier-transform infrared studies of polymer blends. II. Poly(ε-caprolactone)-poly(vinyl chloride) system', *J. Polym. Sci.: Polym. Phys.* **17** 837–50 (1979).

C. Tosi, P. Corradini, A. Valvassori and F. Ciampelli, 'Infrared determination of sequence distributions and randomness in copolymers', *J. Polym,. Sci. C* **22** 1085–92 (1969).

I. V. Kumpanenko and K. S. Kazanskii, 'Analysis of IR-band shapes in studies on the monomer sequence distribution in macromolecules', *J. Polym. Sci. Symp.* **42** 973–80 (1973).

M. Meeks and J. L. Koenig, 'Laser-Raman spectra of vinyl chloride–vinylidene chloride copolymers', *J. Polym. Sci. A2* **9** 717–29 (1971).

6.2

G. Natta, 'Precisely constructed polymers', *Scientific American* **205** No. 2 33–41 (1961).

R. G. Snyder and J. H. Schactschneider, 'Valence force calculations of the vibrational spectra of crystalline isotactic polypropylene and some deuterated polypropylenes', *Spectrochim. Acta* **20** 853–69 (1964).

M. Peraldo and M. Cambini, 'Infrared spectra of syndiotactic polypropylene', *Spectrochim. Acta* **21** 1509–25 (1965).

R. G. Snyder and J. H. Schactschneider, 'Valence force calculations of the vibrational frequencies of two forms of crystalline syndiotactic polypropylene', *Spectrochim. Acta* **21** 1527–42 (1965).

6.3

C. Baker and W. F. Maddams, 'Infrared spectroscopic studies on polyethylene. 1. The measurement of low levels of chain branching', *Makromol. Chemie* **177** 437–48 (1976).

M. A. McRae and W. F. Maddams, 'Infrared spectroscopic studies on polyethylene. 2. The characterisation of specific types of alkyl branches in low-branched ethylene polymer and copolymers', *Makromol. Chemie* **177** 449–59 (1976).

G. S. Park, 'Short chain branching in poly(vinyl chloride)', *J. Vinyl Technol.* **7** 60–4 (1985).

6.4

S. Crawley and I. McNeill, 'Preparation and degradation of head-to-head PVC', *J. Polym. Sci.: Polym., Chem.* **16** 2593–606 (1978).

6.5

C. Baker and W. F. Maddams, 'Infrared spectroscopic studies on polyethylene. 1. The measurement of low levels of chain branching', *Makromol. Chemie* **177** 437–48 (1976).

M. A. McRae and W. F. Maddams, 'Infrared spectroscopic studies on polyethylene. 3. The bromination of unsaturated groups', *Makromol. Chemie* **177** 461–71 (1976).

W. F. Maddams, 'Spectroscopic studies on the characterisation of defects in the molecular structure of poly(vinyl chloride)', *J. Vinyl Technol.*, **7** 65–73 (1985).

6.6

L. R. Schroeder and S. L. Cooper, 'Hydrogen bonding in polyamides', *J. Appl. Phys.* **47** 4310–17 (1976).

V. W. Srichatrapimuk and S. L. Cooper, 'Infrared thermal analysis of polyurethane block polymers', *J. Macromol. Sci. Phys. B* **15** 267–311 (1978).

6.7

F. J. Boerio and J. L. Koenig, 'Raman scattering in crystalline polyethylene', *J. Chem. Phys.* **52** 3425–31 (1970).

M. Glotin and L. Mandelkern, 'A Raman spectroscopic study of the morphological structure of the polyethylenes', *Colloid and Polymer Sci.* **260** 182 92 (1982).

M. E. R. Robinson, D. I. Bower and W. F. Maddams, 'Intermolecular forces and the Raman spectrum of syndiotactic PVC', *Polymer* **17** 355–7 (1976).

W. Frank, H. Schmidt and W. Wulff, 'Complete temperature dependence of the far infrared absorption spectrum of linear polyethylene from 14°K to the melting point', *J. Polym. Sci. Polym. Symp.* **61** 317–26 (1977).

J. F. Rabolt, 'Characterization of lattice disorder in the low temperature phase of irradiated PTFE by vibrational spectroscopy', *J. Polym. Sci.: Polym. Phys.* **21** 1797–805 (1983).

G. Zerbi, F. Ciampelli and V. Zamboni, 'Classification of crystallinity bands in the infrared spectra of polymers', *J. Polym. Sci. C* **7** 141–51 (1964).

6.9

R. F. Schaufele and T. Schimanouchi, 'Longitudinal acoustic vibrations of finite polymethylene chains', *J. Chem. Phys.* **47** 3605–10 (1967).

W. L. Peticolas, G. W. Hibler, J. L. Lippert, A. Peterlin and H. Olf, 'Raman scattering from longitudinal acoustical vibrations of single crystals of polyethylene', *Appl. Phys. Lett.* **18** 87–9 (1971).

B. Fanconi and J. F. Rabolt, 'The determination of longitudinal crystal moduli in polymers by spectroscopic methods', *J. Polym. Sci.: Polym. Phys.* **23** 1201–15 (1985).

6.10

A. Cunningham, I. M. Ward, H. A. Willis and V. Zichy, 'An infrared spectroscopic study of molecular orientation and conformational changes in poly(ethylene terephthalate)', *Polymer* **15** 749–56 (1974).

D. I. Bower, 'Investigation of molecular orientation distributions by polarized Raman scattering and polarized fluorescence', *J. Polym. Sci.: Polym. Phys.* **10** 2135–53 (1972).

D. I. Bower, D. A. Jarvis and I. M. Ward, 'Molecular orientation in biaxially oriented sheets of poly(ethylene terephthalate) I. Characterization of orientation and comparison with models', *J. Polym. Sci. B* **24** 1459–79 (1986).

A note on the use of the indexes

It is not possible to cite in the main index every occurrence of the use of many of the basic concepts introduced in chapters 1–4, and in general only selected examples have been given there. It is hoped that judicious use of the specialised indexes, in conjunction with the main index, will enable others to be readily found. The following abbreviations have been used in the indexes, in addition to more obvious ones:

ATR	attenuated total reflection spectroscopy
f.f.	force field
FTIR	Fourier transform infrared spectroscopy
H-bonding	hydrogen bonding
h-h	head-to-head
h.p.	high pressure
IR	infrared spectroscopy or spectrum
LA(M)	longitudinal acoustic mode
l.d.	low density
PBD	polybutadiene
PE	polyethylene
PED	deuterated polyethylene; potential energy distribution
PEH	ordinary hydrogenated polyethylene
PIB	polyisobutene
PMMA	poly(methyl methacrylate)
POM	polyoxymethylene
PP	polypropylene
PS	polystyrene
PTFE	polytetrafluoroethylene
PVC	poly(vinyl chloride)
PVDC	poly(vinylidene chloride)
R	Raman spectroscopy or spectrum
res. R	resonance Raman spectroscopy or spectrum

Index of spectra illustrated

Index of point groups

Index of group modes

Index of polymers

Main index